Current Topics in Microbiology and Immunology

Volume 429

The review series *Current Topics in Microbiology and Immunology* provides a synthesis of the latest research findings in the areas of molecular immunology, bacteriology and virology. Each timely volume contains a wealth of information on the featured subject. This review series is designed to provide access to up-to-date, often previously unpublished information.

2019 Impact Factor: 3.095., 5-Year Impact Factor: 3.895

2019 Eigenfaktor Score: 0.00081, Article Influence Score: 1.363

2019 Cite Score: 6.0, SNIP: 1.023, h5-Index: 43

More information about this series at http://www.springer.com/series/82

Sho Yamasaki
Editor

C-Type Lectins in Immune Homeostasis

Responsible Series Editor: Shizuo Akira

Springer

Editor
Sho Yamasaki
Department of Molecular Immunology
Research Institute for Microbial
Diseases (RIMD)
Osaka University
Suita, Japan

ISSN 0070-217X ISSN 2196-9965 (electronic)
Current Topics in Microbiology and Immunology
ISBN 978-3-030-62236-7 ISBN 978-3-030-62237-4 (eBook)
https://doi.org/10.1007/978-3-030-62237-4

This Springer imprint is published by the registered company Springer Nature Switzerland AG
The registered company address is: Gewerbestrasse 11, 6330 Cham, Switzerland

Contents

C-Type Lectin Receptors in Phagocytosis 1
Kai Li and David M. Underhill

C-type Lectins in Immunity to Lung Pathogens 19
Benjamin B. A. Raymond, Olivier Neyrolles, and Yoann Rombouts

Signaling C-Type Lectin Receptors in Antifungal Immunity 63
Maxine A. Höft, J. Claire Hoving, and Gordon D. Brown

Recognition of Mycobacteria by Dendritic Cell Immunoactivating Receptor 103
Kenji Toyonaga and Sho Yamasaki

Sensing Tissue Damage by Myeloid C-Type Lectin Receptors 117
Carlos Del Fresno, Francisco J. Cueto, and David Sancho

Structural Aspects of Carbohydrate Recognition Mechanisms of C-Type Lectins 147
Masamichi Nagae and Yoshiki Yamaguchi

Physiological and Pathological Functions of CARD9 Signaling in the Innate Immune System 177
Larsen Vornholz and Jürgen Ruland

C-Type Lectin Receptors in Phagocytosis

Kai Li and David M. Underhill

Contents

1 Introduction 2
2 Mechanisms of Phagocytosis 3
3 C-Type Lectin Receptors 5
3.1 Dectin-1 5
3.2 Dectin-2 7
3.3 Mincle and MCL 8
3.4 DC-SIGN and DC-SIGNR 9
3.5 LSECtin 10
3.6 Lox-1 11
3.7 DNGR-1 11
4 Additional Phagocytosis-Linked C-Type Lectin Domain-Containing Proteins 12
4.1 The Mannose Receptor 12
4.2 Collectins 13
5 Conclusion 14
References 15

Abstract C-type lectin receptors (CLRs) are a family of transmembrane proteins having at least one C-type lectin-like domain (CTLD) on the cell surface and either a short intracellular signaling tail or a transmembrane domain that facilitates interaction with a second protein, often the Fc receptor common gamma chain (FcRγ), that mediates signaling. Many CLRs directly recognize microbial cell walls

K. Li · D. M. Underhill (✉)
F. Widjaja Inflammatory Bowel and Immunobiology Research Institute, Cedars-Sinai Medical Center, Los Angeles, CA 90048, USA
e-mail: David.Underhill@csmc.edu

K. Li · D. M. Underhill
Division of Immunology, Department of Biomedical Sciences, Cedars-Sinai Medical Center, Los Angeles, CA 90048, USA

Current Topics in Microbiology and Immunology (2020) 429: 1–18
https://doi.org/10.1007/82_2020_198

Published Online: 15 February 2020

and influence innate immunity by activating inflammatory and antimicrobial responses in phagocytes. In this review, we examine the contributions of certain CLRs to activation and regulation of phagocytosis in cells such as macrophages, dendritic cells and neutrophils.

1 Introduction

The notion of pattern recognition receptors (PRRs), germline-encoded innate immune receptors recognizing common microbial structures, was articulated by Charles Janeway in the late 1980s (Janeway 1989) and was embodied by the discovery and characterization of Toll-like receptors (TLRs) (Medzhitov et al. 1997). The TLRs comprised a family of PRRs with similar structures and signaling activities, but which recognized different microbial products (e.g., lipids, nucleic acids, proteins). These receptors were generally pro-inflammatory in that they stimulated transcription of inflammatory cytokines and chemokines through the activation of, among others, the NF-κB pathway. Details of TLR signaling pathways continue to be defined, and subtle, and not so subtle, differences in how different TLRs instruct inflammatory immune responses and influence development of adaptive immune responses are studied by investigators worldwide.

In the early 2000s, another family of PRRs emerged with the characterization of the C-type lectin receptors (CLRs), the focus of this volume. Conventionally, the CLRs are a family of transmembrane proteins having at least one C-type lectin-like domain (CTLD) on the cell surface and either a short intracellular signaling tail or a transmembrane domain that facilitates interaction with a second protein, generally the Fc receptor common gamma chain (FcRγ), that mediates signaling. Similarities and differences between the different CLRs will be discussed further below. Interestingly, the emergence of CLRs came not from an effort to identify mechanisms of inflammatory signaling akin to TLRs, but rather from studies aimed at understanding the mechanisms of phagocytosis (Brown and Gordon 2001). In an expression cloning screen aimed at identifying a receptor that could allow macrophages to eat (phagocytose) yeast cell wall particles (zymosan), Brown and colleagues came up with a protein that had previously been called dendritic cell-associated C-type lectin-1 (Dectin-1) (Ariizumi et al. 2000). The investigators went on to discover that this receptor binds to β-glucan, a structural carbohydrate polymer in fungal cell walls, and this classified the receptors a prime example of a PRR.

In this review, we will look at the growing family of C-type lectin receptors in the context of what is currently known about their contributions (or not) to phagocytosis of microbial pathogens by myeloid cells including macrophages, dendritic cells and neutrophils. The reader is directed to other chapters in this volume and elsewhere for deeper discussions of individual CLRs or mechanisms of signaling for the production of inflammatory mediators not associated with phagocytosis.

2 Mechanisms of Phagocytosis

Phagocytosis is an evolutionarily conserved process that mediates the internalization of large particles (>0.5 μm) (Aderem and Underhill 1999). In metazoans, phagocytosis is predominantly performed by a group of specialized cells termed phagocytes, including macrophages, dendritic cells, monocytes and neutrophils. Depending on the nature of the ingested particles and the cell type that engulfs the particle, phagocytosis can lead to diverse biological consequences. For example, macrophages that phagocytose bacteria generally induce pro-inflammatory responses, whereas macrophages that ingest apoptotic cells generate anti-inflammatory responses. Meanwhile, neutrophils and macrophages that phagocytose bacteria elicit direct bactericidal programs and completely degrade the internalized microbe, while dendritic cells tend to preserve antigenic peptides by partially digesting microbial proteins (Savina et al. 2006; Watts 2006). The particle-containing vacuole formed during phagocytosis, termed a phagosome, is a degradative compartment in which microbes are killed, microbial products are degraded and recycled, antigens are recovered, and microbial products released during degradation may be detected.

There are at least two different types of phagocytosis, namely reaching phagocytosis and sinking phagocytosis, which apparently adopt different mechanisms of internalization. Reaching phagocytosis, also known as "zipper phagocytosis," is characterized by cell membrane actively reaching out to the particulate target, forming pseudopodia to complete cargo engulfment. The best-studied model of reaching phagocytosis is Fcγ receptor-mediated phagocytosis. During Fcγ receptor-mediated phagocytosis, binding of IgG-opsonized particle to Fcγ receptors leads to receptor clustering and activation of receptor-proximal Src family kinases (SFKs). The activated SFKs subsequently phosphorylate the dual tyrosine residues within the immunoreceptor tyrosine-based activation motif (ITAM) at the cytoplasmic tail of Fcγ receptors, providing a dock site for Syk recruitment and activation. Activated Syk coordinates lipid modifications and actin remodeling by relaying signals to downstream PI3K, PLCγ, or exchange factors that activate GTPases, culminating in cell membrane protrusion and engulfment of the particle (Freeman and Grinstein 2014; Goodridge et al. 2012). By contrast, sinking phagocytosis triggered by complement receptors generates an inward force effectively submerging the bound particle in the plasma membrane. This process often first requires receptor activation-induced inside-out signaling, which breaks the closed conformation of complement receptors to drive phagocytosis (Luo et al. 2007).

Many types of receptors can contribute to the mechanisms of phagocytosis, somewhat blurring the definition of a "phagocytic receptor" (Fig. 1). First, receptors may mediate direct binding to a microbe or binding to opsonins deposited on microbial surfaces. This binding is required for phagocytosis, and receptors that enhance binding will enhance phagocytosis, even if they do not necessarily trigger internalization themselves. Second, some receptors are sufficient to activate the

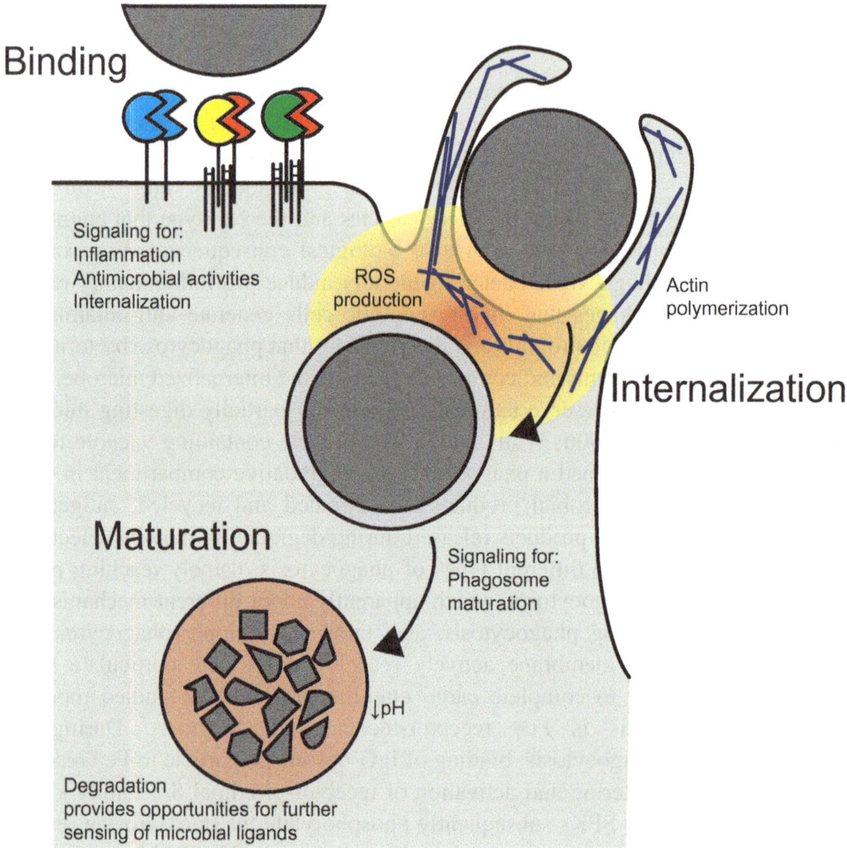

Fig. 1 Mechanisms of phagocytosis. C-type lectin receptors that bind to microbial surfaces may influence phagocytosis in several ways. First, simply by binding to microbial surfaces, they facilitate capture and ingestion of microbes. These receptors may promote inflammatory signals or antimicrobial activities such as the production of reactive oxygen species (ROS). Second, they may directly trigger intracellular signaling that is sufficient, even in the absence of other signals, to trigger membrane reorganization and phagocytosis of the microbe. Finally, the receptors may influence the rate of phagosome maturation or trafficking of intracellular materials to the phagosome

intracellular signaling necessary to coordinate membrane and actin cytoskeletal movements required for microbial engulfment. Receptors falling into this category are generally considered as bona fide phagocytic receptors. Most of the time, this definition can be experimentally translated as receptors that, when ectopically expressed (alone or together with other minimally required signaling adaptors) in an otherwise non-phagocytic cell type, mediate the internalization of phagocytic cargo. Finally, signaling by microbe recognition receptors can influence the speed and nature of phagosome maturation.

3 C-Type Lectin Receptors

Classical C-type lectins are proteins that bind various carbohydrate moieties in a calcium-dependent manner. By now, more than one thousand members within the C-type lectin family have been identified (including those from genome sequence only). Most of the family members are categorized as C-type lectins by the presence of at least one CTLD, which is defined by sequence and structural homology, instead of functional evidence. Thus, many C-type lectins do not retain the carbohydrate-binding capacity or the calcium-dependence for carbohydrate-binding (Zelensky and Gready 2005). Membrane-bound C-type lectins have been of particular interest because they have the potential to directly connect the microenvironment cues with cellular responses such as phagocytosis. Below, we review individual CLRs that have been reported to participate in different phagocytic processes.

3.1 Dectin-1

Dectin-1 (*CLEC7A*) is a prototypic CLR. Murine Dectin-1 is highly expressed on myeloid macrophage, dendritic cells and neutrophils. It is also expressed on a subset of T lymphocytes, albeit at a much lower level (Taylor et al. 2002). Human Dectin-1 is additionally expressed on B cells, eosinophils and mast cells (Willment et al. 2005). Dectin-1 was initially identified as the receptor for β-1,3 glucan, which is a major component of the fungal cell wall. Accordingly, Dectin-1-deficient mice show enhanced fungal dissemination upon intravenous *Candida albicans* infection (Taylor et al. 2007), and humans carrying a *CLEC7A* (Dectin-1) gene variant with an early stop codon mutation are predisposed to mucosal candidiasis (Ferwerda et al. 2009). The extracellular domain of Dectin-1 has the C-type lectin domain which mediates the carbohydrate-binding in a metal ion-independent manner. The intracellular cytoplasmic tail of Dectin-1 harbors an imperfect ITAM motif, termed hemITAM, which is essential for signal transduction following ligand recognition and binding (Kerrigan and Brown 2010).

In the seminal paper that described Dectin-1, it was reported that Dectin-1 can mediate the non-opsonic internalization of heat-killed *Candida albicans* when expressed in normally non-phagocytic NIH3T3 fibroblasts (Brown and Gordon 2001). Based on this root finding, Dectin-1 has long been established as a bona fide phagocytic receptor. Similar to the well-defined Fcγ receptor-mediated phagocytosis, Dectin-1 mediated phagocytosis is actin-dependent and requires Rho GTPase, especially Cdc42 (Herre et al. 2004). Also, the intracellular hemITAM motif is essential for the phagocytic activity of Dectin-1, since an inactivating mutation of the membrane-proximal tyrosine residue abrogates its phagocytic activity in NIH3T3, RAW264.7 and 293T cell overexpressing Dectin-1 (Herre et al. 2004; Underhill et al. 2005). However, unlike the Fcγ receptor, the role of Syk in the

Dectin-1-initiated phagocytosis is controversial. While Syk inhibitors significantly block Dectin-1-mediated phagocytosis of zymosan particles in transduced NIH3T3 fibroblast or 293T cells, it has no effect on zymosan uptake in both RAW264.7 and bone marrow macrophages. Moreover, Syk-knockout bone marrow macrophages show no defect in the internalization of unopsonized zymosan (Herre et al. 2004; Underhill et al. 2005). However, Syk-deficient murine bone marrow dendritic cells or Syk inhibitor-treated human monocyte show reduced phagocytosis of zymosan (Elsori et al. 2011; Rogers et al. 2005), implying a possible cell type-dependent role of Syk in Dectin-1-mediated phagocytosis.

Although the role of a conserved signaling motif and Syk in Dectin-1-mediated phagocytosis has been tested, the connection between these receptor-proximal signaling mechanisms to the classical events of phagocytosis, such as cytoskeleton and membrane remodeling, is still mostly unexplored. It is possible that Dectin-1, as a receptor that directly senses phagocytic cargoes, can on its own generate all the signaling required for particle uptake when activated. On the other hand, Dectin-1 could potentially signal to other more specialized phagocytic receptors, such as integrins, and "outsource" their internalizing activity. In vivo, it is possible that the mechanisms by which Dectin-1 drives phagocytic activity might depend on the availability of other phagocytic receptors in a particular cell type. At least in neutrophils, Dectin-1 signaling has been observed to couple to the Mac-1 (CD11b/CR3) adhesion receptor via vav1/3 and PLCγ/calcium pathways for particle internalization. Dectin-1 activation by zymosan is able to induce Mac-1 activation and Mac-1-deficient neutrophils exhibit impaired internalization of zymosan particles (Li et al. 2011). However, this process seems to be cell type specific, as Mac-1-deficient macrophages show unimpaired phagocytosis of zymosan particles.

Besides mediating particle uptake, the structure formed during the initial phase of Dectin-1-mediated phagocytosis, termed phagocytic synapse, is crucial for Dectin-1-triggered inflammatory responses. Although Dectin-1 binds soluble polymeric β-1,3 glucan with high affinity and single β-glucan polymers could presumably oligomerize Dectin-1, it appears that receptor clustering alone by soluble β-glucan is insufficient to fully activate Dectin-1-dependent signals. IFN-γ-primed bone marrow macrophages do not produce TNF-α or generate reactive oxygen species when stimulated with soluble β-glucan polymers with molecular sizes of 400 kD. By contrast, insoluble β-glucan particles, beads coated with β-glucan, or β-glucan coated on tissue culture plates, can robustly trigger TNF-α production and ROS burst in macrophages, suggesting that activation of Dectin-1 signaling requires its ligand to be present on immobilized surfaces (Goodridge et al. 2011). However, cytoskeletal mobilization, which is required for particle internalization, is not necessary for most parts of Dectin-1 signaling. Pretreating macrophage or dendritic cells with an actin-polymerization inhibitor does not impair glucan particle-induced cytokine production (Brown et al. 2003).

Furthermore, Dectin-1 has also been shown to influence the biological consequences of phagocytosis after particle internalization. In bone marrow-derived dendritic cells and macrophages, Dectin-1-signaling can direct the recruitment of LC3-II, a component of the autophagy pathway to phagosome. Although this process does not affect initial particle phagocytosis, microbe-killing or inflammatory cytokine production, it has an impact on MHC-II molecule recruitment to phagosomes, antigen presentation and ultimately T cell priming (Ma et al. 2012). Thus, Dectin-1 signaling continues to influence phagosome maturation after internalization.

3.2 Dectin-2

Dectin-2 (*CLEC6A*) is another CLR that plays an important role in anti-fungal immune responses. Mice that are deficient in this receptor show increased susceptibility to disseminated *Candida albicans* (Saijo et al. 2010) or *Candida glabrata* infection (Ifrim et al. 2014). The carbohydrate-binding domain of Dectin-2 has the classical carbohydrate-binding EPN motif and can bind high mannose structures in a calcium-dependent manner (McGreal et al. 2006). Dectin-2 deficient dendritic cells are defective in mannan-induced production of inflammatory cytokines (Saijo et al. 2010). Unlike Dectin-1, Dectin-2 does not have the classical ITAM or hemITAM motif in its cytoplasmic tail and instead relies on associating with the ITAM motif-containing Fc receptor gamma chain (FcRγ) for signaling.

To our knowledge, the potential of Dectin-2 to function as a bona fide phagocytic receptor has not been reported. Sato et al. have shown that CHO cells that ectopically express full-length Dectin-2 can bind *Candida* hyphae (Sato et al. 2006). However, whether these cells could proceed to drive the ingestion of mannan-containing particles was not described in their report. As Dectin-2 requires the FcRγ association for both its intracellular signaling and optimal surface expression, FcRγ might need to be concomitantly expressed in non-phagocytic cells to determine if Dectin-2 activation alone is sufficient to mediate phagocytosis. Although the evidence of Dectin-2 being a true phagocytic receptor is lacking, genetic deletion of Dectin-2 in professional phagocytes has been shown to moderately impair the uptake of fungi and the maturation of fungus-containing phagosomes (Haider et al. 2019; Ifrim et al. 2014).

Besides its role in anti-fungal immune responses, Dectin-2 has been shown to be required for suppressing liver metastasis of colon carcinoma, Lewis lung carcinoma and melanoma in mice (Kimura et al. 2016). Interestingly, Dectin-2 seems to selectively control liver metastasis, but not lung metastasis or growth of the primary tumor. The specificity of liver metastasis suppression by Dectin-2 has been largely attributed to the phagocytic activity of Kupffer cells, which are the specialized liver-resident phagocytes directly internalizing live tumor cells. Dectin-2-deficient Kupffer cells show reduced uptake of tumor cells, although Dectin-2 does not appear to directly bind the tumor cells. Moreover, Dectin-2 mediated phagocytosis of tumor cells is only limited to Kupffer cells, because the deficiency of Dectin-2 in bone marrow-derived macrophages or alveolar macrophages does not show any defect in tumor cell uptake. Similar to the earlier mentioned neutrophil-specific cooperation between Dectin-1 and Mac-1, it is conceivable that Dectin-2 could function as a phagocytic co-receptor for a yet to be identified receptor that directly binds live tumor cells.

3.3 Mincle and MCL

Mincle (macrophage-inducible C-type lectin, *CLEC4E*) is, like Dectin-2, a receptor with a very short cytoplasmic tail lacking the classical immunoreceptor-associated motifs for intracellular signaling. It is also thought to rely on coupling to the ITAM motif-containing FcRγ for signaling upon receptor crosslinking and activation. The Mincle receptor is mainly expressed by phagocytic cells in response to different inflammatory signals (Matsumoto et al. 1999). Trehalose-6,6-dimyoclate (TDM, also known as cord factor), which is found in the cell walls of some mycobacterium species, has been identified as the exogenous ligand for Mincle (Ishikawa et al. 2009). In line with the identity of its ligand, Mincle has been shown to play an important role in the immune response against *Mycobacterium bovis* bacillus Calmette–Guerin (BCG) infection in vivo (Behler et al. 2012). Mincle has also been shown to recognize the α-mannose moiety, but not mannan, on the surface of pathogenic fungus *Malassezia* (Yamasaki et al. 2009). Moreover, Mincle is able to sense splice-associated protein 130 (SAP130) endogenously released by necrotic cells and is therefore considered as a DAMP-sensing PRR. Accordingly, Mincle-deficient mice show an impaired inflammatory response upon irradiation-induced cellular damage (Yamasaki et al. 2008).

Another closely related CLR, named MCL (macrophage C-type lectin, Dectin-3, *CLEC4D*), has been considered as an ancillary receptor for both Mincle and Dectin-2 and forms heterodimers with either receptor (Lobato-Pascual et al. 2013; Zhu et al. 2013). Some evidence suggests that MCL could directly bind both TDM (Mincle ligand) (Miyake et al. 2013) and α-mannan (Dectin-2 ligand) (Zhu et al. 2013), synergistically enhancing the response of both receptors. Interestingly, MCL can also enhance surface expression of Mincle, while deficiency of MCL does not seem to impact the surface expression of Dectin-2.

When individually expressed in 293T cells, both Mincle and MCL can mediate uptake of beads coated with antibodies against the corresponding receptor (Lobato-Pascual et al. 2013). The phagocytic activities of these receptors are further increased when the FcRγ adapter is co-expressed with the receptors. Moreover, co-expressing both Mincle and MCL in 293T cells has a synergistic effect on the uptake of beads coating with antibodies against either Mincle or MCL. These data imply Mincle and MCL are true phagocytic receptors. As Mincle or MCL can induce particle uptake in the absence of the ITAM-containing FcRγ adapter, albeit to a lesser degree, it raises some questions as to whether Mincle-induced phagocytosis is dependent on an ITAM motif and Syk.

3.4 DC-SIGN and DC-SIGNR

DC-SIGN (dendritic cell-specific ICAM-3-grabbing nonintegrin, *CLEC4L*) is a single CTLD-containing CLR, and its expression was initially thought to be dendritic cell-restricted, hence the name. Originally identified as a binding molecule of HIV-1 gp120 glycoprotein on DC transporting HIV-1 virus to secondary lymphoid organs to enhance trans-infection (Geijtenbeek et al. 2000a, b), DC-SIGN has since been reported to directly bind a broad range of pathogens, including viruses (HCV, Ebola virus, Dengue viruses, Measles virus), fungi (*Candida albicans*, *Aspergillus fumigatus*), bacteria (Mycobacteria or non-pathogenic *E. coli*) or nematodes (*Schistosoma mansoni*) (Cambi et al. 2003; de Witte et al. 2006; Geijtenbeek et al. 2003; Lozach et al. 2003; Serrano-Gómez et al. 2004; Tassaneetrithep et al. 2003). Minimally, DC-SIGN recognizes high mannose oligosaccharides and fucosylated glycans (van Liempt et al. 2006). DC-SIGNR is a C-type lectin that shows 73% identity to DC-SIGN at the DNA level. In terms of ligand binding, DC-SIGNR greatly overlaps with DC-SIGN by being able to bind mannose-rich moieties (Park et al. 2001).

The intracellular domain of DC-SIGN contains three putative motifs (di-leucine and tri-acidic clusters, hemITAM) that are involved in interactions with intracellular proteins and the cytoskeleton. Given that the extracellular domain of DC-SIGN can directly bind pathogens with phagocytosable sizes, it is conceivable that this receptor could mediate phagocytosis. Indeed, several lines of evidence support the phagocytic potential of this receptor. There are at least four studies that used K562 cells ectopically expressing DC-SIGN as a model to assess phagocytic functions of DC-SIGN in response to *Aspergillus fumigatus*, *Candida albicans*, *Mycobacterium tuberculosis* or beads coated with Man-LAM (the major mycobacterial ligand for DC-SIGN), respectively (Azad et al. 2008; Cambi et al. 2003; Geijtenbeek et al. 2003; Serrano-Gómez et al. 2004). All these studies demonstrated that non-phagocytic K562 cells can engulf particles when transfected to express DC-SIGN. Furthermore, one study showed that K562 cells expressing a DC-SIGN mutant lacking the tri-acidic cluster show markedly reduced phagocytosis of Man-LAM-coated beads (Azad et al. 2008). Similar approaches have also employed HeLa cells, where ectopic expression of DC-SIGN permits HeLa cells to adhere and internalize non-pathogenic *E. coli* (Zhang 2006). Besides cell line models, the potential role of DC-SIGN in phagocytosis has also been tested in primary cells. In monocyte-derived dendritic cells, phagocytosed *Aspergillus fumigatus*, *Candida albicans* and mycobacteria colocalize with DC-SIGN on phagosomes. Furthermore, a blocking antibody against DC-SIGN inhibits binding (or internalization) of *Aspergillus fumigatus*, *Candida albicans* and mycobacteria to primary dendritic cells.

Notably, the phagocytic capacity of the murine DC-SIGNR homolog, named DC-SIGNR1 (*CLEC4M*), has also been studied (Taylor et al. 2004). DC-SIGNR1-transduced NIH3T3 fibroblasts gain the ability to bind zymosan particles and heat-killed *Candida albicans*. However, unlike Dectin-1-transduced

NIH3T3, DC-SIGNR1-transduced NIH3T3 cells cannot internalize the bound fungal particles, implying that DC-SIGNR facilitates phagocytosis by increasing particle binding but requires other receptors, such as Dectin-1, to complete phagocytosis.

3.5 LSECtin

LSECtin (Liver and lymph node sinusoidal endothelial cell C-type lectin, *CLEC4G*) is a small type II transmembrane protein that is encoded by a gene within the "DC-SIGN cluster." It is mainly expressed on liver and lymph node sinusoidal endothelial cells, and tissue-resident macrophages. The CTLD domain of LSECtin contains calcium-binding sites as well as an EPN motif associated with binding to mannose/N-acetylglucosamine. It is thus predicted to be a calcium-dependent carbohydrate-binding CLR. Indeed, mannose has been reported to be a monosaccharide ligand for LSEctin (Liu et al. 2004). Further, glycan array analysis shows that LSECtin can specifically bind glycoproteins with GlcNAcβ1-2Man termini (Powlesland et al. 2008). In line with the fact that such glycan moieties are present on the glycoprotein of Ebola virus, LSECtin has been shown to bind Ebola virus particles and mediate Ebola virus infection-induced inflammatory responses (Zhao et al. 2016). Besides exogenous ligands, LSECtin has also been reported to bind apoptotic cells. The precise nature of the LSECtin ligand present on dead cells is still undetermined, even though the possibility of phosphatidylserine has been excluded (Yang et al. 2018).

The potential function of LSECtin in phagocytosis has been investigated in the context of dead cell clearance. Li et al. have observed that LSECtin-deficient mice show impaired intestinal epithelium recovery upon gut barrier disruption by DSS treatment (Yang et al. 2018). They attributed this dysfunction to the defective clearance of dead cells by macrophages. Detailed analyses show that ex vivo peritoneal macrophages or Kupffer cells from LSECtin-deficient mice exhibit reduced uptake of apoptotic thymocytes, but not latex beads or bacteria, indicating that LSECtin is necessary for optimal internalization of specific ligand-containing particles. Furthermore, ectopic expression of LSECtin in non-phagocytic NIH3T3 cells permits the internalization of apoptotic thymocytes, supporting the idea that LSECtin alone is sufficient to drive phagocytosis of dead cells. However, further studies using alternative non-phagocytic cell lines and more robust phagocytosis assays are required to claim LSECtin as a true phagocytic receptor, as empty vector-transduced NIH3T3 also internalized a significant number of dead cells in the study.

3.6 Lox-1

LOX-1 (Lectin-like oxidized low-density lipoprotein receptor-1, *CLEC8A*) is predominantly expressed on the surface of endothelial cells, macrophages and smooth muscle cells in atherosclerotic lesions of humans. LOX-1 is the major receptor for oxidized low-density lipoprotein (OxLDL) and has been shown to bind a broad range of endogenous and exogenous ligands, including oxidized erythrocytes, apoptotic cells, activated platelets and bacteria (Oka et al. 1998; Shimaoka et al. 2001).

OxLDL can block the engulfment of aged and apoptotic cells by bovine aortic endothelial cells (BAE). Given that LOX-1 is the major receptor for OxLDL on BAE, it was predicted to mediate phagocytosis of BAE. In line with this hypothesis, CHO cells stably transformed to express bovine LOX-1 gain the ability to bind and internalize aged RBC and apoptotic cells by recognizing phosphatidylserine (PS) exposed on those cells. As PS is known to be procoagulant, it is proposed that increased OxLDL promotes coagulation in atherosclerosis by blocking prompt removal of PS-containing cell via LOX-1 (Oka et al. 1998).

3.7 DNGR-1

DNGR-1 (*CLEC9A*) is a danger-sensing receptor that specifically binds F-actin exposed by damaged or necrotic cells (Zhang et al. 2012). The expression of DNGR-1 is restricted to $CD8\alpha^+$ subset of dendritic cells. Genetic tracing of DNGR-1 expression history delineates cells of the DC lineage. Therefore, DNGR-1 is now widely used as a marker of DCs (Schraml et al. 2013).

The intracellular domain of DNGR-1 contains a tyrosine-based motif that resembles the hemITAM of Dectin-1, which is important for Dectin-1-triggered phagocytosis. To see if DNGR-1 is thus capable of mediating phagocytosis, Huysamen et al. constructed a chimeric receptor containing the extracellular domain of Dectin-1 coupled to the intracellular part of DNGR-1 (Huysamen et al. 2008). While this chimeric receptor can mediate the binding to the defined Dectin-1 ligand zymosan when expressed on non-phagocytic NIH3T3 fibroblasts, it does not induce particle uptake, suggesting that DNGR-1 is not a phagocytic receptor. Meanwhile, this also implies that other biological events during Dectin-1 activation, besides ligand-induced clustering of its extracellular domain and hemITAM-mediated signaling, are required for phagocytosis as well. Although unable to trigger phagocytosis, DNGR-1 has been shown to impact the functional consequences of dead cell phagocytosis by diverting necrotic cell cargo to favor cross-presentation to $CD8^+$ T cells, playing an important role in anti-viral CTL responses (Iborra et al. 2012).

4 Additional Phagocytosis-Linked C-Type Lectin Domain-Containing Proteins

It is arguably not entirely fair to credit, as we have done in the introduction to this review, the "discovery" of the family of CLRs including Dectin-1 as a second example of PRRs after TLRs or as a "discovery" of the role of CLRs in phagocytosis. We think it is fair to characterize how they were introduced to and discussed in the scientific community at the time this way. However, it is important to acknowledge that C-type lectin domain-containing receptors and soluble proteins had been implicated in phagocytosis and recognition of microbial cell walls long before the Dectin-related CLR family was identified.

4.1 *The Mannose Receptor*

The macrophage mannose receptor (CD206, MRC1) was first characterized in the late 1970s and was cloned in the 1990s (Alan et al. 1990; Stahl et al. 1978). The receptor is a type I transmembrane protein with a short 45 amino acid cytoplasmic tail. The extracellular domain of the receptor consists of eight C-type lectin carbohydrate recognition domains (CRDs), which bind α-mannans found on cell walls of many microbes, a short amino-terminal cysteine-rich region, which binds glycoproteins bearing sulfated sugars that terminate in sulfated N-Acetyl-D-galactosamines, and a fibronectin type II repeat. The receptor is expressed broadly by myeloid phagocytes and is particularly upregulated on these cells during the resolution of inflammation. The MR serves as a homeostatic receptor by binding and endocytosing unwanted high mannose N-linked glycoproteins as well as pituitary hormones from the circulation (Allavena et al. 2004).

The MR directly binds a wide variety of microorganisms, including bacteria, fungi, viruses, and parasites and appears sufficient to trigger phagocytosis of large microbes. Expression of human mannose receptors in normally non-phagocytic COS cells is enough to mediate the internalization of zymosan. Importantly, expression of a tailless mutant mannose receptor-mediated zymosan binding but not internalization (Alan et al. 1990), and expression of a chimeric receptor consisting of the extracellular domain of FcγRI fused to the transmembrane and intracellular domain of the mannose receptor facilitated uptake of IgG-opsonized particles (Kruskal et al. 1992). While these observations suggested that signals initiated by mannose receptor ligation are sufficient to induce particle internalization, the molecular mechanisms by which mannose receptors activate downstream signals for particle internalization have until recently remained largely undescribed.

In 2017, a new evaluation of the mechanisms involved in mannose receptor engagement of *Mycobacterium tuberculosis* found that expression of human mannose receptor at the cell surface is dependent on association with the FcRγ-chain (Rajaram et al. 2017). Typically, an association of a receptor with

FcRγ-chain requires a charged amino acid in the receptor's transmembrane domain, and such a residue is present in the human mannose receptor but not in the mouse, suggesting that there may be differences in how the receptor functions in mice and humans. In human cells, the investigators found that tyrosine phosphorylation of the mannose receptor intracellular tail facilitates the initial recruitment of Grb2 that facilitates activation of the Rac/Cdc42/PAK complex that is essential for phagocytosis. Subsequently, Grb2 recruits SHP-1 to newly formed phagosomes, and this causes the phagosomes to mature more slowly. Whether or how the FcRγ-chain contributes to signaling in this context is not yet clear.

4.2 Collectins

The collectins, C-type lectins with collagenous regions, are soluble molecules and not transmembrane receptors, which are generally the focus of this review (Casals et al. 2019). However, they are secreted pattern recognition proteins that bind directly to microbes via their C-type lectin domains and enhance phagocytosis. Collectins recognize carbohydrates on the surface of microorganisms (mainly mannose and N-acetyl-glucosamine), while they do not usually detect carbohydrates that are exposed on typical mammalian glycoproteins, such as galactose and sialic acid.

Perhaps the most well-known collectin is the mannose-binding lectin (MBL), a serum protein produced in the liver. The MBL protein (also sometimes called mannan-binding protein (MBP)) consists of a couple of amino-terminal collagen-like domains, a short neck region, and a carboxy-terminal C-type lectin domain, and the assembled protein is a bouquet of three to six trimers (Casals et al. 2019). It binds to carbohydrates including mannose, N-acetylmannosamine, N-acetylglucosamine, glucose and fucose on the cell walls of many kinds of microbes. As a multimer, binding strongly promotes aggregation of target microbes. Proteins associated with the collagenous regions active complement deposition (the lectin pathway) on microbes leading to killing and opsonization for phagocytosis by complement receptors. MBL can also act directly as an opsonic receptor through binding by to collectin receptors (e.g., C1qR) that can trigger phagocytosis (Van De Wetering et al. 2004).

Two other well-known collectins are surfactant proteins A (SPA) and D (SPD). SPA and SPD are secreted in the lung together with other surfactants. Unlike other surfactants, however, the main functions of SPA and SPD are in host defense. The proteins are assembled as multimers with mature SPA being an octodecamer resembling a bouquet, and SPD being a dodecameric cruciform. As multimers presenting many lectin domain binding sites, the proteins are particularly effective at aggregating targets including bacteria, fungi and viruses. SPA and SPD can promote phagocytosis, similar to MBL, through interaction with collectin receptors.

Another collectin, Collectin-12 (also known as collectin placenta 1), has been shown to have both a membrane-bound form and a secreted soluble form. As a cell surface-expressed receptor, Collectin-12 was often defined as a scavenger receptor C-type lectin that is mainly found in vascular endothelial cells and is expressed at a particularly high level in umbilical cord endothelial cells (Ohtani et al. 2001). Like many other scavenger receptors, Collectin-12 can bind a broad spectrum of ligands, including gram-positive and gram-negative bacteria, yeast and OxLDL. Notably, CHO cell and human umbilical artery endothelial cells, both of which are normally non-phagocytic, can ingest zymosan particles when transfected to express human Collectin-12 (Jang et al. 2009), indicating hCollectin-12 is potentially a phagocytic receptor. Besides residing on the cell surface, Collectin-12 can also be shed as a soluble protein, which can opsonize microorganisms such as *Aspergillus fumigatus* through its CRD domain and activate the alternative complement pathway, similar to other soluble collectins (Ma et al. 2015).

5 Conclusion

Soluble C-type lectins have been known to be essential for pathogen opsonization and subsequent phagocytosis by immune cells for more than half a century. Additionally, in the past two decades, membrane-bound CLRs have increasingly become recognized as a major class of PRR that are important in innate immune responses, involved in pathogen recognition, phagocytosis, inflammatory signaling and antigen processing. Although the phagocytic potential of CLRs has been recognized since the establishment of the field, data on the specific contributions of all the different CLRs to phagocytosis and phagocytosis-related processes is scattered and often difficult to interpret and compare. There are many more CLRs than we have discussed here, but their possible roles in phagocytosis are not yet clear. We propose that systematic investigations into the following aspects of CLR-mediated phagocytosis are needed in future studies: (A) The ability of CLR to facilitate surface binding (the affinity and avidity of the ligand-CLR interaction) and assessment of how ligand binding contributes to the whole phagocytosis process (e.g., by generating chimeric receptors with the ligand-binding extracellular domain fused to the transmembrane and intracellular part of a bona fide phagocytic CLR like Dectin-1). (B) The sufficiency of CLRs to direct phagocytosis (e.g., by determining the phagocytic capacity in non-phagocytic cell lines ectopically expressing the CLR in question). (C) The contribution of CLR to the functional consequences of phagocytosis after particle internalization (e.g., ROS production and antigen processing). We believe that this framework would provide clear approaches to comparing the functions and mechanisms of CLRs in phagocytosis and host-pathogen interactions.

References

Aderem A, Underhill DM (1999) Mechanisms of phagocytosis in macrophages. Annu Rev Immunol 17:593–623

Alan R, Ezekowitz B, Sastry K et al (1990) Molecular characterization of the human macrophage mannose receptor: demonstration of multiple carbohydrate recognition-like domains and phagocytosis of yeasts in cos-1 cells. J Exp Med 172:1785–1794

Allavena P, Chieppa M, Monti P et al (2004) From pattern recognition receptor to regulator of homeostasis: the double-faced macrophage mannose receptor. Crit Rev Immunol 24:179–192

Ariizumi K, Shen GL, Shikano S et al (2000) Identification of a novel, dendritic cell-associated molecule, dectin-1, by subtractive cDNA cloning. J Biol Chem 275:20157–20167

Azad AK, Torrelles JB, Schlesinger LS (2008) Mutation in the DC-SIGN cytoplasmic triacidic cluster motif markedly attenuates receptor activity for phagocytosis and endocytosis of mannose-containing ligands by human myeloid cells. J Leukoc Biol 84:1594–1603

Behler F, Steinwede K, Balboa L et al (2012) Role of mincle in alveolar macrophage-dependent innate immunity against mycobacterial infections in mice. J Immunol 189:3121–3129

Brown GD, Gordon S (2001) A new receptor for β-glucans. Nature 413:36–37

Brown GD, Herre J, Williams DL et al (2003) Dectin-1 mediates the biological effects of β-glucans. J Exp Med 197:1119–1124

Cambi A, Gijzen K, de Vries IJM et al (2003) The C-type lectin DC-SIGN (CD209) is an antigen-uptake receptor for *Candida albicans* on dendritic cells. Eur J Immunol 33:532–538

Casals C, García-Fojeda B, Minutti CM (2019) Soluble defense collagens: sweeping up immune threats. Mol Immunol

de Witte L, Abt M, Schneider-Schaulies S et al (2006) Measles virus targets DC-SIGN to enhance dendritic cell infection. J Virol 80:3477–3486

Elsori DH, Yakubenko VP, Roome T et al (2011) Protein kinase Cδ is a critical component of Dectin-1 signaling in primary human monocytes. J Leukoc Biol 90:599–611

Ferwerda B, Ferwerda G, Plantinga TS et al (2009) Human dectin-1 deficiency and mucocutaneous fungal infections. N Engl J Med 361:1760–1767

Freeman SA, Grinstein S (2014) Phagocytosis: receptors, signal integration, and the cytoskeleton. Immunol Rev 262:193–215

Geijtenbeek TBH, Kwon DS, Torensma R et al (2000a) DC-SIGN, a dendritic cell-specific HIV-1-binding protein that enhances trans-infection of T cells. Cell 100:587–597

Geijtenbeek TBH, Torensma R, Van Vliet SJ et al (2000b) Identification of DC-SIGN, a novel dendritic cell-specific ICAM-3 receptor that supports primary immune responses. Cell 100:575–585

Geijtenbeek TBH, Van Vliet SJ, Koppel EA et al (2003) Mycobacteria target DC-SIGN to suppress dendritic cell function. J Exp Med 197:7–17

Goodridge HS, Reyes CN, Becker CA et al (2011) Activation of the innate immune receptor Dectin-1 upon formation of a "phagocytic synapse". Nature 472:471–475

Goodridge HS, Underhill DM, Touret N (2012) Mechanisms of Fc receptor and Dectin-1 activation for phagocytosis. Traffic 13:1062–1071

Haider M, Dambuza IM, Asamaphan P et al (2019) The pattern recognition receptors dectin-2, mincle, and FcRγ impact the dynamics of phagocytosis of Candida, Saccharomyces, Malassezia, and Mucor species. PLoS ONE 14:e0220867

Herre J, Marshall ASJ, Caron E et al (2004) Dectin-1 uses novel mechanisms for yeast phagocytosis in macrophages. Blood 104:4038–4045

Huysamen C, Willment JA, Dennehy KM et al (2008) CLEC9A is a novel activation C-type lectin-like receptor expressed on BDCA3 + dendritic cells and a subset of monocytes. J Biol Chem 283:16693–16701

Iborra S, Izquierdo HM, Martínez-López M et al (2012) The DC receptor DNGR-1 mediates cross-priming of CTLs during vaccinia virus infection in mice. J Clin Invest 122:1628–1643

Ifrim DC, Bain JM, Reid DM et al (2014) Role of dectin-2 for host defense against systemic infection with *Candida glabrata*. Infect Immun 82:1064–1073

Ishikawa E, Ishikawa T, Morita YS et al (2009) Direct recognition of the mycobacterial glycolipid, trehalose dimycolate, by C-type lectin Mincle. J Exp Med 206:2879–2888

Janeway CA (1989) Approaching the asymptote? Evolution and revolution in immunology. Cold Spring Harb Symp Quant Biol 54:1–13

Jang SJ, Ohtani K, Fukuoh A et al (2009) Scavenger receptor collectin placenta 1 (CL-P1) predominantly mediates zymosan phagocytosis by human vascular endothelial cells. J Biol Chem 284:3956–3965

Kerrigan AM, Brown GD (2010) Syk-coupled C-type lectin receptors that mediate cellular activation via single tyrosine based activation motifs. Immunol Rev 234:335–352

Kimura Y, Inoue A, Hangai S et al (2016) The innate immune receptor Dectin-2 mediates the phagocytosis of cancer cells by Kupffer cells for the suppression of liver metastasis. Proc Natl Acad Sci USA 113:14097–14102

Kruskal BA, Sastry K, Warner AB et al (1992) Phagocytic chimeric receptors require both transmembrane and cytoplasmic domains from the mannose receptor. J Exp Med 176:1673–1680

Li X, Utomo A, Cullere X et al (2011) The β-glucan receptor dectin-1 activates the integrin Mac-1 in neutrophils via vav protein signaling to promote *Candida albicans* clearance. Cell Host Microbe 10:603–615

Liu W, Tang L, Zhang G et al (2004) Characterization of a novel C-type lectin-like gene, LSECtin: demonstration of carbohydrate binding and expression in sinusoidal endothelial cells of liver and lymph node. J Biol Chem 279:18748–18758

Lobato-Pascual A, Saether PC, Fossum S et al (2013) Mincle, the receptor for mycobacterial cord factor, forms a functional receptor complex with MCL and FcεRI-γ. Eur J Immunol 43:3167–3174

Lozach PY, Lortat-Jacob H, Lacroix De, de Lavalette A et al (2003) DC-SIGN and L-SIGN are high affinity binding receptors for hepatitis C virus glycoprotein E2. J Biol Chem 278:20358–20366

Luo B-H, Carman CV, Springer TA (2007) Structural basis of integrin regulation and signaling. Annu Rev Immunol 25:619–647

Ma J, Becker C, Lowell CA et al (2012) Dectin-1-triggered recruitment of light chain 3 protein to phagosomes facilitates major histocompatibility complex class II presentation of fungal-derived antigens. J Biol Chem 287:34149–34156

Ma YJ, Hein E, Munthe-Fog L et al (2015) Soluble collectin-12 (CL-12) is a pattern recognition molecule initiating complement activation via the alternative pathway. J Immunol 195:3365–3373

Matsumoto M, Tanaka T, Kaisho T et al (1999) A novel LPS-inducible C-type lectin is a transcriptional target of NF-IL6 in macrophages. J Immunol 163:5039–5048

McGreal EP, Rosas M, Brown GD et al (2006) The carbohydrate-recognition domain of Dectin-2 is a C-type lectin with specificity for high mannose. Glycobiology 16:422–430

Medzhitov R, Preston-Hurlburt P, Janeway CA (1997) A human homologue of the Drosophila Toll protein signals activation of adaptive immunity. Nature 388:394–397

Miyake Y, Toyonaga K, Mori D et al (2013) C-type Lectin MCL Is an FcRγ-coupled receptor that mediates the adjuvanticity of mycobacterial cord factor. Immunity 38:1050–1062

Ohtani K, Suzuki Y, Eda S et al (2001) The membrane-type collectin CL-P1 is a scavenger receptor on vascular endothelial cells. J Biol Chem 276:44222–44228

Oka K, Sawamura T, Kikuta KI et al (1998) Lectin-like oxidized low-density lipoprotein receptor 1 mediates phagocytosis of aged/apoptotic cells in endothelial cells. Proc Natl Acad Sci USA 95:9535–9540

Park CG, Takahara K, Umemoto E et al (2001) Five mouse homologues of the human dendritic cell C-type lectin, DC-SIGN. Int Immunol 13:1283–1290

Powlesland AS, Fisch T, Taylor ME et al (2008) A novel mechanism for LSECtin binding to Ebola virus surface glycoprotein through truncated glycans. J Biol Chem 283:593–602

Rajaram MVS, Arnett E, Azad AK et al (2017) M. tuberculosis-initiated human mannose receptor signaling regulates macrophage recognition and vesicle trafficking by FcRγ-Chain, Grb2, and SHP-1. Cell Rep 21:126–140

Rogers NC, Slack EC, Edwards AD et al (2005) Syk-dependent cytokine induction by dectin-1 reveals a novel pattern recognition pathway for C type lectins. Immunity 22:507–517

Saijo S, Ikeda S, Yamabe K et al (2010) Dectin-2 recognition of α-mannans and induction of Th17 cell differentiation is essential for host defense against *Candida albicans*. Immunity 32:681–691

Sato K, Yang XL, Yudate T et al (2006) Dectin-2 is a pattern recognition receptor for fungi that couples with the Fc receptor γ chain to induce innate immune responses. J Biol Chem 281:38854–38866

Savina A, Jancic C, Hugues S et al (2006) NOX_2 controls phagosomal ph to regulate antigen processing during crosspresentation by dendritic cells. Cell 126:205–218

Schraml BU, Van Blijswijk J, Zelenay S et al (2013) Genetic tracing via DNGR-1 expression history defines dendritic cells as a hematopoietic lineage. Cell 154:843–858

Serrano-Gómez D, Domínguez-Soto A, Ancochea J et al (2004) Dendritic cell-specific intercellular adhesion molecule 3-grabbing nonintegrin mediates binding and internalization of *Aspergillus fumigatus* conidia by dendritic cells and macrophages. J Immunol 173:5635–5643

Shimaoka T, Kume N, Minami M et al (2001) LOX-1 supports adhesion of gram-positive and gram-negative bacteria. J Immunol 166:5108–5114

Stahl PD, Rodman JS, Miller MJ et al (1978) Evidence for receptor-mediated binding of glycoproteins, glycoconjugates, and lysosomal glycosidases by alveolar macrophages. Proc Natl Acad Sci USA 75:1399–1403

Tassaneetrithep B, Burgess TH, Granelli-Piperno A et al (2003) DC-SIGN (CD209) mediates dengue virus infection of human dendritic cells. J Exp Med 197:823–829

Taylor PR, Brown GD, Herre J et al (2004) The role of SIGNR1 and the β-Glucan receptor (Dectin-1) in the nonopsonic recognition of yeast by specific macrophages. J Immunol 172:1157–1162

Taylor PR, Brown GD, Reid DM et al (2002) The β-Glucan receptor, Dectin-1, is predominantly expressed on the surface of cells of the monocyte/macrophage and neutrophil lineages. J Immunol 169:3876–3882

Taylor PR, Tsoni SV, Willment JA et al (2007) Dectin-1 is required for β-glucan recognition and control of fungal infection. Nat Immunol 8:31–38

Underhill DM, Rossnagle E, Lowell CA et al (2005) Dectin-1 activates Syk tyrosine kinase in a dynamic subset of macrophages for reactive oxygen production. Blood 106:2543–2550

Van De Wetering JK, Van Golde LMG, Batenburg JJ (2004) Collectins: players of the innate immune system. Eur J Biochem 271:1229–1249

van Liempt E, Bank CMC, Mehta P et al (2006) Specificity of DC-SIGN for mannose- and fucose-containing glycans. FEBS Lett 580:6123–6131

Watts C (2006) Phagosome neutrality in host defense. Cell 126:17–19

Willment JA, Marshall AS, Reid DM et al (2005) The human β-glucan receptor is widely expressed and functionally equivalent to murine Dectin-1 on primary cells. Eur J Immunol 35:1539–1547

Yamasaki S, Ishikawa E, Sakuma M et al (2008) Mincle is an ITAM-coupled activating receptor that senses damaged cells. Nat Immunol 9:1179–1188

Yamasaki S, Matsumoto M, Takeuchi O et al (2009) C-type lectin Mincle is an activating receptor for pathogenic fungus, *Malassezia*. Proc Natl Acad Sci USA 106:1897–1902

Yang Z, Li Q, Wang X et al (2018) C-type lectin receptor LSECtin-mediated apoptotic cell clearance by macrophages directs intestinal repair in experimental colitis. Proc Natl Acad Sci U S A 115:11054–11059

Zelensky AN, Gready JE (2005) The C-type lectin-like domain superfamily. FEBS J 272:6179–6217

Zhang JG, Czabotar PE, Policheni AN et al (2012) The dendritic cell receptor Clec9A binds damaged cells via exposed Actin Filaments. Immunity 36:646–657
Zhang P (2006) DC-SIGN (CD209) recognition of *Neisseria gonorrhoeae* is circumvented by lipooligosaccharide variation. J Leukoc Biol 79:731–738
Zhao D, Han X, Zheng X et al (2016) The Myeloid LSECtin Is a DAP12-coupled receptor that is crucial for inflammatory response induced by Ebola Virus Glycoprotein. PLoS Pathog 12: e1005487
Zhu LL, Zhao XQ, Jiang C et al (2013) C-type lectin receptors dectin-3 and dectin-2 form a heterodimeric pattern-recognition receptor for host defense against fungal infection. Immunity 39:324–334

C-type Lectins in Immunity to Lung Pathogens

Benjamin B. A. Raymond, Olivier Neyrolles and Yoann Rombouts

Contents

1 Introduction: C-Type Lectin Receptors and Their Importance to Lung Homeostasis 20
2 *Aspergillus fumigatus* 26
3 Pneumocystis 30
4 *Cryptococcus neoformans* 32
5 *Mycobacterium tuberculosis* 35
6 Pneumonia-Causing Bacteria 41
6.1 Streptococcus pneumoniae 41
6.2 Klebsiella pneumoniae 44
6.3 Haemophilus influenzae 46
7 Influenza Virus 47
8 Concluding Remarks 49
References 50

Abstract The respiratory tract is tasked with responding to a constant and vast influx of foreign agents. It acts as an important first line of defense in the innate immune system and as such plays a crucial role in preventing the entry of invading pathogens. While physical barriers like the mucociliary escalator exert their effects through the clearance of these pathogens, diverse and dynamic cellular mechanisms exist for the activation of the innate immune response through the recognition of pathogen-associated molecular patterns (PAMPs). These PAMPs are recognized by pattern recognition receptors (PRRs) that are expressed on a number of myeloid cells such as dendritic cells, macrophages, and neutrophils found in the respiratory tract. C-type lectin receptors (CLRs) are PRRs that play a pivotal role in the innate

B. B. A. Raymond (✉) · O. Neyrolles · Y. Rombouts
Institut de Pharmacologie et de Biologie Structurale, IPBS, Université de Toulouse, CNRS, UPS, 205 Route de Narbonne, 31400 Toulouse, France
e-mail: benjamin.raymond@ipbs.fr

Current Topics in Microbiology and Immunology (2020) 429: 19–62
https://doi.org/10.1007/82_2020_197

Published Online: 15 February 2020

immune response and its regulation to a variety of respiratory pathogens such as viruses, bacteria, and fungi. This chapter will describe the function of both activating and inhibiting myeloid CLRs in the recognition of a number of important respiratory pathogens as well as the signaling events initiated by these receptors.

1 Introduction: C-Type Lectin Receptors and Their Importance to Lung Homeostasis

Historically, pulmonary infections like influenza and tuberculosis have been some of the most devastating diseases to affect humankind. Thus, the lungs are under constant assault by foreign agents, whether they be pathogens or allergens, and as such have developed many mechanisms to combat these foes. As part of its first line of defense, the lung possesses a number of physical barriers such as the mucociliary clearance system that, through the action of mucins and cilia, expels invading pathogens from the airways. At the cellular level, lung epithelial cells, as well as alveolar macrophages (AMs) and dendritic cells (DCs) in intimate contact with the lung epithelium, represent a first barrier in the defense against a wide range of inhaled foreign agents by phagocytosing them and/or by secreting effector molecules such as surfactant proteins, cytokines, or antimicrobial peptides. Activation of these cells not only recruits other innate immune cells (neutrophils, monocytes, eosinophils, NK cells) but also contributes to the development and activation of the adaptive immune response. A considerably diverse group of proteins, termed pattern recognition receptors (PRRs), have evolved for the specific purpose of sensing conserved molecular motifs found on pathogens. The most well-studied groups of PRRs are the toll-like receptors (TLRs), NOD-like receptors (NLRs), RIG-I-like receptors (RLRs), and C-type lectins.

The superfamily of C-type lectins comprises more than 1000 members which can be either transmembrane or soluble proteins (Brown et al. 2018). C-type lectins are characterized by the presence of at least one C-type lectin-like domain. C-type lectins play a particularly important role in health and homeostasis as they can recognize both exogenous (non-self) ligands and endogenous (self) ligands. Initially, C-type lectins were characterized for their capacity to bind carbohydrates in a Ca^{2+}-dependent manner. However, it is now clear that many C-type lectins lack the motifs required for Ca^{2+}-dependent binding of carbohydrates and can recognize a broad range of ligands including lipids, proteins (e.g., F-actin in dead or dying cells) as well as organic and inorganic molecules (e.g., uric acid crystals and ice) (Ahrens et al. 2012; Neumann et al. 2014).

In the context of pulmonary infection, C-type lectins interact, in most cases, with carbohydrate moieties from invading bacteria, fungi, and viruses in order to sense the influx of these microbes from the environment. While soluble C-type lectins in the lung are primarily secreted by epithelial cells, membrane-bound C-type lectins, also named C-type lectin receptors (CLR), are highly expressed by innate immune cells. For example, and as will be described later in this chapter, AMs, DCs, and neutrophils express numerous CLRs that play important roles in the recognition and killing of pathogenic organisms. For an in-depth view of the expression profile of CLRs in cells of the lung, see Fig. 1. Thus, the function(s) of CLRs are diverse and can play opposing roles in the regulation of the immune response, especially in the lung.

Broadly, CLRs can be grouped according to the signaling motif present in their cytoplasmic tails as follows: (i) immunoreceptor tyrosine-based activating motif (ITAM)-coupled CLRs, (ii) hemi-ITAM-bearing CLRs, (iii) immunoreceptor tyrosine-based inhibitory motif (ITIM)-containing CLRs, and (iv) CLRs that do not possess an ITAM/ITIM signaling motif (Del Fresno et al. 2018). Briefly, ITAM-coupled CLRs such as Dectin-2 (*Clec6a*), macrophage inducible Ca^{2+}-dependent lectin receptor (Mincle/*Clec4e*), and dendritic cell activating receptor (DCAR/*Clec4b1*) associate with ITAM (YxxI/Lx_{6-12}YxxI/L)-containing adaptor proteins, such as Fc receptor γ (FcRγ) chain or DNAX-activation protein 12 (DAP12) (Del Fresno et al. 2018). Hemi-ITAM-bearing CLRs, such as Dectin-1 (*Clec7a*), contain a single YxxI/Lx_{6-12}YxxI/L motif in their cytoplasmic tail and finally, ITIM-containing CLRs namely, dendritic cell immunoreceptor (DCIR/*clec4a*) contain a single S/I/V/LxYxxI/V/L motif in their cytoplasmic tail (Del Fresno et al. 2018). Additionally, soluble CLRs, termed collectins, do not typically exhibit the inherent ability to transduce signals (Kuroki et al. 2007), instead playing prominent roles as opsonins that enhance recognition and phagocytosis of lung pathogens by PRRs (Kuroki et al. 2007). For an in-depth description of the CLRs described in this chapter, see Fig. 1 and Table 1. Generally, ITAM and hemi-ITAM CLRs are involved in mediating activating signals through phosphorylation of spleen tyrosine kinase (Syk). Depending upon CLR, Syk can generally (shown in detail in Fig. 1) transduce its signal through the CARD9/Bcl10/Malt-1 complex to activate NF-κB for the production of proinflammatory cytokines (Brown et al. 2018), while the less-studied ITIM CLRs are believed to transduce inhibitory signals through the recruitment of tyrosine phosphatases, such as Src homology region 2 domain-containing phosphatases (SHPs), which can inhibit STAT dimerization, and therefore, subsequent activation of interferon-stimulated genes (ISGs). Finally, CLRs that do not associate or possess an ITAM/ITIM, like dendritic cell-specific ICAM-grabbing non-integrin (DC-SIGN/*Clec4l*), and Langerin, instead possess internalization motifs in their cytoplasmic tails that are believed to mediate endocytosis of bound ligands.

While CLRs have largely been described for their role in recognizing fungi, this chapter will expand upon this to detail the major fungal, bacterial, and viral respiratory pathogens that CLRs recognize and/or mediate innate immunity against and the known involvement of these CLRs in disease progression.

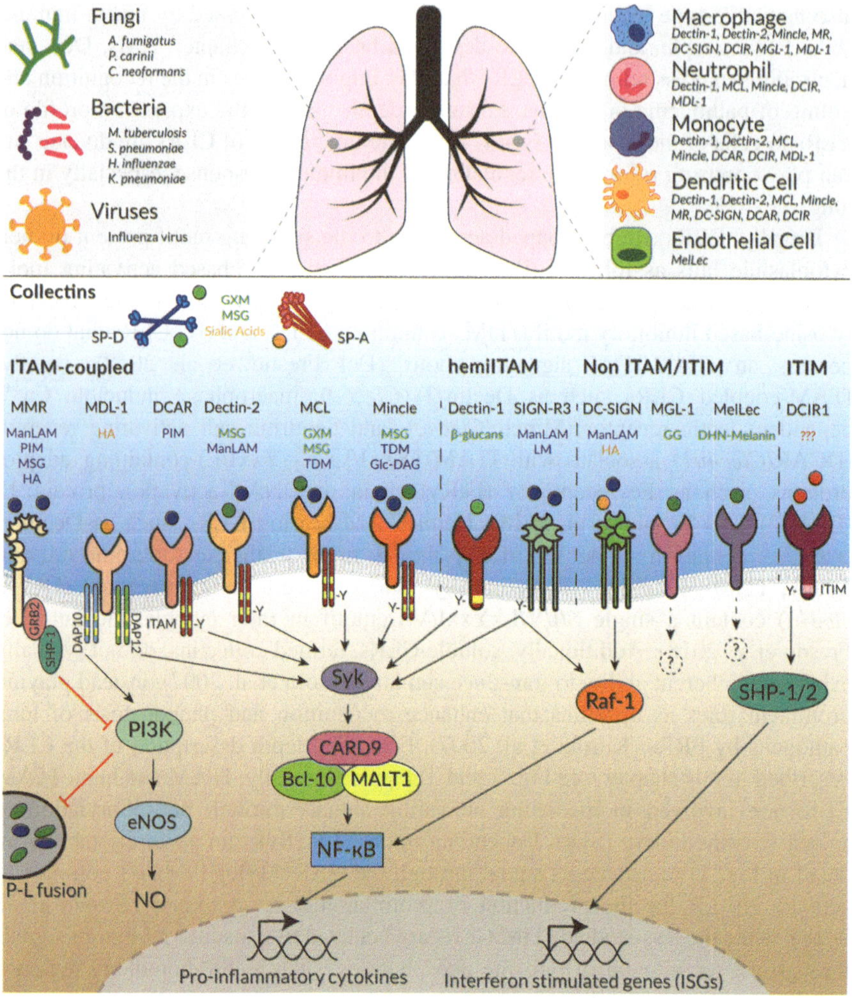

Fig. 1 Expression, ligand recognition, and known signaling pathways of the major CLRs (from human and mouse), as well as the pathogens that engage them, that are described in this chapter. CLRs are grouped according to (i) association with an ITAM-containing adaptor molecule (ITAM-coupled), (ii) presence of an ITAM in their cytoplasmic tail (HemiITAM), (iii) no ITAM/ITIM, or (iv) presence of an ITIM in their cytoplasmic tail. Ligands are color coded to depict the pathogen class that they belong to; fungi (green), bacteria (blue), or virus (yellow) that are also depicted in the top left panel of the figure. Broad expression patterns of CLRs discussed in this chapter on myeloid and endothelial cells are shown in the top right panel. Specific murine CLRs such as MDL-1, DCAR, SIGN-R3, MGL-1, and DCIR1 are included as the majority of the literature has focused on these murine homologs (otherwise, each CLR is meant to represent a broad overview of both human and murine orthologues). SIGN-R3 and DC-SIGN are depicted in their tetrameric form as ligand binding efficiency is enhanced by the formation of tetramers through tandem repeat regions present in their transmembrane domains. Unknown signaling pathways are depicted by question marks. While MMR has been shown to associate with FcRγ, hence its inclusion as an ITAM-coupled CLR, its signaling pathway in this regard has not been studied in depth and thus is not shown. Phagosome–lysosome fusion has been abbreviated to P-L fusion

Table 1 Description of the major CLRs described in this review

	CLR name	Gene names (notable murine homologs)	Associated respiratory pathogen(s)	Known pathogenic ligand(s)	Major consequence of knockout in mouse models
ITAM	Mannose receptor (**MMR**)	MRC1/*Clec13d*	*C. neoformans* *M. tuberculosis* *P. carinii* *S. pneumoniae*	ManLAM PIM MSG/gpA MPs PLY Hemagglutinin	↑ pulmonary infiltration of leukocytes (*P. carinii*) ↑ susceptibility to *C. neoformans*
	Dectin-2	*Clec6a*	*A. fumigatus* *C. neoformans* *M. tuberculosis* *P. carinii* *S. pneumoniae*	MSG/gpA ManLAM	↑ production of Th2 cytokines (*C. neoformans*) ↑ pulmonary bacterial burden (*M. avium* complex) ↑ susceptibility to *S. pneumoniae*
	Macrophage C-type lectin (**MCL**)	Dectin-3/*Clec4d*/ CLECSF8	*C. neoformans* *M. tuberculosis*	TDM GXM MSG/gpA	↑ pulmonary fungal burden and susceptibility (*C. neoformans*)
	Macrophage inducible C-type lectin (**Mincle**)	*Clec4e*	*M. tuberculosis* *P. carinii* *S. pneumoniae*	TDM Glc-DAG MSG/gpA	↑ pulmonary fungal burden (*P. murina*) ↑ pulmonary bacterial burden (Mtb & *S. pneumoniae*)
	Myeloid DAP12-associating lectin (**MDL-1**)	*Clec5a*	Influenza A Virus	Hemagglutinin	↑ survival against IAV
	Dendritic cell activating receptor (**DCAR**)	*Clec4b1* (only found in mouse)	*M. tuberculosis*	PIM	↑ pulmonary bacterial burden (Mtb)

(continued)

Table 1 (continued)

	CLR name	Gene names (notable murine homologs)	Associated respiratory pathogen(s)	Known pathogenic ligand(s)	Major consequence of knockout in mouse models
hemiITAM	Dectin-1	*Clec7a*	*A. fumigatus* *C. neoformans* *H. influenzae* *M. tuberculosis* *P. carinii*	Beta-glucan	↑ susceptibility to *A. fumigatus* ↓ pulmonary bacterial burden (Mtb)
ITIM	Dendritic cell immunoreceptor (**DCIR**)	*Clec4a* (*Clec4a1*, *Clec4a2*, *Clec4a3* & *Clec4a4*)	*M. tuberculosis*	Unknown	↓ pulmonary bacterial burden (Mtb) ↑ pulmonary infiltration of leukocytes (Mtb)
Non-ITAM/ ITIM	Dendritic Cell-Specific Intercellular adhesion molecule-3-Grabbing Non-integrin (**DC-SIGN**)	CD209/*Clec4l* (SIGN-R1 & SIGN-R3)	*A. fumigatus* Influenza A Virus *M. tuberculosis*	ManLAM Hemagglutinin MSG/gpA Galactomannan	↑ susceptibility to IAV
	Macrophage galactose-type lectin (**MGL**)	CD301/*Clec10a* (MGL-1/CD301a & MGL-2/CD301b)	*A. fumigatus* Influenza A Virus	Galactosaminogalactan	↑ susceptibility to *K. pneumoniae*
	Melanin-sensing C-type lectin receptor (**MelLec**)	*Clec1a*	*A. fumigatus*	DHN-melanin	↑ susceptibility to *A. fumigatus* (systemic model)
	Langerin	CD207/*Clec4k*	Influenza A Virus	Hemagglutinin	NA

(continued)

Table 1 (continued)

	CLR name	Gene names (notable murine homologs)	Associated respiratory pathogen(s)	Known pathogenic ligand(s)	Major consequence of knockout in mouse models
Soluble	SP-A & SP-D	SFTPA1/*COLEC4* SFTPD/*COLEC7*	*A. fumigatus* *C. neoformans* *P. carinii* *H. influenzae* Influenza A Virus *S. pneumoniae*	GXM Sialic acids MSG/gpA	↓ pulmonary fungal burden (SP-D; *C. neoformans*) ↓ bacterial clearance (*S. pneumoniae*) ↑ susceptibility to *H. influenzae* (SP-A) ↓ viral clearance (SP-A & SP-D)

2 *Aspergillus fumigatus*

Aspergillus fumigatus primarily causes invasive pulmonary aspergillosis, a severe fungal respiratory disease that largely affects immunocompromised individuals, particularly those with cystic fibrosis or undergoing transplantation, which can be fatal if not treated (Kousha et al. 2011). *A. fumigatus* accounts for 65% of *Aspergillus* spp-causing pulmonary infections (Warris 2014). Aspergillosis can be present in both an acute and chronic form, with the latter being the most common. In healthy individuals, the asexual spores, termed conidia, are transmitted through inhalation and are typically eliminated by antifungal mechanisms involving epithelial cells and immune cells (i.e., alveolar and interstitial macrophages, neutrophils, monocytes, dendritic cells, and $CD4^+$ T cells) (Lionakis et al. 2017). CLRs, expressed on or secreted by innate immune cells, are involved in the initial recognition of these conidia and the activation of cellular and innate immune mechanisms.

The most studied CLR in *Aspergillus* spp. infection is Dectin-1 that recognizes the β-(1,3)-D-glucan, a glucose polymer constituting about 50% of the dry weight of the fungal cell wall (Beauvais and Latge 2001; Brown and Gordon 2001). Dectin-1 has been shown to preferentially bind to "swollen conidia," a phenotypically distinct form to the inhaled "resting conidia," with minimal binding to germinating hyphae forms observed (Steele et al. 2005). Hohl et al. further explored this concept by demonstrating that β-(1,3)-D-glucans become surface-exposed during the switch from resting to swollen conidia and that Dectin-1 is recruited to the phagosomal membrane in AMs surrounding ingested swollen but not resting conidia (Hohl et al. 2005). These studies indicate that Dectin-1 plays an important role in early detection of *A. fumigatus* before germination. Indeed, Dectin-1 has proven to be essential for mounting an efficient immune response to *A. fumigatus*. Upon ligand binding, the ITAM-like motif of Dectin-1 undergoes tyrosine phosphorylation by Src kinases and initiates both SYK-dependent and SYK-independent signaling pathways. The latter triggers phagocytosis, regulates phagolysosomal and autophagosome maturation (Kyrmizi et al. 2013; Mansour et al. 2013), activates the inflammasome (Gross et al. 2009), and induces the production of inflammatory mediators (cytokines, chemokines, reactive oxygen species, and lipid mediators) (Drummond et al. 2011; Kerrigan and Brown 2011). Infection of wild-type (WT) mice with *A. fumigatus* is characterized by intense inflammation in the lung, accompanied by a rapid production of proinflammatory cytokines (TNF-α, IL-1α, IL-1β) and chemokines (CCL3/MIP-1α, CCL4/MIP-1β, CXCL1/KC) by AMs in lung homogenates (Werner et al. 2009; Steele et al. 2005). These inflammatory mediators, in particular TNF-α, CCL3, and CXCL1, allow the recruitment of neutrophils into the lung alveoli that interact with and kill the fungi. In Dectin-$1^{-/-}$ mice infected with *A. fumigatus*, the production of these cytokines/chemokines by AMs and the recruitment of neutrophils are strongly impaired (Werner et al. 2009). Also, neutrophils isolated from Dectin-$1^{-/-}$ mice fail to produce reactive oxygen species (ROS) in response to *A. fumigatus* infection that correlates with an inability of these cells to kill the pathogen (Werner et al. 2009). Consequently, mice lacking Dectin-1 expression exhibited a heightened susceptibility to *A. fumigatus* infection, with over 80% mortality observed within 5 days (Werner et al. 2009). Interestingly, Werner et al. also observed that IL-23 and IL-17 production was reduced in Dectin-$1^{-/-}$ mice at 1–2 days post-infection with *A. fumigatus*, and when IL-17A was neutralized in WT mice during infection, a significant increase in fungal pulmonary burden was observed (Werner et al. 2009). While IL-23 stimulates the production of IL-17A, the latter signals especially on lung epithelial cells to promote an innate-like acute immune response characterized by the production of cytokines promoting myeloid inflammation (i.e., IL-6, G-CSF), neutrophil-attracting chemokines (CXCL1, CXCL2, CXCL8), and antimicrobial peptides (McGeachy et al. 2019). IL-17 was shown to be produced by various immune cell population including γδ T cells, natural killer T cells, innate lymphoid cells, ($CD4^+$) T helper 17 cells (Th17), as well as some myeloid–lineage cells (especially neutrophils and eosinophils). At the early stage of *A. fumigatus* infection, Dectin-1 mediates IL-23 production by lung cells (most likely eosinophils) which trigger the secretion of IL-17A by neutrophils (Muntoni et al. 1993; Werner et al. 2011). Additionally, it is known that the development of the Th17 response, peaking at one week following intratracheal inoculation of mice with *A. fumigatus* conidia, is important for the clearance of the pathogen (Rivera et al. 2011). Dectin-1 on DCs plays a crucial role in this process by enhancing the production of cytokines (e.g., IL-23, IL-6, and IL-1β) required for the differentiation of Th17 cells, while simultaneously downregulating the secretion of the Th1-associated cytokine IL-12 (Dennehy et al. 2009). Collectively, this is indicative of a protective role of Dectin-1 by regulating the innate and adaptive immune responses during *A. fumigatus* infection.

In addition to Dectin-1, several other C-type lectins have been involved in the recognition of *A. fumigatus* including Dectin-2, DC-SIGN (Serrano-Gomez et al. 2004, 2005), Mincle (Zhao et al. 2017; Lin et al. 2017), Lox1 (Jiang et al. 2019; He et al. 2016; Gao et al. 2016), CD23 (Guo et al. 2018), MelLec (Stappers et al. 2018) and the soluble mannose-binding lectin (MBL) (Hogaboam et al. 2004; Neth et al. 2000) and the lung surfactant proteins (SP) SP-A and SP-D.

While Dectin-1 is involved in binding β-glucan on *A. fumigatus* conidia, the ITAM-coupled CLR Dectin-2 (*Clec6a*) has been shown to play a prominent role in the recognition of hyphal forms of *A. fumigatus* by plasmacytoid DCs (pDCs) (Loures et al. 2015). Competition binding assays demonstrated that mannans, but not laminarin, could inhibit the binding of *A. fumigatus* hyphae by human pDCs (Loures et al. 2015). Additionally, blocking antibodies specific for Dectin-2 could not only decrease antifungal immunity by pDCs, but effectively reduced the production of proinflammatory cytokines, such as TNF-α and IFN-α, in response to *A. fumigatus* infection (Loures et al. 2015). Interestingly, *A. fumigatus* hyphae, and not conidia, were found to activate Dectin-2 transduced reporter cells, indicating that Dectin-2 initiates signaling through the ITAM-containing coupled FcRγ (Loures et al. 2015). Furthermore, these pDCs were shown to be capable of producing extracellular traps; however, whether or not their production was mediated by Dectin-2 has not been demonstrated (Sun et al. 2013). In addition to pDCs, human AMs have been shown to express high levels of Dectin-2 after infection with *A. fumigatus* (Sun et al. 2013). A subsequent study has established that both hyphae and conidia could upregulate the expression of Dectin-2 in THP-1 macrophages and that this correlated with the production of proinflammatory cytokines such as IL-1β, IL-10, IL-23p19, and TNF-α (Sun et al. 2014). Moreover, the production of these cytokines was diminished when Dectin-2 expression was silenced in these cells, indicating that these observations were due to Dectin-2 upregulation and subsequent activation, further evidenced by an increase in the phosphorylation of Syk (Sun et al. 2014). Additionally, Sun et al. demonstrated that in the absence of Dectin-2, NF-κB activation was diminished (Sun et al. 2014). Furthermore, it was shown that human macrophages derived from THP-1 cell lines mediated the killing of *A. fumigatus* via phagocytosis and ROS production in a Dectin-2/Syk-dependent manner (Sun et al. 2014). Together, these studies suggest a prominent role for Dectin-2 in antifungal defenses in the lung.

Recently, the newly discovered CLR melanin-sensing C-type lectin receptor (MelLec/*Clec1a*) has been implicated in the immune response against *A. fumigatus*. MelLec is expressed mainly by non-myeloid cells, specifically by $CD31^{+}$ $EpCAM^{-}$ endothelial cells in mice (Stappers et al. 2018). MelLec specifically recognizes 1, 8-dihydroxynaphthalene (DHN)-melanin found on *A. fumigatus* conidia but not swollen conidia (Stappers et al. 2018). The removal of the hydrophobic rodlet layer that covers the surface of conidia leads to a drastic increase in binding of MelLec to *A. fumigatus* cells, suggesting that DHN-melanin is mainly hidden under this layer (Stappers et al. 2018). The susceptibility profile of $MelLec^{-/-}$ mice to *A. fumigatus* challenge is largely dependent upon the immune status of the mouse as well as the route of administration of conidia. Healthy $MelLec^{-/-}$ mice intratracheally infected with *A. fumigatus* did not exhibit any differences in survival compared with WT mice, but did display a reduced infiltration of neutrophils to the lung shortly after challenge (Stappers et al. 2018). Immunocompromised $MelLec^{-/-}$ mice infected intravenously with *A. fumigatus* displayed a higher susceptibility to infection, characterized by increased fungal burdens in a number of tissue sites (Stappers et al. 2018).

Some other lesser studied membrane-bound CLRs such as CD23 (*Clec4j*) and macrophage galactose-type lectin-1 (MGL-1) also play a role in immunity against *A. fumigatus*. MGL-1 binds to the *A. fumigatus* polysaccharide galactosaminogalactan (GG) that is present on the cell wall of germinating conidia and secreted from mycelium (Fontaine et al. 2011). While the implications of this remain to be seen, GG appears to promote aspergillosis and so MGL-1 may inhibit its immunomodulatory effects (Fontaine et al. 2011). CD23 is a type II transmembrane CLR that is mainly expressed on B cells but also on a variety of hematopoietic cells (monocytes, activated macrophages, follicular dendritic cells) and is known to bind IgE, CD21, CD11b, and CD11c (Bajorath and Aruffo 1996; Wurzburg et al. 2006).

$CD23^{-/-}$ mice are more susceptible to *A. fumigatus* infection, as are CD23/Rag1 double knockout (KO) mice indicating that the observed phenotype was due to CD23 expressed on innate immune cells rather than B or T cells (Guo et al. 2018). CD23 directly binds to *A. fumigatus*, and the engagement of CD23 by α-mannan and β-glucan leads to NF-κB activation through FcRγ in macrophages (Guo et al. 2018). These findings indicate the potentially important roles of understudied CLRs in immunity to fungal pathogens such as *A. fumigatus*.

The collectins are a family of soluble C-type lectin with a collagen-like domain that bind to pathogens and facilitate their clearance via phagocytosis, opsonization, or complement activation (Carreto-Binaghi et al. 2016). Within this family are the surfactant proteins A and D (SP-A and SP-D, respectively) that are produced by large alveolar cells (also known as type II pneumocytes) within the lung (Johansson et al. 1994). In particular, SP-A and SP-D have been well described for their roles in antifungal immunity through their binding to *A. fumigatus* conidia thereby causing agglutination and immobilization of the fungus (Madan et al. 1997a). SP-A and SP-D specifically recognize two glycoproteins from *A. fumigatus*, gp55 and gp45 (Madan et al. 1997b). However, it remains unknown whether other ligands exist. SP-A and SP-D can act as chemoattractants for neutrophils as well as opsonins, increasing the rate of phagocytosis and killing of *A. fumigatus* by AMs and neutrophils (Madan et al. 1997a). Furthermore, mice that were administered SP-A and SP-D exhibited a 60–80% increase in survival after a fatal challenge with *A. fumigatus* (Madan et al. 2001). Interestingly, mice deficient for SP-A were more resistant to challenge with *A. fumigatus* than WT mice, while SP-D-deficient mice were more susceptible (Madan et al. 2010). These studies suggest that while surfactant proteins do play a role in the *A. fumigatus* killing *in vitro*, their function(s) in vivo are more complex. In addition to its role as an opsonin, like SP-A and SP-D, another collection called mannan-binding lectin (MBL) can activate the complement pathway through the action of MBL-associated serine proteases (MASPs) (Matsushita and Fujita 1992). MBL binds strongly to *A. fumigatus* (Neth et al. 2000) and plays a protective role in immunity against the pathogen. Mice that were administered recombinant MBL (rMBL) exhibited significantly higher survival rates and lower pulmonary fungal loads than control groups in a murine model of invasive pulmonary aspergillosis (Kaur et al. 2007). These mice also had higher levels of proinflammatory cytokines such as TNF-α, IL-1β, and IFN-γ (Kaur et al. 2007). The concentration of rMBL correlated with the deposition of complement component 4b (C4b) on *A. fumigatus* conidia and hyphae which was not seen in cells not incubated with rMBL (Kaur et al. 2007). Furthermore, a higher level of engulfment of MBL-bound *A. fumigatus* conidia and oxidative burst by PMNs was observed; however, the latter was only true in the presence of MBL-deficient serum (Kaur et al. 2007). These findings indicate that the antimicrobial activities of MBL can occur in both complement-dependent and -independent mechanisms.

3 Pneumocystis

Pneumocystis pneumonia (PCP) is a fungal respiratory disease primarily caused by the opportunistic pathogen *Pneumocystis jirovecii,* previously referred to as *Pneumocystis carinii* (Sokulska et al. 2015). In actuality, *P. jirovecii* is a human-specific pathogen, whereas *P. carinii and Pneumocystis murina* infect rats and mice, respectively. PCP is mainly seen in immunocompromised patients such as those infected with HIV or undergoing chemotherapy and can become fatal if left untreated (Sokulska et al. 2015). There are two main morphological forms of all *Pneumocystsis* spp.: an asexual proliferative trophozoite form and a reproductive cyst form (Sokulska et al. 2015). Transmission of this pathogen occurs via the airborne route through the inhalation of these cysts (Sokulska et al. 2015). The cell wall of *Pneumocystis* spp. is decorated with a number of molecules that are recognized by ITAM-coupled CLRs including the most abundant cell surface glycoprotein (MSG/gpA) as well as β-(1,3)- and β-(1,6)-glucans (Matsumoto et al. 1989).

The macrophage mannose receptor 1 (MMR/*Mrc1*) is among the first CLRs characterized as a *Pneumocystis* receptor. MMR is a type I transmembrane CLR that is highly expressed on AMs and as its name suggests has an affinity for mannans. Ezekowitz et al. demonstrated that MMR mediates the uptake of *P. carinii* by both human and rat AMs (Ezekowitz et al. 1991). This MMR-dependent recognition and phagocytosis of *P. carinii* by AMs can be inhibited by competition with α-mannans, demonstrating that the interaction is carbohydrate-dependent. Subsequently, the mannose-rich MSG/gpA on the cell surface of *P. carinii* was identified as the main ligand for MMR, although the exact glycan epitope recognized remains unknown (Ezekowitz et al. 1991; Sassi et al. 2018). Recognition and phagocytosis of non-opsonized *P. carinii* by human AMs triggers the production of matrix metallopeptidase 9 (MMP-9) and neutrophil-chemoattractant CXCL8 (IL-8), but not that of common proinflammatory cytokines such as TNF-α, IL-6, or IL-1β. In fact, it has been shown that MMR on AMs contributes to the production of CXCL8 when it binds to non-opsonized *P. carinii,* inhibiting the release of proinflammatory cytokines, such as TNF-α (Zhang et al. 2005). Infection of macrophages with *P. carinii* also induces cell surface shedding of MMR, and soluble MMR was detected in cell-free alveolar fluid from infected humans (Fraser et al. 2000). This shed soluble form of MMR has been shown to bind *P. carinii* and reduces phagocytosis of the organism, acting as a so-called anti-opsonin (Fraser et al. 2000). In vivo studies of $MMR^{-/-}$ mice infected with *P. carinii* highlighted no differences in fungal loads when compared to WT mice. However, following depletion of $CD4^{+}$ T cells, they did exhibit a greater level of phagocyte infiltration and pulmonary pathology (Swain et al. 2003). Based on these data, it appears that there is a level of redundancy between CLRs, with the reduced phenotype observed in vivo.

In addition to MMR, Mincle, in particular, has been shown to specifically bind to MSG/gpA as well as to whole cyst forms of the fungus (Kottom et al. 2017).

Interestingly, bone marrow-derived macrophages (BMDMs) from Mincle$^{-/-}$ mice exhibited little binding to whole *P. carinii* (Stringer et al. 2002), thus demonstrating that Mincle is a major receptor for *P. carinii* in BMDMs (Kottom et al. 2017). It was shown that binding of *P. carinii* initiated signaling through Mincle-coupled FcRγ, as well as Syk phosphorylation (Kottom et al. 2017). However, in macrophages silenced for the expression of Mincle, Syk phosphorylation, albeit significantly lower, was still detected after addition of *P. carinii,* suggesting that other ITAM-coupled CLRs may also play a role (Kottom et al. 2017). Expression of Mincle is upregulated in mice in response to infection with *P. carinii* (Kottom et al. 2017, 2018). When Mincle$^{-/-}$ mice (immune suppressed by CD4^{+} T cell depletion) were infected with *P. murina*, a significantly higher burden of the organism was detected, and albeit no differences in mortality were noted (Kottom et al. 2017). Compared to WT mice, Mincle$^{-/-}$ mice infected with *P. murina* produce significantly more IL-1ra, IL-1α, and IL-1β, but less IL-17 (Kottom et al. 2017). Interestingly, the expression levels of other CLRs such as Dectin-1, Dectin-2, and Dectin-3 (also known as macrophage C-type lectin/MCL) in the lungs of Mincle$^{-/-}$ mice infected with *P. murina* were found to be significantly higher than WT mice, indicating that there is a possible compensatory effect in the absence of Mincle (Kottom et al. 2017).

Dectin-2 has been shown to interact with both MSG/gpA and the cell wall β-glucan of *Pneumocystis* spp. (Kottom et al. 2018, 2019). *P. carinii* cell wall β-glucans were able to induce Syk phosphorylation in AMs through Dectin-2, which was notably diminished in cells not expressing Dectin-2 (Kottom et al. 2018). However, a complete abolishment of Syk phosphorylation was not observed in Dectin-2$^{-/-}$ cells, indicating here also that other CLRs, such as Mincle and Dectin-1, may also be involved in signaling (Kottom et al. 2018). Of note, a significant decrease in proinflammatory cytokines, such as TNF-α and IL-6, was observed in Dectin-2$^{-/-}$ AMs stimulated with *P. carinii* cell wall β-glucans or MSG/gpA (Kottom et al. 2018). However, in contrast to the results obtained in MMR$^{-/-}$or Mincle$^{-/-}$ mice, Dectin-2 deficiency does not significantly impair *P. carinii* clearance in mice immune suppressed by CD4 depletion (Kottom et al. 2018). Likewise, in striking contrast to Mincle-deficient animals, Dectin-2$^{-/-}$ mice infected with *P. carinii* displayed a lower level of expression for other PRRs, for example, Dectin-1, Mincle, TLR2, and complement receptor 3 (CR3) (Kottom et al. 2018). While these results do not fully explain the functional role of Dectin-2 in the innate immune response to *P. carinii* infection, it does clearly demonstrate that there is interplay between the expression and function of various ITAM-coupled CLRs. Furthermore, these results indicate that CLRs, like Dectin-2 and Mincle, may share redundant functions in response to pulmonary pathogens such as *Pneumocystis*.

The surfactant proteins SP-A and SP-D, MCL/Dectin-3, and DC-SIGN have also been found to interact with *Pneumocystis* MSG/gpA, but their role in immune defense against this pathogen has not been studied (Kottom et al. 2019; McCormack et al. 1997; Vuk-Pavlovic et al. 2001). SP-A and SP-D, in particular, might be important because their expression is significantly increased during PCP (Atochina et al. 2001; Phelps et al. 1996; Qu et al. 2001).

Vassallo et al. demonstrated that β-glucans on the *P. carinii* cell wall induce potent TNF-α production in the lungs of mice after intratracheal challenge, which was partially driven by AMs (Vassallo et al. 2000). Dectin-1 has been shown to bind to *P. carinii* cysts and opsonize these cells (Rapaka et al. 2007) and more specifically, Dectin-1 interacts with *Pneumocystis* cell wall β-glucans (Saijo et al. 2007). An earlier study has demonstrated that Dectin-1 is required for phagocytosis and killing of non-opsonized *P. carinii* by AMs (Steele et al. 2003). Accordingly, it has been shown that the production of ROS by AMs infected with *P. carinii* cysts is dependent upon Dectin-1 (Saijo et al. 2007). In mice infected with *P. carinii*, Dectin-1 plays an important role during the early phase of infection. Indeed, as compared to WT mice, Dectin-1$^{-/-}$ animals showed twice as many *P. carinii* cysts in their lung at 1 or 2 weeks after intranasal administration of the pathogen (Saijo et al. 2007). However, both WT and Dectin-1$^{-/-}$ mice eventually cleared the *P. carinii* cysts at 3 weeks post-infection: a clearance mediated by adaptive immune cells. These data suggest that in the absence of an intact acquired immune system, such as in HIV-infected patients, Dectin-1 is required for protection against *Pneumocystis*. Accordingly, hydrodynamic injection of plasmid DNA or administration of an adenoviral vector expressing Dectin-1 fused to the Fc portion of mouse IgG1 (Dectin-1: mIgG1 Fc) in mice lacking B and T cells (SCID and Rag1$^{-/-}$ deficient mice) prior to infection enhanced the clearance of *Pneumocystis* organisms compared to mice that were not administered Dectin-1 (Rapaka et al. 2007; Ricks et al. 2013). Mechanistically, the recombinant Dectin-1: mIgG1 Fc binds to *P. carinii* cysts and increases the killing of the pathogen by macrophages in an FcγRII/III-dependent manner (Rapaka et al. 2007). Aside from its role in activating macrophages, *Pneumocystis* cell wall β-glucans also stimulate the production of CXCL2 by alveolar epithelial cells leading to the recruitment of neutrophils, which are seen as detrimental during severe PCP due to their involvement in the development of respiratory failure (Hahn et al. 2003).

4 *Cryptococcus neoformans*

Cryptococcus neoformans is an opportunistic fungal pathogen that can colonize a number of sites and primarily causes pneumonia and meningitis in immunocompromised patients. Unfortunately, *C. neoformans* is perhaps best known for being the leading fungal-associated cause of death in HIV/AIDS patients (Denning 2016). *C. neoformans* is found ubiquitously within the environment in a yeast-like form and primarily enters the body via inhalation of spores or small yeast cells. The clearance of the organism is highly dependent upon the concerted actions of the initial innate immune response and adaptive immunity (Heung 2017). During *C. neoformans* infection, Th1 immune responses are beneficial and promote fungal clearance, whereas Th2 immunity worsens disease (Wiesner et al. 2016). Th17 cells have been associated with both protection and increased disease, depending upon the model (Mukaremera and Nielsen 2017).

Like other fungal species, *C. neoformans* has a cell wall composed of chitin, chitosan, α-linked and β-linked glucans, and heavily mannosylated proteins. However, *C. neoformans* is unique among fungal pathogens for its ability to produce a characteristic polysaccharide capsule surrounding its cell wall. The capsule consists of two polysaccharides, glucuronoxylomannan (90% of the capsule mass) and glucuronoxylomannogalactan (around 10% of the capsule mass), with trace amount of mannosylated proteins (<1%). Based on antigenic differences in the polysaccharide capsules, *C. neoformans* has been classified into three serotypes A, D, and AD hybrid. The serotypes B and C have been attributed to a third variety of *C. neoformans* that is now considered a distinct species called *Cryptococcus gattii.*

Initial work has demonstrated that the phagocytosis of acapsular mutants of *C. neoformans* by murine macrophages can be inhibited by soluble mannans and β-(1,3)-glucans, thereby suggesting an important role of mannose and β-glucan receptors in this process (Cross and Bancroft 1995). Using blocking antibodies, the contribution of MMR/*Mrc1* to the recognition and ingestion of acapsular mutants of *C. neoformans* was confirmed in human DCs (Syme et al. 2002). However, neither the binding nor uptake of fully encapsulated *C. neoformans* and *C. gattii* by human and murine phagocytes are mediated by MMR, except maybe in human blood monocyte-derived macrophages differentiated with GM-CSF (Lim et al. 2018; Dan et al. 2008). Thus, it has been proposed that the capsular polysaccharides of C. *neoformans* conceal the cell wall components and prevent their recognition by CLRs. Aside from phagocytosis, MMR was shown to be crucial for the activation of adaptive immunity during *C. neoformans* infection. Using cell lines transfected with human CLRs, Mansour et al. showed that MMR and DC-SIGN strongly bind to cell wall mannoproteins (MP) of *C. neoformans* (Mansour et al. 2002, 2006). In human and mouse, *C. neoformans* infection induces a protective Th1 immune response including MP-specific $CD4^+$ T cells (Rohatgi and Pirofski 2015). During *C. neoformans* infection, the activation of adaptive immunity by DCs was shown to be dependent upon mannose receptors, such as MMR and DC-SIGN, for the efficient uptake and processing of MP antigens (Mansour et al. 2006). Consequently, $MMR^{-/-}$ mice infected with *C. neoformans* exhibit less MP-specific $CD4^+$ T cells, a higher pulmonary fungal load, and died at a faster rate compared with WT mice (Dan et al. 2008).

In addition to MMR and DC-SIGN, Dectin-2 is also a good candidate to interact with the abundant cell wall mannose structures of *Cryptococcus* spp. (McGreal et al. 2006). However, no study has been able to demonstrate the contribution of Dectin-2 in the binding and phagocytosis of cryptococcal spores by primary murine macrophages and DCs (Lim et al. 2018; Nakamura et al. 2015; Walsh et al. 2017). One can speculate that the capsule of *C. neoformans* may inhibit its recognition by Dectin-2. After *C. neoformans* infection, Dectin-$2^{-/-}$ mice exhibit a pronounced increase in the secretion of anti-inflammatory cytokine IL-10, Th2 cytokines (IL-4, IL-5, and IL-13) and mucus (Nakamura et al. 2015). However, Dectin-$2^{-/-}$ animals display no difference in their survival or in their ability to clear the pathogen from the lungs compared with WT mice (Walsh et al. 2017; Nakamura et al. 2015).

The role of the archetype β-glucan receptor, Dectin-1, in the phagocytosis of *C. neoformans* remains elusive. While some studies have reported that Dectin-1 binds to encapsulated *C. neoformans* spores and contributes to the phagocytosis of the fungi by macrophages and DCs (Lim et al. 2018; Giles et al. 2009), another study reports no involvement of this lectin (Walsh et al. 2017). In addition to the presence of capsular polysaccharides, the low amount of β-(1,3)-linked glucans (relative to β-(1,6)-glucans) in the *C. neoformans* cell wall could be unfavorable for the binding of Dectin-1 (Lim et al. 2018; Giles et al. 2009; Walsh et al. 2017; Nakamura et al. 2007). In mouse models, Dectin-1$^{-/-}$ animals infected with *C. neoformans* exhibit no differences in survival compared to WT counterparts (Nakamura et al. 2007).

While Dectin-1 and Dectin-2 do not appear to be required to trigger an effective immune response against C. *neoformans* in vivo, their downstream signaling molecule CARD9 seems essential (Huang et al. 2018; Yamamoto et al. 2014). This suggests that an ITAM-signaling CLR other than Dectin-1 and Dectin-2, or a concerted action of ITAM-signaling CLRs, is needed for phagocytosis and immunity against *Crytococcus* spp. The efficacy of CLR-mediated phagocytosis of *Cryptococcus* spp. and antifungal immunity may also depend on the type of capsular polyssaccharides and therefore on serotypes. In this context, MCL (i.e., Dectin-3) has been shown to play a crucial role in the initial immune response to some serotypes of *Cryptococcus* strains in mice. Capuzano et al. demonstrated that MCL$^{-/-}$ mice infected with *C. neoformans* strain H99 (serotype A) displayed no significant changes in neither survival nor pulmonary fungal burden when compared to WT mice (Campuzano et al. 2017). Consequently, they observed no difference in the infiltration of leukocytes to the lung during infection between WT and MCL$^{-/-}$ mice (Campuzano et al. 2017). They also found no differences in the production of cytokines in the lungs of WT and MCL$^{-/-}$ mice, nor in the ability of macrophages and DCs from these mice to internalize *C. neoformans* cells (Campuzano et al. 2017). Nonetheless, they did report that infection with the less virulent strain 52D of *C. neoformans* (serotype D) resulted in 20% reduction in the survival rate of MCL$^{-/-}$ mice (Campuzano et al. 2017). While the authors of this initial study have concluded that MCL does not contribute to immunity against *Cryptococcus* spp., a recent report showed that this lectin is essential for the early immune defense against specific serotypes of *Cryptococcus* spp. Thus, Huang et al. showed that a recombinant soluble MCL binds to the capsular glucuronoxylomannan (GXM) from all serotypes of *C. neoformans* and *C. gattii*. Nonetheless, only the GXM of the *C. neoformans* serotype AD and *C. gattii* serotype D induce a NF-κB and ERK-mediated production of proinflammatory cytokines (i.e., TNF-α, IL-6, IL-12p40) in murine BMDMs. Importantly, the secretion of these proinflammatory cytokines is dependent on the presence of MCL and CARD9 as demonstrated by the use of MCL$^{-/-}$ and CARD9$^{-/-}$ macrophages. Not only did MCL$^{-/-}$ mice exhibit a higher pulmonary fungal burden than WT mice after 3 days of *C. gattii* (serotype B) infection, but all infected mice died within 45 days, whereas 70% of WT mice persisted for at least 65 days post-infection (Huang et al. 2018). Likewise, as compared to WT mice, MCL$^{-/-}$ mice display a higher pulmonary fungal burden at 3 days post-infection with *C. neoformans* serotype AD but only after infection with the serotypes C and D. AMs are particularly important for the initial response to *Cryptococcus* spp. infection in the lung. In agreement, compared with WT mice, MCL$^{-/-}$ animals displayed a significant reduction in the number of AMs present in the lung after *C. gattii* serotype B and *C. neoformans* serotype AD infections. Furthermore, AMs from MCL$^{-/-}$ mice displayed a marked decrease in their ability to kill *C. gattii* serotype B and *C. neoformans* serotype AD in vitro when compared to WT cells (Huang et al. 2018).

While SP-A and SP-D were both shown to bind to *C. neoformans*, they only bound to acapsulated forms of the fungi (Walenkamp et al. 1999; Schelenz et al. 1995). Additionally, SP-D but not SP-A was capable of causing agglutination of *C. neoformans* cells, and this aggregation could be inhibited by the addition of GXM (Schelenz et al. 1995). This would suggest that capsular GXM plays a role in subverting SP-D-mediated agglutination of *C. neoformans* cells (Schelenz et al. 1995). Notably, SP-D$^{-/-}$ mice infected with *C. neoformans* displayed a decreased pulmonary fungal load and longer survival times, suggesting that SP-D renders mice more susceptible to infection (Geunes-Boyer et al.2012). At the cellular level, this was partly explained by the fact that SP-D-bound *C. neoformans* cells were more resistant to ROS production by macrophages (Geunes-Boyer et al. 2012). SP-A$^{-/-}$ mice, on the other hand, did not display any differences in pulmonary fungal loads or cytokine production compared to WT mice after infection with *C. neoformans* (Giles et al. 2007). This once again demonstrates the convoluted roles played by surfactant proteins in pulmonary fungal infection.

From these data, it is evident that CLRs have an important role in innate immunity against *Cryptococcus*; however, due to the marked differences between the numerous strains and serotypes, the conceptualization of experiments is crucial to properly elucidate their roles.

5 *Mycobacterium tuberculosis*

Mycobacterium tuberculosis is the causative agent of tuberculosis, a chronic respiratory disease that has persisted throughout humanity for millennia. In 2018 alone, it was reported that tuberculosis was responsible for 1.5 million deaths (Harding 2019). The persistence of tuberculosis can be explained by the growing number of cases of extensively drug-resistant tuberculosis and the lack of fully efficacious vaccines. *M. tuberculosis* is transmitted via microaerophilic droplets that are inhaled into the lung and enter the lower respiratory tract. Once the bacteria enter the alveolar space, they are encountered by AMs (Philips and Ernst 2012), which act as the first line of defense, and are believed to be important in determining the outcome of infection. These cells possess numerous cellular mechanisms to eliminate the intracellular mycobacteria, such as the generation of phagolysosomes and reactive oxygen and nitrogen species, and *M. tuberculosis* in-turn, possesses numerous mechanisms to subvert these defenses, utilizing these cells as a replicative niche (Mayer-Barber and Sher 2015). Other myeloid cell subsets recruited into the lung (e.g. neutrophils, monocytes, interstitial macrophages, and DCs) are also important regulators of the immune response to *M. tuberculosis* (Srivastava et al. 2014). Activation or infection of these cells leads to the production of various cytokines, chemokines, and lipid mediators that help fight the infection, but also trigger tissue pathology. Among cytokines, TNF-α, IL-12p70, and IFN-γ have proven to be essential for the control of *M. tuberculosis* growth. TNF-α promotes the production of reactive oxygen and nitrogen species by macrophages, and as consequence, the intracellular killing of *M. tuberculosis*, IL-12p70 produced by mature DCs is essential for the priming of *M. tuberculosis*-specific $CD4^+$ T helper 1 (Th1) cells in the lymph node. Following expansion and migration within the lung, Th1 cells respond to *M. tuberculosis* antigens by secreting IFN-γ which activates, in concert with TNF-α, the microbicidal functions of macrophages. The recognition of *M. tuberculosis*-associated PAMPs by PRRs is crucial for the host–pathogen interplay described above. The involved PRRs include TLRs (i.e., TLR-2, -4, and -9) but also several C-type lectins which interact with the unique cell envelope-associated (glyco) lipids and (lipo)polysaccharides produced by *M. tuberculosis* (Brightbill et al. 1999; Bafica et al. 2005; Means et al. 1999; von Meyenn et al. 2006; Jones et al. 2001; Underhill et al. 1999; Lugo-Villarino et al. 2011; Goyal et al. 2016).

MMR is probably the first CLR that has been found to interact with *M. tuberculosis* through binding to mycobacterial mannose-capped lipoarabinomannan (ManLAM), lipomannan (LM), and phosphatidylinositol mannosides (PIMs) (Goyal et al. 2016; Schlesinger et al. 1994, 1996; Schlesinger 1993). The binding of *M. tuberculosis* to MMR on human phagocytes is thought to provide a favorable outcome for the pathogen. Indeed, the interaction of MMR on human macrophages with ManLAM of *M. tuberculosis* has been shown to limit phagosome–lysosomal fusion and to promote the intracellular survival of the pathogen (Kang et al. 2005; Astarie-Dequeker et al. 1999). In human DCs, the engagement of MMR by ManLAM decreases the production of IL-12p70 (induced by TLR4 agonists), thereby supporting an inhibitory role of MMR in the polarization of Th1 effector cells (Nigou et al. 2001; Chieppa et al. 2003). Interestingly, $MMR^{-/-}$ mice infected with *M. tuberculosis* did not display any differences in survival and pulmonary bacterial loads (at 4 and 12 weeks post-infection) compared to WT mice. This suggests that MMR may play a redundant role in the wider scope of *M. tuberculosis* disease progression (Court et al. 2010; Schlesinger et al. 1994, 1996; Schlesinger 1993). The divergence of results between the murine and human contexts may also be explained by species-specific differences in the MMR signaling pathway. Recently, Rajaram et al. demonstrated that the engagement of human MMR by *M. tuberculosis* results in the phosphorylation of SHP-1 and its recruitment to *M. tuberculosis*-containing phagosomes (Rajaram et al. 2017). Furthermore, the inhibition or knockdown of SHP-1 in human monocyte-derived macrophages infected with *M. tuberculosis* resulted in an increase in phagosome-lysosome fusion, as well as a marked reduction in bacterial growth. This observation was shown to occur through the inhibition of the Class III phosphoinositide 3-kinase (PI3K) Vps34 by SHP-1 that subsequently led to a reduction in phagosome–lysosome fusion (Rajaram et al. 2017). To elucidate the differences in mouse and human MMR signaling, transgenic mice expressing human MMR (hMMR mice) were constructed and tested for their ability to transduce signals in BMDMs and AMs following infection with *M. tuberculosis*. BMDMs and AMs from hMMR mice exhibited an increase in SHP-1 phosphorylation following infection, in addition to an increase in *M. tuberculosis* growth compared with cells from WT mice (Rajaram et al. 2017). Collectively, these data indicate that in human macrophages, MMR signals through SHP-1 to inhibit phagosome–lysosome fusion and that this mechanism does not seem to exist in murine macrophage. This finding may explain why previous studies have not observed differences between WT and $MMR^{-/-}$ mice after infection with *M. tuberculosis*.

DC-SIGN is a type II transmembrane CLR that does not possess an ITAM or ITIM in its cytoplasmic tail. In the lung, it is mainly expressed on subsets of DCs, AMs, or alternatively activated macrophages (Tailleux et al. 2003, 2005; Lugo-Villarino et al. 2018). Its cytoplasmic tail contains internalization motifs that have been shown to be involved in endocytosis and subsequent antigen presentation to $CD4^{+}$ T cells (Engering et al. 2002; Geijtenbeek et al. 2000). There are eight murine homologs of human DC-SIGN, but none of them displays identical functions (e.g. glycan specificity, internalization, intracellular trafficking, and signaling) to those of human DC-SIGN (Garcia-Vallejo and van Kooyk 2013; Park et al. 2001). Therefore, for the purpose of this chapter, we will primarily focus on studies pertaining to human DC-SIGN. DC-SIGN has an affinity for fucosylated glycans and mannosylated glycoconjugates (Mitchell et al. 2001; Geijtenbeek et al. 2000). *M. tuberculosis* has been shown to bind to HeLa cells expressing DC-SIGN up to 25 times more than cells that did not (Tailleux et al. 2003). Additionally, when using blocking antibodies against DC-SIGN, MMR, or CR3, only antibodies against DC-SIGN significantly inhibited the binding of *M. tuberculosis* to human blood monocyte-derived DCs (Tailleux et al. 2003). Notably, while the same pattern of inhibition was not observed in human blood monocyte-derived macrophages (Tailleux et al. 2003), subsequent work demonstrated that DC-SIGN is a major receptor for *M. tuberculosis* in human AMs taken from patients with TB (Tailleux et al. 2005). Furthermore, it has been established that DC-SIGN recognizes several glycoconjugates of *M. tuberculosis* including ManLAM (Tailleux et al. 2003; Geijtenbeek et al. 2003; Pitarque et al. 2005; Maeda et al. 2003), mannose-capped arabinomannan, LM, mannosylated glycoproteins (Pitarque et al. 2005), PIM (Driessen et al. 2009), α-glucan (Geurtsen et al. 2009), but also non-glycosylated proteins such as chaperone protein DnaK and glyceraldehyde-3 phosphate dehydrogenase (Carroll et al. 2010), to name a few. The binding of ManLAM to DC-SIGN has been shown to induce the production of the immunosuppressive cytokine IL-10, but only when cells were simultaneously stimulated through TLR4 (Geijtenbeek et al. 2003). Due to the differences in signaling and internalization motifs between DC-SIGN, and notable murine homologs such as SIGN-R1 and SIGN-R3, Schaefer et al. created human DC-SIGN transgenic mice (DC-SIGN mice) (Schaefer et al. 2008). Bone marrow-derived dendritic cells (BMDCs) from WT to DC-SIGN mice showed no difference in IL-10 production after infection with *M. bovis* BCG (Schaefer et al. 2008). However, a decrease in IL-12p40 was observed in BMDCs from DC-SIGN mice when compared with WT mice, which subsequently translated into a decrease in IFNγ secretion by $CD4^+$ T cells primed by these cells in co-culture experiments (Schaefer et al. 2008). While the number of $CD4^+$ T cells in the lungs of WT and DC-SIGN mice infected with *M. tuberculosis* did not differ significantly, there was a significant reduction in $CD8^+$ T cells in the lungs of DC-SIGN mice (Schaefer et al. 2008). Notably, DC-SIGN mice exhibited reduced lung pathology and prolonged survival times after challenge with a high dose (2000 CFU) of *M. tuberculosis* when compared with WT mice (Schaefer et al. 2008). These findings would indicate that human DC-SIGN differs in quite a few ways to its murine homologs and that it may have evolved to dampen the immune response to lessen tissue damage upon infection with *M. tuberculosis*. These types of studies need to be taken into consideration when using mouse models to study CLRs in the context of *M. tuberculosis* especially when specific orthologues do not exist.

The theme of crosstalk between CLRs and other PRRs, in response to *M. tuberculosis*, is not distinct to DC-SIGN and has also been demonstrated between MMR and TLR4 (Nigou et al. 2001), SIGN-R3 and TLR2 (Tanne et al. 2009) and even between MMR, DC-SIGN, and Dectin-1 (Lugo-Villarino et al. 2018; Zenaro et al. 2009). On the crosstalk between DC-SIGN and Dectin-1, it was shown that in the context of *M. tuberculosis*, DC-SIGN may negatively act on Dectin-1-driven production of proinflammatory cytokines in human, IL-4 activated, monocyte-derived macrophages (Lugo-Villarino et al. 2018).

Despite the absence of β-glucans in the mycobacterial cell envelop, Dectin-1 has been shown to interact with several mycobacteria species. Initially, Yadav and Schorey demonstrated that the production of proinflammatory cytokines (e.g. TNF-α, IL-6) in murine BMDMs infected with non-pathogenic (*M. smegmatis*) or facultative-pathogenic mycobacterial species (*M. bovis* BCG) was Dectin-1-dependent and that this effect was abolished in the absence of TLR2 (Yadav and Schorey 2006). However, this phenotype could not be reproduced with the virulent *M. tuberculosis* strain H37Rv (Yadav and Schorey 2006). In contrast, a later study firmly established that Dectin-1 does interact with pathogenic mycobacteria including *M. tuberculosis* strain H37Rv (Rothfuchs et al. 2007); however, the specific ligand remains unknown. Additionally, the expression of Dectin-1 in human A549 airway epithelial cells has been shown to be induced by *M. tuberculosis* in a TLR2-dependent manner and leads to phagocytosis and the production of proinflammatory cytokines (Lee et al. 2009). From a functional point of view, the interaction of *M. tuberculosis* with Dectin-1 on murine and human DCs has been shown to regulate the production of proinflammatory cytokines and the priming of T cells by DCs. Thus, using Dectin-1 blocking antibodies, Rothfuchs et al. found that activation of Dectin-1 is required for the production of IL-12p40 and IL-12p70 by murine splenic DCs following infection with *M. tuberculosis* (Rothfuchs et al. 2007). Using a similar methodology, others found that *M. tuberculosis* triggers the production of TNF-α, IL-6, IL-1β, and IL-23 by human blood monocyte-derived DCs through the engagement of Dectin-1 (Zenaro et al. 2009). In agreement with the involvement of these cytokines in the polarization of T cells into Th1/Th17, the stimulation of Dectin-1 enables DCs to instruct the generation of Th1/Th17 cells. Interestingly, concomitant engagement of MMR or DC-SIGN inhibits the Dectin-1-induced Th17 response, while promoting the Th1 response to eliminate *M. tuberculosis* (Zenaro et al. 2009). In vivo, Marakalala et al. demonstrated that Dectin-$1^{-/-}$ mice infected with *M. tuberculosis* strain H37Rv exhibited a reduced bacterial burden in the lung compared to WT mice; however, no difference in the production of proinflammatory cytokines or the overall survival of the mice was observed (Marakalala et al. 2011). A tendency toward a reduction in IL-10 production in Dectin-$1^{-/-}$ mice infected with *M. tuberculosis* was deemed to partially explain the reduction in bacterial burden in the lung (Marakalala et al. 2011); however, more in-depth studies will be needed to confirm this.

Dectin-2 can bind directly to *M. tuberculosis*, specifically recognizing ManLAM (Yonekawa et al. 2014), and is the sole *M. tuberculosis*-derived ligand for this receptor (Decout et al. 2018). ManLAM stimulation of BMDCs produced a robust proinflammatory response that was abolished in BMDCs from Dectin-$2^{-/-}$ mice, indicating that ManLAM mediates the inflammatory response in a Dectin-2-dependent manner (Yonekawa et al. 2014). Interestingly, the production of immunosuppressive cytokines such as IL-10 was also diminished in BMDCs from the same mice (Yonekawa et al. 2014). Given DC-SIGN and MMR are also known to bind ManLAM, it was examined to see if this phenotype could be explained by a potential crosstalk between Dectin-2 and MMR, or a murine homolog of DC-SIGN, SIGN-R1. However, neither SIGN-R1 nor MMR were found to play a role in the observed phenotype (Yonekawa et al. 2014). Dectin-$2^{-/-}$ mice that were infected with *Mycobacterium avium* complex displayed an increase in bacterial burden as well as lung pathology compared with WT mice (Yonekawa et al. 2014), suggesting that while it has not been shown for *M. tuberculosis*, Dectin-2 may play a role in the clearance of the pathogen.

Trehalose-6,6′-dimycolate (TDM, also known as cord factor) is a mycobacterial cell wall glycolipid that has been studied extensively as an immunomostimulatory molecule (Ishikawa et al. 2009; Schoenen et al. 2010; Werninghaus et al. 2009). It was simultaneously observed that TDM is a ligand for Mincle (Ishikawa et al. 2009; Schoenen et al. 2010). Additional mycobacterial ligands for Mincle such as glucose monomycolate, diacyl-trehalose, and triacyl-trehalose have also been identified (Decout et al. 2017). While there have been contradicting findings regarding the outcome of Mincle$^{-/-}$ mice infected with *M. tuberculosis* (Heitmann et al. 2013), there is some data to suggest that mice deficient for Mincle are more susceptible to infection with *M. tuberculosis* (Behler et al. 2012; Lee et al. 2012). The in vitro effects of Mincle deficiency have proven less contradictory, as TDM-dependent activation of Mincle in murine BMDMs was demonstrated to be important for the production of proinflammatory cytokines as well as nitric oxide (Ishikawa et al. 2009).

Additionally, neutrophil recruitment to the lung has been shown to occur in a Mincle-dependent manner and was enhanced through crosstalk with TLR2 (Lee et al. 2012). This study, in particular, found that Mincle played a protective role against *M. tuberculosis* in vivo, demonstrating higher bacterial loads in the lungs of Mincle$^{-/-}$ mice. Like Mincle, MCL can also recognize TDM (Miyake et al. 2013), and it was further demonstrated that these two CLRs, through FcRγ, form a functional complex with one another (Lobato-Pascual et al. 2013). Additionally, TDM-induced activation of MCL leads to Mincle expression in BMDMs that occurs through CARD9/Bcl10/Malt-1-dependent activation of NF-κB (Zhao et al. 2014). There is also data that suggests that Mincle is crucial for surface expression of MCL in BMDCs (Miyake et al. 2015), and while this association is known, its role in *M. tuberculosis* immunity remains unclear. MCL$^{-/-}$ mice are more susceptible to infection with *M. tuberculosis*, whereby they had a 20% reduction in survival rates compared with WT mice within 6 weeks post-infection (Wilson et al. 2015). While no differences were observed at earlier time points, MCL$^{-/-}$ mice infected with *M. tuberculosis* displayed increased bacterial loads, proinflammatory cytokines, and pulmonary lesions at 4 months post-infection (Wilson et al. 2015). MCL$^{-/-}$ mice infected with *M. bovis* BCG exhibited less association between mycobacteria and pulmonary leukocytes suggesting that MCL plays an important role in their binding to these cells (Wilson et al. 2015). Notably, data mining experiments of publicly available datasets showed that polymorphisms in the rs4304840 allele were associated with susceptibility to TB, with further experiments demonstrating that this polymorphism affected the transmembrane region of MCL, reducing its surface expression (Wilson et al. 2015). These findings warrant further interrogation as they reveal an important role for MCL, specifically the incidence of polymorphisms, in anti-TB immunity.

Recently, DCAR has been added to the growing list of CLRs that recognize *M. tuberculosis*. It was shown that DCAR recognizes the mycobacterial envelope glycolipid, acylated phosphatidylinositol dimannosides (AcPIM2 and Ac2PIM2) (Toyonaga et al. 2016). Notably, DCAR is only found in mouse, with blood DC antigen 2 (BDCA-2) being the closest homolog in human (Kanazawa et al. 2003). DCAR is an FcRγ-coupled activating receptor expressed on monocyte-derived cells that can transduce signal upon engagement by *M. tuberculosis* (Toyonaga et al. 2016). DCAR$^{-/-}$ mice infected intraperitoneally with *Mycobacterium bovis* exhibited reduced infiltration of monocyte-derived inflammatory cells, as well as a reduction in IFNγ and IL-12, indicative of a protective role of DCAR during mycobacterial infection. Furthermore, bacterial loads in the lung were significantly higher in DCAR$^{-/-}$ mice infected with *M. bovis* compared with WT mice. These results were also recapitulated using *M. tuberculosis* (Toyonaga et al. 2016), supporting the role of DCAR in promoting the Th1 response that is crucial for the clearance of *M. tuberculosis*.

While studies pertaining to ITIM-containing CLRs are limited, they present an interesting view of a subset of CLRs that while opposing, potentially work in concert with ITAM-containing CLRs to balance the innate immune response. DCIR is currently the only ITIM-bearing CLR shown to play a role in immunity to lung pathogens and has recently been studied for its role in regulating the type I IFN response to infection with *M. tuberculosis* (Troegeler et al. 2017). It was shown that mice deficient for a murine homolog of DCIR (mDCIR1) had a defect in type I IFN signaling, attributed to a reduction in STAT phosphorylation (Troegeler et al. 2017). Additionally, infected mDCIR1$^{-/-}$ mice exhibited enhanced bacterial killing, which was evidenced by a reduced bacterial load in the lung as well as a higher level of lung infiltration by leukocytes (Troegeler et al. 2017). This effect was attributed to an increase in IL-12 production in the lungs, likely due to the observed reduction in type I IFN signaling (Troegeler et al. 2017). These data would indicate that even though it is thought of as an inhibitory receptor, DCIR's role in signaling may be more complex. Additionally, unlike other CLRs, the ligand(s) of DCIR remain unknown. The literature describing the role of soluble CLRs in immunity to *M. tuberculosis* is diverse and not within the scope of this chapter. For a detailed review on the role of these lectins in immunity to *M. tuberculosis*, see Goyal et al.

When considering the role that CLRs play in the recognition of *M. tuberculosis* and the subsequent innate immune mechanisms that are activated in response to this, a number of important points need to be highlighted. In terms of ligand recognition, the biological significance of these interactions is oftentimes complicated, evidenced by DC-SIGN being able to recognize non-TB complex ManLAMs but not to the cells themselves (Pitarque et al. 2005). This would indicate that while a wide range of ligands for CLRs have been identified, the biological significance of these interactions must be further explored. Additionally, due to the constant observation of crosstalk between CLRs and other PRRs in the recognition of *M. tuberculosis*, there are perhaps complex synergies that have not yet been uncovered and need to be considered. Finally, in addition to the inherent resistance of C57BL/6 mice to infection with *M. tuberculosis* (Chackerian et al. 2002), knockouts of numerous CLRs in mice do not routinely exhibit the same phenotypes as their ex vivo counterparts post-challenge, demonstrating the complexities that exist when using these mouse models. Despite these complexities, however, CLRs have overwhelmingly been shown to play a crucial role in immunity against *M. tuberculosis* and with the further development of more appropriate models to study host-pathogen interactions, these mechanisms may be further dissected.

6 Pneumonia-Causing Bacteria

6.1 *Streptococcus pneumoniae*

S. pneumoniae is a gram-positive bacterium and the leading cause of community-acquired pneumonia (Brown 2012). *Pneumococcal pneumonia*, which mostly affects children and the elderly, is known to disseminate and cause invasive disease (van der Poll and Opal 2009). Upon inhalation and colonization of the upper respiratory tract, *S. pneumoniae* uses numerous mechanisms to subvert the host defenses, such as capsular polysaccharide and lipoteichoic acid, which have been shown to protect it from the lytic effects of neutrophils extracellular traps (Wartha et al. 2007). Based on the chemical composition of, and antigenic differences in the capsular polysaccharide, more than 97 pneumococcal capsular serotypes of *Streptococcus pneumoniae* have been identified with differing potential to cause invasive pneumococcal disease.

Among CLRs, Dectin-2, SIGN-R1, MMR and MGL-1, have been identified as receptors for the capsular polysaccharides of specific *S. pneumoniae* serotypes (McGreal et al. 2006; Miyasaka et al. 2013; Zamze et al. 2002; Saba et al. 2009; Akahori et al. 2016; Kang et al. 2004). The involvement of Dectin-2 in immunity against *S. pneumoniae* was first established in mice immunized with a 23-valent pneumococcal polysaccharide vaccine (PPSV), which is effective against 23 different pneumococcal capsular types. Upon PPSV immunization, WT mice produce protective antibodies, including IgM and IgG, targeting the pneumococcal polysaccharides. However, the quantity of anti-pneumococcal polysaccharide antibodies was dramatically reduced in Dectin-$2^{-/-}$ mice immunized with PPSV (Miyasaka et al. 2013). Mechanistically, Dectin-2 is required in vitro for the production of IL-12 by DCs stimulated by PPSV and in vivo for the synthesis of IFN-γ by NKT cells, which are most probably activated by IL-12 secreted by DCs. As NKT-derived IFN-γ contributes to antibody production induced by PPV immunization, it explains why Dectin-$2^{-/-}$ mice have less anti-pneumococcal polysaccharide antibodies compared to WT mice (Miyasaka et al. 2013). In a follow-up study, the role of Dectin-2 in the early immune response against a clinically relevant strain of *S. pneumoniae* (serotype 3) was investigated. Upon infection with serotype 3 of *S. pneumoniae*, Dectin-$2^{-/-}$ mice exhibit a heightened mortality compared to WT mice (Akahori et al. 2016). This effect was attributed to the twofold decrease in the ability of Dectin-$2^{-/-}$ neutrophils to phagocytose *S. pneumoniae* cells and to the reduced production of serotype-specific IgG against capsular polysaccharide, which is known to be important for opsonic phagocytosis (Akahori et al. 2016).

The receptor SIGN-R1, a mouse homolog of human DC-SIGN, has been involved in the uptake of *S. pneumoniae* capsular polysaccharides (serotypes 3, 14, 23, and 2) by marginal zone macrophages in the spleen (Kang et al. 2004; Koppel et al. 2005, 2008). This uptake triggers the production of anti-pneumococcal IgM by marginal zone B cells (Koppel et al. 2005, 2008). As a consequence, mice lacking SIGN-R1 are significantly more susceptible to lung infection with *S. pneumoniae* (serotype 3) than WT mice as revealed by the higher bacterial load and the enhanced lung inflammation (higher level of proinflammatory cytokines/chemokines, more edema, and recruited neutrophils) at 48 h post-infection (Koppel et al. 2005).

As mentioned above, MMR was initially found to bind *S. pneumoniae* capsular polysaccharides (McGreal et al. 2006; Zamze et al. 2002). Nevertheless, the cholesterol-dependent cytolysin pneumolysin (PLY) of *S. pneumoniae* was recently identified as a novel MMR ligand. Interaction of active PLY with MMR mediates the uptake of *S. pneumoniae* by human blood-monocyte-derived DCs and inhibits the fusion of pathogen-containing vacuoles with lysosomes. Binding of PLY to MMR upregulates the expression of the suppressor of cytokine signaling 1

(SOCS1) and as such downregulates the synthesis of proinflammatory cytokines (specifically TNF-α, IL-1β, IL-12) by DCs infected with non-encapsulated *S. pneumoniae*. Consequently, MMR-silenced DCs infected with *S. pneumoniae* or stimulated with recombinant PLY elicited higher IFN-y and lower IL-4 from naïve T helper cells. The interaction of PLY, with MMR in human DCs and BMDMs is also needed for the induction of regulatory T cells (Tregs) (expressing Foxp3 and IL-10). Together, these data suggest an important role of MMR in the induction of T cell tolerance during pneumococci infection. In vivo, compared to controls, mice treated with an anti-MMR antibody had higher levels of TNF-α and IL-12, with lower levels of IL-10, in the bronchoalveolar lavage fluid (BALF) at 6 h post-infection with *S. pneumoniae*. In agreement with this increased inflammation, the anti-MMR antibody treatment reduced bacterial numbers in the airway at 6 h post-infection with non-encapsulated *S. pneumoniae*. Likewise, MMR$^{-/-}$ mice displays higher levels of proinflammatory cytokines (TNF-α, IL-6) but reduced levels of anti-inflammatory IL-10 and TGF-β in the nasopharynx at 6 h and 1 day post-infection with *S. pneumoniae*. Consequently, MMR$^{-/-}$ mice had a significantly lower bacterial count in the nasopharynx at 7 and 14 days after infection. Altogether, these data suggest that the interaction of *S. pneumoniae* PLY with MMR on airway macrophages and DCs dampens the cytokine responses and promotes T cell tolerance, thereby allowing the establishment of intracellular residency by pneumococci.

The role of Mincle in the innate immune response against *S. pneumoniae* seems to be somewhat limited from earlier research, where it was found that macrophages and neutrophils from the lungs of Mincle$^{-/-}$ mice displayed no difference in their ability to phagocytose or kill *S. pneumoniae* cells (Rabes et al. 2015). Moreover, Mincle$^{-/-}$ mice were equally susceptible to *S. pneumoniae* infection, with no differences in cytokine production, survival rate or BALF bacterial load (Rabes et al. 2015). However, in a subsequent study, AMs and neutrophils have been observed to have significantly higher expression of Mincle after infection with *S. pneumoniae* (Behler-Janbeck et al. 2016). Originally, Mincle had been shown to bind to *S. pneumoniae* (Rabes et al. 2015), and following studies have demonstrated that it specifically recognizes the glycolipid glucosyl-diacylglycerol (Glc-DAG), which is able to stimulate Mincle reporter cells to comparable levels as TDM derived from *Mycobacterium Tuberculosis* (Behler-Janbeck et al. 2016). Similarly, Mincle directly engages *Streptococcus pyogenes* (group A streptococcus) through its lipoteichoic acid anchor glycolipid monoglucosyldiacylglycerol that induces proinflammatory cytokine production in BMDCs (Imai et al. 2018). Additionally, it has been shown that Glc-DAG simulation initiated Mincle-dependent proinflammatory cytokine release in murine AMs (Behler-Janbeck et al. 2016). Consistent with previous studies, Behler-Janbeck et al. found that Mincle$^{-/-}$ mice infected with *S. pneumoniae* showed no difference in survival percentage when compared with WT mice, but they did exhibit higher bacterial load in the BALF as well as increased lung pathology (Behler-Janbeck et al. 2016). However, this was not attributed to altered leukocyte infiltration or the ability of these cells to phagocytose and kill the bacteria (Behler-Janbeck et al. 2016). Furthermore, Mincle$^{-/-}$ mice infected with *S. pneumoniae* had a significantly altered profile of pro- and anti-inflammatory cytokines, such that their production was prolonged over the course of challenge, whereas WT mice exhibited a "peak" and then "diminishment" of these cytokines, concurrent with the clearance of disease (Behler-Janbeck et al. 2016).

The surfactant proteins have also been shown to play a role in the innate immune response against *S. pneumoniae*. SP-A binding to *S. pneumoniae* has been shown to enhance phagocytosis through the upregulation of scavenger receptor-A at the surface of AMs (Kuronuma et al. 2004). SP-D also has the ability to bind and agglutinate *S. pneumoniae* cells; however, it does not enhance neutrophil-mediated killing of the pathogen (Jounblat et al. 2004). A subsequent study by the same group demonstrated that SP-$D^{-/-}$ mice infected with *S. pneumoniae*, displayed a diminished ability to clear the bacteria from the lungs compared with WT mice, with an overall decrease in the number of infiltrating neutrophils seen at 48 h post-infection (Jounblat et al. 2005).

6.2 *Klebsiella pneumoniae*

K. pneumoniae is a gram-negative bacterium that comprises the normal flora of the skin and gut and upon inhalation is capable of causing serious pneumonia that can lead to sepsis. One of the main virulence factors of *K. pneumoniae* is its capsule, with unencapsulated strains significantly less virulent than encapsulated strains (Paczosa and Mecsas 2016). Notably, *K. pneumoniae* is described for its ability to vary the expression of its capsular polysaccharide through phase variation, adding to its virulence (Favre-Bonte et al. 1999).

The initial innate immune response against *K. pneumoniae* is important for preventing the establishment of infection and phagocytosis of the bacteria by AMs is a key event. Early studies have demonstrated that *K. pneumoniae* strains which contain Manα-(1,2/3)-Man or Rhaα-(1,2/3)-Rha sequences in their capsular polysaccharide are recognized by the macrophage MMR or with lung SP-A, thus participating in the attachment, phagocytosis, and killing of the bacteria by the phagocytes (Kabha et al. 1997; Keisari et al. 1997; Athamna et al. 1991). A follow-up study demonstrated that while encapsulated strains of *K. pneumoniae* expressing dimannose were readily recognized by MMR and induced the production of IL-6, IL-12, and TNF-α, unencapsulated forms were not (Keisari et al. 2001). A contrasting study, however, demonstrated that MMR could only bind to LPS and not capsular polysaccharide from *K. pneumoniae* (Zamze et al. 2002), suggesting that the findings of Keisari et al. could be attributed to another CLR expressed by macrophages. In contrast to MMR and SP-A, SP-D binds unencapsulated *K. pneumoniae* strains, through the recognition of the mannose-rich O-polysaccharides from LPS (Sahly et al. 2002), and only agglutinating these unencapsulated forms of the bacteria leading to their phagocytosis by AMs (Ofek et al. 2001). Encapsulated strains, however, were not readily agglutinated

(Ofek et al. 2001), indicating that phase variation of capsular polysaccharide is an important virulence mechanism for avoiding SP-D-mediated opsonization. SP-D-bound unencapsulated strains of *K. pneumoniae* cells induced robust induction of proinflammatory cytokines (Keisari et al. 2001) demonstrating that SP-D may have evolved as a mechanism to combat phase variation in *K. pneumoniae*.

Besides phagocytosis by macrophages, neutrophils, in particular, are crucial for *K. pneumoniae* clearance and the prevention of sepsis in mice (Balamayooran et al. 2012). However, hyperinflammation and exacerbated recruitment of neutrophils can cause fatal lung damages. Sharma et al. demonstrated that Mincle$^{-/-}$ mice infected with a sublethal dose of *K. pneumoniae* were more susceptible to infection than WT mice and display significantly higher bacterial loads in the lung and systemic bacterial dissemination within a few days post-infection (Sharma et al. 2014). Consequently, all Mincle$^{-/-}$ mice succumbed to infection at 6 days post-infection, while the majority of infected WT mice cleared the infection by day 5. Mincle$^{-/-}$ mice infected with *K. pneumoniae* present a significant increase of TNF-α, IL-1β, IFN-γ, CCL3 and IL-6 in the lungs compared with WT mice (Sharma et al. 2014). This hyperinflammatory response was also accompanied by an increase in neutrophil accumulation in the lungs of Mincle$^{-/-}$ mice, due especially to an increase in production of the neutrophil-recruiting chemokines CXCL2 and CXCL6 (Sharma et al. 2014). While higher neutrophil recruitment should promote killing of the bacteria, Sharma et al. found that neutrophils from Mincle$^{-/-}$ mice are deficient in their ability to phagocytose non-opsonized *K. pneumoniae* cells and to produce NETs (Sharma et al. 2014). A subsequent study demonstrated that Mincle plays an important role in NET formation by neutrophils in a murine pneumonic sepsis model of *K. pneumoniae* infection. Mincle$^{-/-}$ mice infected with *K. pneumoniae* exhibited a significant reduction in the proportion of neutrophils with NETs, but displayed a similar level of ROS production (Sharma et al. 2017). Furthermore, this Mincle-mediated NET production by neutrophils was shown to occur through the activation of autophagy; neutrophils from Mincle$^{-/-}$ mice were unable to effectively activate autophagy through processing of LC3 (Sharma et al. 2017). Interestingly, this effect was not due to canonical CARD9/Bcl10/Malt-1 signaling, but to Mincle-mediated activation of mTOR (Sharma et al. 2017). Syk has been shown to regulate mTOR signaling (Carnevale et al. 2013; Leseux et al. 2006), and so this may explain the observed phenotype in this study as Mincle$^{-/-}$ neutrophils displayed higher levels of phosphorylated Syk after stimulation with phorbol myristate acetate (PMA) (Sharma et al. 2017). These findings suggest that Mincle plays a protective role against *K. pneumoniae* through bacterial clearance by neutrophils.

MCL is a relatively poorly understood CLR that is expressed mainly by peripheral blood neutrophils and monocytes (Graham et al. 2012). MCL has been shown to induce signaling through Syk in addition to being able to induce respiratory burst in macrophages (Graham et al. 2012). MCL can also form a heterodimeric receptor complex with Dectin-2 that results in high-affinity recognition of fungal wall compounds such as α-mannans and for the effective triggering of innate immunity (Zhu et al. 2013). MCL $^{-/-}$ mice are highly susceptible to *K. pneumoniae* infection, with an observed increase in pulmonary bacterial burden as well as hyperinflammation compared with WT mice (Steichen et al. 2013). Similarly to Mincle$^{-/-}$ mice, this hyperinflammatory response in MCL$^{-/-}$ mice was characterized by an increase in proinflammatory cytokines such as TNF-α, IFN-γ, and IL-6, severe lung pathology and increased accumulation of neutrophils (Steichen et al. 2013). CXCL1 and CXCL6 levels were also increased, suggesting that MCL plays an important role in regulating neutrophil turnover that is important for resolving inflammation to prevent further lung damage (Steichen et al. 2013).

MGL-1 is a type II transmembrane protein whose expression is increased in the lungs of *K. pneumoniae*-infected mice (Jondle et al. 2016). MGL-1$^{-/-}$ mice were more susceptible to infection with *K. pneumoniae* compared with WT mice, but did not have any differences in pulmonary bacterial burden (Jondle et al. 2016). Similarly to the observed phenotype in Mincle$^{-/-}$ and MCL$^{-/-}$ mice, MGL-1$^{-/-}$ mice infected with *K. pneumoniae* displayed a hyperinflammatory response and increased neutrophil accumulation and infiltration in the lung (Jondle et al. 2016). Collectively, it is clear that membrane-bound CLRs play an important role in regulating the innate immune response against *K. pneumoniae* infection for both clearance and resolution of inflammation.

6.3 *Haemophilus influenzae*

H. influenzae is a gram-negative bacterium that is largely responsible for opportunistic infections in immunocompromised patients. It was originally shown that Dectin-1 was important for the activation of eosinophils and the entry of unencapsulated strains, often referred to as nontypeable *Haemophilus influenzae* (NTHi) into human monocytic and epithelial cells (Ahren et al. 2003, 2001). The gene expression of Dectin-1 on in vitro cultured human primary small airway epithelial and normal bronchial epithelial cells has been shown to be upregulated upon stimulation of TLR3, IFNAR, and IFNGR, while surface expression of the CLR is only upregulated after stimulation with the TLR3 agonist poly(I·C) (Heyl et al. 2014). Importantly, Dectin-1 is expressed on cells within the bronchial and alveolar epithelium and pleura of the lung (Heyl et al. 2014). Infection of healthy human bronchial epithelial cells with NTHi was found to elicit a comparable level of IL-6 and IL-8 production to that noted upon activation of Dectin-1 by its agonistic ligand, zymosan. Additionally, this response could be inhibited by antagonistic compounds against Dectin-1 (Heyl et al. 2014). While Dectin-1 does not appear to be an attachment site for NTHi, affinity pull-down experiments have demonstrated that Dectin-1 does recognize components of the bacterium (Heyl et al. 2014). Interestingly, the entry of NTHI into the well characterized cellular model for respiratory infections, A549 lung carcinoma cells, was shown to be dependent upon the expression of Dectin-1 (Heyl et al. 2014).

SP-A and SP-D have both been shown to play a role in innate immunity to *H. influenzae* to some degree. SP-A or SP-D deficient mice infected with *H. influenzae* displayed increased immune cell recruitment and inflammation, and

SP-A$^{-/-}$ mice were more susceptible to infection (LeVine et al. 2000). Notably, AMs from SP-A$^{-/-}$ mice were less adept at killing *H. influenzae*, which was shown to be attributed to a decrease in ROS production (LeVine et al. 2000).

While the literature is relatively limited pertaining to the involvement of CLRs in bacterial pneumonia, when compared with fungi, the studies presented here point to a fundamental role that these receptors play in the recognition and clearance of both gram-positive and gram-negative pathogens.

7 Influenza Virus

Viral respiratory infections are responsible for numerous epidemics and pandemics (Iwasaki and Pillai 2014). Due to their intimate association with host cells, it is important to understand the complex interactions that occur during the different stages following the inhalation of viral particles. Influenza virus is an RNA virus that causes influenza, a mild to severe respiratory disease that usually only causes death in immunocompromised people such as children and the elderly. The IAV envelope contains two main glycoproteins, the hemagglutinin (HA) and the neuraminidase (NA), which bind to sialic acids linked to glycoproteins and glycolipids at the host cell surface. HA mediates the attachment of IAV to cells, triggers viral entry and contributes to the fusion of the virus with endosomes. HA binding specificity is one of the most important factors that determine host tissues tropism, including interspecies adaptation and transmissibility. NA is an enzyme that cleaves sialic acids and releases HA-based attachment. Specifically, NA promotes viral infiltration into the protective mucosa and enables the release of progeny virions from the cell surface. Based on the 18 known HA and 11 NA glycoproteins, IAVs can be further classified into subtypes (e.g., H1N1). Besides the contribution of viral envelop glycoproteins, several CLRs have also been implicated in IAV recognition and internalization into cells. The involvement of CLRs in IAV attachment and cell entry has often been studied in sialic acid deficient cells lines (e.g., Lec2 cells derived from CHO cells) in order to bypass the HA-mediated binding of virus to the cells. Whenever possible, the results were validated in primary cells, such as macrophages and DCs, through the use of, for instance, blocking antibodies and gene silencing approaches. MMR and MGL-1 were among the first CLRs to be identified as being implicated in the endocytosis of some IAV strains into mouse macrophages (Reading et al. 2000; Ng et al. 2014; Upham et al. 2010). Of the human CLRs, DC-SIGN (CD209), L-SIGN (DC-SIGNR and CD209L), as well as Langerin (CD207), have been shown to act both as an attachment site and as an endocytic receptor for influenza A virus (IAV) (Hillaire et al. 2013; Gillespie et al. 2016; Londrigan et al. 2011; Wang et al. 2008; Ng et al. 2016). MMR, DC-SIGN, L-SIGN, and Langerin are mannose-specific lectins that bind to mannose-rich glycans present on HA and NA. As such, their binding to IAVs depends on the type and degree of glycosylation of the envelop glycoproteins, which vary among virus strains and subtypes. Importantly, these lectins differ with respect to their tissue distribution implying that their contribution to IAV recognition and entry is cell-type specific. As previously mentioned, MMR and DC-SIGN are mainly expressed in the lung by human AMs and subpopulations of DCs (Tailleux et al. 2003, 2005; Lugo-Villarino et al. 2018). L-SIGN is expressed by alveolar epithelial cells and endothelial cells, whereas Langerin is expressed by Langerhans cells residing in the upper respiratory tract (van der Vlist et al. 2011; Jeffers et al. 2004).

In addition to their role as binding and endocytic receptors, some CLRs have also been shown to have an important role in driving anti-viral immunity. A recent study by Palomino-Segura et al. demonstrated that in a murine model for influenza, inflammatory DCs expressing SIGN-R1 are recruited to the site of infection and bind to influenza virus (Palomino-Segura et al. 2019). Specifically, it was shown that SIGN-R1 recognizes N-glycans on the HA glycoprotein (Palomino-Segura et al. 2019). In order to investigate whether SIGN-R1 mediates signaling upon recognition of influenza virus, SIGN-R1$^{-/-}$ mice were examined for chemokine production. When compared to WT mice, SIGN-R1$^{-/-}$ mice were shown to produce less chemokines, such as CCL5, CXCL9, and CXCL10, and inflammatory DCs were shown to be responsible for the production of these chemokines (Palomino-Segura et al. 2019). To further explore these findings, Palomino-Segura et al. demonstrated that this reduction in chemokines led to a reduction in the recruitment of natural killer (NK) cells, which are vital for viral clearance (Palomino-Segura et al. 2019). Finally, it was eloquently shown that SIGN-R1$^{-/-}$ mice infected with influenza displayed lower survival rates, higher viral titers, and a higher percentage of weight loss, which was all comparable to the effects of NK cell depletion as well as inflammatory DCs depletion (Palomino-Segura et al. 2019). Collectively, these results demonstrate that SIGN-R1-expressing inflammatory DCs recognize viral HA resulting in the production of chemokines that recruit NK cells to the site of infection and are necessary for viral clearance.

Myeloid DAP12-associating lectin (MDL)-1 (*Clec5a*) is type II transmembrane receptor mainly expressed by monocytes, macrophages, and neutrophils (Bakker et al. 1999). Upon ligand recognition, its short intracellular domain recruits the transmembrane adaptor proteins DAP-10 and DAP-12 (10-KDa and 12-kDa DNAX-activating proteins) (Bakker et al. 1999; Inui et al. 2009). While DAP-12 contains an ITAM motif and activates the Syk and ZAP-70 tyrosine kinase pathways, DAP10 signaling is thought to be mediated by phosphatidylinositol 3 kinase (PI3K) (Bakker et al. 1999; Inui et al. 2009; Lanier et al. 1998; Cheung et al. 2011). MDL-1 is purported to be a CLR that mainly recognizes mannose and fucose moieties from *Flavivirus* (Chen et al. 2008). To decipher the role played by MDL-1 in immunity to influenza, it was demonstrated that MDL-1 could bind to HA glycoproteins derived from an H5N1 isolate of IAV (Teng et al. 2017). Although MDL-1 plays no role in viral uptake in human monocyte-derived macrophages, it induces the production of proinflammatory cytokine (e.g. TNF-α, IFNα, and IL-6) in a Syk-dependent manner (Teng et al. 2017). This effect is independent of IAV subtypes, indicating that MDL-1 interacts with a range of different IAV subtypes. These results were recapitulated in MDL-1$^{-/-}$ mice infected with influenza, which displayed lower levels of proinflammatory cytokines and reduced immune cell infiltration into the lung (Teng et al. 2017). Interestingly, in a low-dose challenge, MDL-1$^{-/-}$ mice exhibited a lower production of cytokines as well as a better survival than WT mice, indicating that an overzealous immune response may be detrimental during early stages of infection (Teng et al. 2017).

The collectins SP-A and MBL bind to influenza virus and effectively inhibit virus-mediated agglutination of red blood cells, with MBL specifically binding to neuraminidase (Malhotra et al. 1994). The interaction between SP-A and IAV is believed to occur via the recognition of terminal sialic acids on N-linked glycans (Benne et al. 1997). Furthermore, SP-A was shown to enhance opsonic phagocytosis of IAV by AMs (Benne et al. 1997), and SP-A$^{-/-}$ mice had a significantly reduced ability to clear IAV compared with WT mice (Hawgood et al. 2004). Subsequent studies have demonstrated that cross-linked trimeric forms of SP-D could also bind IAV and induce viral agglutination (Tecle et al. 2008). In vivo studies using SP-D$^{-/-}$ mice infected with IAV demonstrated that mice deficient for SP-D exhibited reduced viral clearance and increased production of proinflammatory cytokines in the lung compared with WT mice (LeVine et al. 2001). Neutrophils from SP-D$^{-/-}$ mice possessed a reduced level of myeloperoxidase activity against IAV that is crucial for the anti-viral response (LeVine et al. 2001). SP-D levels within the lung are increased upon IAV infection (LeVine et al. 2001), exemplifying its role in the early innate immune response against this virus.

8 Concluding Remarks

From this chapter, it is clear that CLRs play a prominent role in innate immunity against pulmonary pathogens such as fungi, bacteria, and viruses. This not only encompasses the recognition of these pathogens, but the initiation of numerous innate immunity mechanisms namely, phagocytosis, complement activation, and cytokine production. For example, the archetype β-glucan receptor Dectin-1 modulates immunity to *A. fumigatus* and plays a largely protective role against infection, with Dectin-1$^{-/-}$ exhibiting heightened mortality. While this is one example from the rich pool of literature detailing the role of activating CLRs and how they respond to a diverse array of respiratory pathogens, there is limited data on so-called inhibitory receptors like DCIR and Clec12a, with only the former being shown to play a role in immunity against *M. tuberculosis*. It was shown that the deletion of Dcir1 in mice conferred a stronger ability to eliminate *M. tuberculosis* from the lung, likely due to a diminished type I IFN response. Following this, it is important to note that a large portion of our understanding of CLRs and their role in innate immunity is based upon in vivo mouse work, and in some cases, their pattern of expression, ligand specificities, and orthologues can differ greatly between mouse and human. This is most apparent for DC-SIGN, whose closest murine homologs exhibit differing patterns of expressions, signaling pathways, and ligand recognition. In these cases, the creation of humanized mouse models is a good tool to dissect the shared/differing functions of these CLRs.

The main themes ascertained from the current research concerning the regulation of the innate immune response to respiratory pathogens by CLRs are complex with many likely redundant processes. This is demonstrated by numerous findings, such as the role of MMR during *P. carinii* infection, where knockouts in specific genes do not oftentimes yield robust phenotypes in vivo. Additionally, it is important to note that CLRs exhibit crosstalk between each other, and therefore, single gene knockout studies are not the best means to elucidate these responses. A prime example of this is the inhibition of Dectin-1-driven Th17 responses by crosstalk between MMR and DC-SIGN. To further complicate things, pathogens often engage CLRs with redundant functions, for instance *M. tuberculosis* and *C. neoformans* that trigger multiple ITAM-coupled CLRs that result in downstream NF-κB activation. Thus, a better understanding of the detailed signaling pathways of CLRs will be beneficial for the development of host-directed therapies that can better promote innate immunity against respiratory pathogens.

Acknowledgements We did not receive specific funding for this work. The Neyrolles laboratory is supported by the Centre National de la Recherche Scientifique (CNRS), the Université Paul Sabatier-Université de Toulouse, the Ministère de l'Enseignement supérieur, de la Recherche et de l'Innovation, the European Union (H2020), the Agence Nationale de la Recherche (ANR), the Fondation Bettencourt Schueller, and the Fondation pour la Recherche Médicale (FRM). BBAR is a fellow of the Marie Skłodowska-Curie action funded by the European Commission. We would like to thank Dr. Jérome Nigou and Dr. Geanncarlo Lugo-Villarino for editing this chapter as well as Dr. Emma Dawson for assistance with the illustration.

References

Ahren IL, Eriksson E, Egesten A, Riesbeck K (2003) Nontypeable Haemophilus influenzae activates human eosinophils through beta-glucan receptors. Am J Respir Cell Mol Biol 29:598–605

Ahren IL, Williams DL, Rice PJ, Forsgren A, Riesbeck K (2001) The importance of a beta-glucan receptor in the nonopsonic entry of nontypeable Haemophilus influenzae into human monocytic and epithelial cells. J Infect Dis 184:150–158

Ahrens S, Zelenay S, Sancho D, Hanc P, Kjaer S, Feest C, Fletcher G, Durkin C, Postigo A, Skehel M, Batista F, Thompson B, Way M, E Sousa CR, Schulz O (2012) F-actin is an evolutionarily conserved damage-associated molecular pattern recognized by DNGR-1, a receptor for dead cells. Immunity 36:635–645

Akahori Y, Miyasaka T, Toyama M, Matsumoto I, Miyahara A, Zong T, Ishii K, Kinjo Y, Miyazaki Y, Saijo S, Iwakura Y, Kawakami K (2016) Dectin-2-dependent host defense in mice infected with serotype 3 *Streptococcus pneumoniae*. BMC Immunol 17:1

Astarie-Dequeker C, N'Diaye EN, le Cabec V, Rittig MG, Prandi J, Maridonneau-Parini I (1999) The mannose receptor mediates uptake of pathogenic and nonpathogenic mycobacteria and bypasses bactericidal responses in human macrophages. Infect Immun 67:469–747

Athamna A, Ofek I, Keisari Y, Markowitz S, Dutton GG, Sharon N (1991) Lectinophagocytosis of encapsulated *Klebsiella pneumoniae* mediated by surface lectins of guinea pig alveolar macrophages and human monocyte-derived macrophages. Infect Immun 59:1673–1682

Atochina EN, Beck JM, Scanlon ST, Preston AM, Beers MF (2001) Pneumocystis carinii pneumonia alters expression and distribution of lung collectins SP-A and SP-D. J Lab Clin Med 137:429–439

Bafica A, Scanga CA, Feng CG, Leifer C, Cheever A, Sher A (2005) TLR9 regulates Th1 responses and cooperates with TLR2 in mediating optimal resistance to *Mycobacterium tuberculosis*. J Exp Med 202:1715–1724

Bajorath J, Aruffo A (1996) Structure-based modeling of the ligand binding domain of the human cell surface receptor CD23 and comparison of two independently derived molecular models. Protein Sci 5:240–247

Bakker AB, Baker E, Sutherland GR, Phillips JH, Lanier LL (1999) Myeloid DAP12-associating lectin (MDL)-1 is a cell surface receptor involved in the activation of myeloid cells. Proc Natl Acad Sci U S A 96:9792–9796

Balamayooran G, Batra S, Theivanthiran B, Cai S, Pacher P, Jeyaseelan S (2012) Intrapulmonary G-CSF rescues neutrophil recruitment to the lung and neutrophil release to blood in Gram-negative bacterial infection in MCP-1-/- mice. J Immunol 189:5849–5859

Beauvais A, Latge JP (2001) Membrane and cell wall targets in *Aspergillus fumigatus*. Drug Resist Updat 4:38–49

Behler-Janbeck F, Takano T, Maus R, Stolper J, Jonigk D, Tarres MT, Fuehner T, Prasse A, Welte T, Timmer MS, Stocker BL (2016) C-type lectin Mincle recognizes glucosyl-diacylglycerol of *Streptococcus pneumoniae* and plays a protective role in pneumococcal pneumonia. PLoS Pathog 12(12):e1006038

Behler F, Steinwede K, Balboa L, Ueberberg B, Maus R, Kirchhof G, Yamasaki S, Welte T, Maus UA (2012) Role of Mincle in alveolar macrophage-dependent innate immunity against mycobacterial infections in mice. J Immunol 189:3121–3129

Benne CA, Benaissa-Trouw B, van Strijp JA, Kraaijeveld CA, van Iwaarden JF (1997) Surfactant protein A, but not surfactant protein D, is an opsonin for influenza A virus phagocytosis by rat alveolar macrophages. Eur J Immunol 27:886–890

Brightbill HD, Libraty DH, Krutzik SR, Yang RB, Belisle JT, Bleharski JR, Maitland M, Norgard MV, Plevy SE, Smale ST, Brennan PJ, Bloom BR, Godowski PJ, Modlin RL (1999) Host defense mechanisms triggered by microbial lipoproteins through toll-like receptors. Science 285:732–736

Brown GD, Gordon S (2001) Immune recognition. A new receptor for beta-glucans. Nature 413:36–37

Brown GD, Willment JA, Whitehead L (2018) C-type lectins in immunity and homeostasis. Nat Rev Immunol 18:374–389

Brown JS (2012) Community-acquired pneumonia. Clin Med (Lond) 12:538–543

Campuzano A, Castro-Lopez N, Wozniak KL, Wager CML, Wormley FL Jr (2017) Dectin-3 is not required for protection against *Cryptococcus neoformans* infection. PLoS ONE 12(1): e0169347

Carnevale J, Ross L, Puissant A, Banerji V, Stone RM, Deangelo DJ, Ross KN, Stegmaier K (2013) SYK regulates mTOR signaling in AML. Leukemia 27:2118–2128

Carreto-Binaghi LE, Taylor ML (2016) Surfactant proteins, SP-A and SP-D, in respiratory fungal infections: their role in the inflammatory response. Respir Res 17(1):66

Carroll MV, Sim RB, Bigi F, Jakel A, Antrobus R, Mitchell DA (2010) Identification of four novel DC-SIGN ligands on *Mycobacterium* bovis BCG. Protein Cell 1:859–870

Chackerian AA, Alt JM, Perera TV, Dascher CC, Behar SM (2002) Dissemination of *Mycobacterium tuberculosis* is influenced by host factors and precedes the initiation of T-cell immunity. Infect Immun 70:4501–4509

Chen ST, Lin YL, Huang MT, Wu MF, Cheng SC, Lei HY, Lee CK, Chiou TW, Wong CH, Hsieh SL (2008) CLEC5A is critical for dengue-virus-induced lethal disease. Nature 453:672–676

Cheung R, Shen F, Phillips JH, McGeachy MJ, Cua DJ, Heyworth PG, Pierce RH (2011) Activation of MDL-1 (CLEC5A) on immature myeloid cells triggers lethal shock in mice. J Clin Invest 121:4446–4461

Chieppa M, Bianchi G, Doni A, del Prete A, Sironi M, Laskarin G, Monti P, Piemonti L, Biondi A, Mantovani A, Introna M, Allavena P (2003) Cross-linking of the mannose receptor on monocyte-derived dendritic cells activates an anti-inflammatory immunosuppressive program. J Immunol 171:4552–4560

Court N, Vasseur V, Vacher R, Fremond C, Shebzukhov Y, Yeremeev VV, Maillet I, Nedospasov SA, Gordon S, Fallon PG, Suzuki H, Ryffel B, Quesniaux VF (2010) Partial redundancy of the pattern recognition receptors, scavenger receptors, and C-type lectins for the long-term control of *Mycobacterium tuberculosis* infection. J Immunol 184:7057–7070

Cross CE, Bancroft GJ (1995) Ingestion of acapsular *Cryptococcus neoformans* occurs via mannose and beta-glucan receptors, resulting in cytokine production and increased phagocytosis of the encapsulated form. Infect Immun 63:2604–2611

Dan JM, Kelly RM, Lee CK, Levitz SM (2008) Role of the mannose receptor in a murine model of *Cryptococcus neoformans* infection. Infect Immun 76:2362–2367

Decout A, Silva-Gomes S, Drocourt D, Barbe S, Andre I, Cueto FJ, Lioux T, Sancho D, Perouzel E, Vercellone A, Prandi J, Gilleron M, Tiraby G, Nigou J (2017) Rational design of adjuvants targeting the C-type lectin Mincle. Proc Natl Acad Sci U S A 114:2675–2680

Decout A, Silva-Gomes S, Drocourt D, Blattes E, Riviere M, Prandi J, Larrouy-Maumus G, Caminade AM, Hamasur B, Kallenius G, Kaur D, Dobos KM, Lucas M, Sutcliffe IC, Besra GS, Appelmelk BJ, Gilleron M, Jackson M, Vercellone A, Tiraby G, Nigou J (2018) Deciphering the molecular basis of mycobacteria and lipoglycan recognition by the C-type lectin Dectin-2. Sci Rep 8:16840

del Fresno C, Iborra S, Saz-Leal P, Martinez-Lopez M, Sancho D (2018) Flexible signaling of myeloid C-type lectin receptors in immunity and inflammation. Front Immunol 9:804

Dennehy KM, Willment JA, Williams DL, Brown GD (2009) Reciprocal regulation of IL-23 and IL-12 following co-activation of Dectin-1 and TLR signaling pathways. Eur J Immunol 39:1379–1386

Denning DW (2016) Minimizing fungal disease deaths will allow the UNAIDS target of reducing annual AIDS deaths below 500,000 by 2020 to be realized. Philos Trans R Soc Lond B Biol Sci 371

Driessen NN, Ummels R, Maaskant JJ, Gurcha SS, Besra GS, Ainge GD, Larsen DS, Painter GF, Vandenbroucke-Grauls CM, Geurtsen J, Appelmelk BJ (2009) Role of phosphatidylinositol mannosides in the interaction between mycobacteria and DC-SIGN. Infect Immun 77:4538–4547

Drummond RA, Saijo S, Iwakura Y, Brown GD (2011) The role of Syk/CARD9 coupled C-type lectins in antifungal immunity. Eur J Immunol 41:276–281

Engering A, Geijtenbeek TB, van Vliet SJ, Wijers M, van Liempt E, Demaurex N, Lanzavecchia A, Fransen J, Figdor CG, Piguet V, van Kooyk Y (2002) The dendritic cell-specific adhesion receptor DC-SIGN internalizes antigen for presentation to T cells. J Immunol 168:2118–2126

Ezekowitz RA, Williams DJ, Koziel H, Armstrong MY, Warner A, Richards FF, Rose RM (1991) Uptake of Pneumocystis carinii mediated by the macrophage mannose receptor. Nature 351:155–158

Favre-Bonte S, Joly B, Forestier C (1999) Consequences of reduction of *Klebsiella pneumoniae* capsule expression on interactions of this bacterium with epithelial cells. Infect Immun 67:554–561

Fontaine T, Delangle A, Simenel C, Coddeville B, van Vliet SJ, van Kooyk Y, Bozza S, Moretti S, Schwarz F, Trichot C, Aebi M, Delepierre M, Elbim C, Romani L, Latge JP (2011) Galactosaminogalactan, a new immunosuppressive polysaccharide of *Aspergillus fumigatus*. PLoS Pathog 7:e1002372

Fraser IP, Takahashi K, Koziel H, Fardin B, Harmsen A, Ezekowitz RA (2000) Pneumocystis carinii enhances soluble mannose receptor production by macrophages. Microbes Infect 2:1305–1310

Gao X, Zhao G, Li C, Lin J, Jiang N, Wang Q, Hu L, Xu Q, Peng X, He K, Zhu G (2016) LOX-1 and TLR4 affect each other and regulate the generation of ROS in A. fumigatus keratitis. Int Immunopharmacol 40:392–399

Garcia-Vallejo JJ, van Kooyk Y (2013) The physiological role of DC-SIGN: a tale of mice and men. Trends Immunol 34:482–486

Geijtenbeek TB, Torensma R, van Vliet SJ, van Duijnhoven GC, Adema GJ, van Kooyk Y, Figdor CG (2000) Identification of DC-SIGN, a novel dendritic cell-specific ICAM-3 receptor that supports primary immune responses. Cell 100:575–585

Geijtenbeek TB, van Vliet SJ, Koppel EA, Sanchez-Hernandez M, Vandenbroucke-Grauls CM, Appelmelk B, van Kooyk Y (2003) Mycobacteria target DC-SIGN to suppress dendritic cell function. J Exp Med 197:7–17

Geunes-Boyer S, Beers MF, Perfect JR, Heitman J, Wright JR (2012) Surfactant protein D facilitates *Cryptococcus neoformans* infection. Infect Immun 80:2444–2453

Geurtsen J, Chedammi S, Mesters J, Cot M, Driessen NN, Sambou T, Kakutani R, Ummels R, Maaskant J, Takata H, Baba O, Terashima T, Bovin N, Vandenbroucke-Grauls CM, Nigou J, Puzo G, Lemassu A, Daffe M, Appelmelk BJ (2009) Identification of mycobacterial alpha-glucan as a novel ligand for DC-SIGN: involvement of mycobacterial capsular polysaccharides in host immune modulation. J Immunol 183:5221–5231

Giles SS, Dagenais TR, Botts MR, Keller NP, Hull CM (2009) Elucidating the pathogenesis of spores from the human fungal pathogen *Cryptococcus neoformans*. Infect Immun 77:3491–3500

Giles SS, Zaas AK, Reidy MF, Perfect JR, Wright JR (2007) *Cryptococcus neoformans* is resistant to surfactant protein A mediated host defense mechanisms. PLoS ONE 2:e1370

Gillespie L, Roosendahl P, Ng WC, Brooks AG, Reading PC, Londrigan SL (2016) Endocytic function is critical for influenza a virus infection via DC-SIGN and L-SIGN. Sci Rep 6:19428

Goyal S, Klassert TE, Slevogt H (2016) C-type lectin receptors in tuberculosis: what we know. Med Microbiol Immunol 205:513–535

Graham LM, Gupta V, Schafer G, Reid DM, Kimberg M, Dennehy KM, Hornsell WG, Guler R, Campanero-Rhodes MA, Palma AS, Feizi T, Kim SK, Sobieszczuk P, Willment JA, Brown GD (2012) The C-type lectin receptor CLECSF8 (CLEC4D) is expressed by myeloid cells and triggers cellular activation through Syk kinase. J Biol Chem 287:25964–25974

Gross O, Poeck H, Bscheider M, Dostert C, Hannesschlager N, Endres S, Hartmann G, Tardivel A, Schweighoffer E, Tybulewicz V, Mocsai A, Tschopp J, Ruland J (2009) Syk kinase signalling couples to the Nlrp3 inflammasome for anti-fungal host defence. Nature 459:433–436

Guo Y, Chang Q, Cheng L, Xiong S, Jia X, Lin X, Zhao X (2018) C-Type Lectin Receptor CD23 Is Required for Host Defense against *Candida albicans* and *Aspergillus fumigatus* Infection. J Immunol 201:2427–2440

Hahn PY, Evans SE, Kottom TJ, Standing JE, Pagano RE, Limper AH (2003) Pneumocystis carinii cell wall beta-glucan induces release of macrophage inflammatory protein-2 from alveolar epithelial cells via a lactosylceramide-mediated mechanism. J Biol Chem 278:2043–2050

Harding E (2019) WHO global progress report on tuberculosis elimination. Lancet Respir Med

Hawgood S, Brown C, Edmondson J, Stumbaugh A, Allen L, Goerke J, Clark H, Poulain F (2004) Pulmonary collectins modulate strain-specific influenza a virus infection and host responses. J Virol 78:8565–8572

He K, Yue LH, Zhao GQ, Li C, Lin J, Jiang N, Wang Q, Xu Q, Peng XD, Hu LT, Zhang J (2016) The role of LOX-1 on innate immunity against *Aspergillus keratitis* in mice. Int J Ophthalmol 9:1245–1250

Heitmann L, Schoenen H, Ehlers S, Lang R, Holscher C (2013) Mincle is not essential for controlling *Mycobacterium tuberculosis* infection. Immunobiology 218:506–516

Heung LJ (2017) Innate immune responses to *Cryptococcus*. J Fungi (Basel) 3

Heyl KA, Klassert TE, Heinrich A, Muller MM, Klaile E, Dienemann H, Grunewald C, Bals R, Singer BB, Slevogt H (2014) Dectin-1 is expressed in human lung and mediates the proinflammatory immune response to nontypeable *Haemophilus* influenzae. MBio 5:e01492–e01514

Hillaire ML, Nieuwkoop NJ, Boon AC, de Mutsert G, Vogelzang-van Trierum SE, Fouchier RA, Osterhaus AD, Rimmelzwaan GF (2013) Binding of DC-SIGN to the hemagglutinin of influenza A viruses supports virus replication in DC-SIGN expressing cells. PLoS ONE 8(2): e56164

Hogaboam CM, Takahashi K, Ezekowitz RA, Kunkel SL, Schuh JM (2004) Mannose-binding lectin deficiency alters the development of fungal asthma: effects on airway response, inflammation, and cytokine profile. J Leukoc Biol 75:805–814

Hohl TM, van Epps HL, Rivera A, Morgan LA, Chen PL, Feldmesser M, Pamer EG (2005) *Aspergillus fumigatus* triggers inflammatory responses by stage-specific beta-glucan display. PLoS Pathog 1:e30

Huang HR, Li F, Han H, Xu X, Li N, Wang S, Xu JF, Jia XM (2018) Dectin-3 Recognizes Glucuronoxylomannan of Cryptococcus neoformans Serotype AD and *Cryptococcus gattii* Serotype B to Initiate Host Defense Against *Cryptococcosis*. Front Immunol 9:1781

Imai T, Matsumura T, Mayer-Lambertz S, Wells CA, Ishikawa E, Butcher SK, Barnett TC, Walker MJ, Imamura A, Ishida H, Ikebe T, Miyamoto T, Ato M, Ohga S, Lepenies B, van Sorge NM, Yamasaki S (2018) Lipoteichoic acid anchor triggers Mincle to drive protective immunity against invasive group A Streptococcus infection. Proc Natl Acad Sci U S A 115: E10662–E10671

Inui M, Kikuchi Y, Aoki N, Endo S, Maeda T, Sugahara-Tobinai A, Fujimura S, Nakamura A, Kumanogoh A, Colonna M, Takai T (2009) Signal adaptor DAP10 associates with MDL-1 and triggers osteoclastogenesis in cooperation with DAP12. Proc Natl Acad Sci U S A 106:4816–4821

Ishikawa E, Ishikawa T, Morita YS, Toyonaga K, Yamada H, Takeuchi O, Kinoshita T, Akira S, Yoshikai Y, Yamasaki S (2009) Direct recognition of the mycobacterial glycolipid, trehalose dimycolate, by C-type lectin Mincle. J Exp Med 206:2879–2888

Iwasaki A, Pillai PS (2014) Innate immunity to influenza virus infection. Nat Rev Immunol 14:315–328

Jeffers SA, Tusell SM, Gillim-Ross L, Hemmila EM, Achenbach JE, Babcock GJ, Thomas WD, Thackray LB, Young MD, Mason RJ, Ambrosino DM (2004) CD209L (L-SIGN) is a receptor for severe acute respiratory syndrome coronavirus. Proc Natl Acad Sci 101(44):15748–15753

Jiang JQ, Li C, Cui CX, Ma YN, Zhao GQ, Peng XD, Xu Q, Wang Q, Zhu GQ, Li CY (2019) Inhibition of LOX-1 alleviates the proinflammatory effects of high-mobility group box 1 in *Aspergillus fumigatus* keratitis. Int J Ophthalmol 12:898–903

Johansson J, Curstedt T, Robertson B (1994) The proteins of the surfactant system. Eur Respir J 7:372–391

Jondle CN, Sharma A, Simonson TJ, Larson B, Mishra BB, Sharma J (2016) Macrophage galactose-type lectin-1 deficiency is associated with increased Neutrophilia and Hyperinflammation in Gram-Negative Pneumonia. J Immunol 196:3088–3096

Jones BW, Means TK, Heldwein KA, Keen MA, Hill PJ, Belisle JT, Fenton MJ (2001) Different Toll-like receptor agonists induce distinct macrophage responses. J Leukoc Biol 69:1036–1044

Jounblat R, Clark H, Eggleton P, Hawgood S, Andrew PW, Kadioglu A (2005) The role of surfactant protein D in the colonisation of the respiratory tract and onset of bacteraemia during pneumococcal pneumonia. Respir Res 6:126

Jounblat R, Kadioglu A, Iannelli F, Pozzi G, Eggleton P, Andrew PW (2004) Binding and agglutination of *Streptococcus pneumoniae* by human surfactant protein D (SP-D) vary between strains, but SP-D fails to enhance killing by neutrophils. Infect Immun 72:709–716

Kabha K, Schmegner J, Keisari Y, Parolis H, Schlepper-Schaeffer J, Ofek I (1997) SP-A enhances phagocytosis of Klebsiella by interaction with capsular polysaccharides and alveolar macrophages. Am J Physiol 272:L344–L352

Kanazawa N, Tashiro K, Inaba K, Miyachi Y (2003) Dendritic cell immunoactivating receptor, a novel C-type lectin immunoreceptor, acts as an activating receptor through association with Fc receptor gamma chain. J Biol Chem 278:32645–32652

Kang PB, Azad AK, Torrelles JB, Kaufman TM, Beharka A, Tibesar E, Desjardin LE, Schlesinger LS (2005) The human macrophage mannose receptor directs *Mycobacterium tuberculosis* lipoarabinomannan-mediated phagosome biogenesis. J Exp Med 202:987–999

Kang YS, Kim JY, Bruening SA, Pack M, Charalambous A, Pritsker A, Moran TM, Loeffler JM, Steinman RM, Park CG (2004) The C-type lectin SIGN-R1 mediates uptake of the capsular polysaccharide of *Streptococcus pneumoniae* in the marginal zone of mouse spleen. Proc Natl Acad Sci U S A 101:215–220

Kaur S, Gupta VK, Thiel S, Sarma PU, Madan T (2007) Protective role of mannan-binding lectin in a murine model of invasive pulmonary aspergillosis. Clin Exp Immunol 148:382–389

Keisari Y, Kabha K, Nissimov L, Schlepper-Schafer J, Ofek I (1997) Phagocyte-bacteria interactions. Adv Dent Res 11:43–49

Keisari Y, Wang H, Mesika A, Matatov R, Nissimov L, Crouch E, Ofek I (2001) Surfactant protein D-coated Klebsiella pneumoniae stimulates cytokine production in mononuclear phagocytes. J Leukoc Biol 70:135–141

Kerrigan AM, Brown GD (2011) Syk-coupled C-type lectins in immunity. Trends Immunol 32:151–156

Koppel EA, Litjens M, van den Berg VC, van Kooyk Y, Geijtenbeek TB (2008) Interaction of SIGNR1 expressed by marginal zone macrophages with marginal zone B cells is essential to early IgM responses against *Streptococcus pneumoniae.* Mol Immunol 45:2881–2887

Koppel EA, Wieland CW, van den Berg VC, Litjens M, Florquin S, van Kooyk Y, van der Poll T, Geijtenbeek TB (2005) Specific ICAM-3 grabbing nonintegrin-related 1 (SIGNR1) expressed by marginal zone macrophages is essential for defense against pulmonary *Streptococcus pneumoniae* infection. Eur J Immunol 35:2962–2969

Kottom TJ, Hebrink DM, Jenson PE, Marsolek PL, Wuthrich M, Wang H, Klein B, Yamasaki S, Limper AH (2018) Dectin-2 is a C-type lectin receptor that recognizes Pneumocystis and participates in innate immune responses. Am J Respir Cell Mol Biol 58:232–240

Kottom TJ, Hebrink DM, Jenson PE, Nandakumar V, Wuthrich M, Wang H, Klein B, Yamasaki S, Lepenies B, Limper AH (2017) The interaction of Pneumocystis with the C-type lectin receptor mincle exerts a significant role in host defense against infection. J Immunol 198:3515–3525

Kottom TJ, Hebrink DM, Monteiro JT, Lepenies B, Carmona EM, Wuethrich M, Santo Dias LD, Limper AH (2019) Myeloid C-type lectin receptors that recognize fungal mannans interact with Pneumocystis organisms and major surface glycoprotein. J Med Microbiol 68(11):1649–1654

Kousha M, Tadi R, Soubani AO (2011) Pulmonary aspergillosis: a clinical review. Eur Respir Rev 20:156–174

Kuroki Y, Takahashi M, Nishitani C (2007) Pulmonary collectins in innate immunity of the lung. Cell Microbiol 9:1871–1879

Kuronuma K, Sano H, Kato K, Kudo K, Hyakushima N, Yokota S, Takahashi H, Fujii N, Suzuki H, Kodama T, Abe S, Kuroki Y (2004) Pulmonary surfactant protein A augments the phagocytosis of *Streptococcus pneumoniae* by alveolar macrophages through a casein kinase 2-dependent increase of cell surface localization of scavenger receptor A. J Biol Chem 279:21421–21430

Kyrmizi I, Gresnigt MS, Akoumianaki T, Samonis G, Sidiropoulos P, Boumpas D, Netea MG, van de Veerdonk FL, Kontoyiannis DP, Chamilos G (2013) Corticosteroids block autophagy protein recruitment in *Aspergillus fumigatus* phagosomes via targeting dectin-1/Syk kinase signaling. J Immunol 191:1287–1299

Lanier LL, Corliss BC, Wu J, Leong C, Phillips JH (1998) Immunoreceptor DAP12 bearing a tyrosine-based activation motif is involved in activating NK cells. Nature 391:703–707

Lee HM, Yuk JM, Shin DM, Jo EK (2009) Dectin-1 is inducible and plays an essential role for mycobacteria-induced innate immune responses in airway epithelial cells. J Clin Immunol 29:795–805

Lee WB, Kang JS, Yan JJ, Lee MS, Jeon BY, Cho SN, Kim YJ (2012) Neutrophils promote mycobacterial trehalose dimycolate-induced lung inflammation via the mincle pathway. PLoS Pathog 8:e1002614
Leseux L, Hamdi SM, Al Saati T, Capilla F, Recher C, Laurent G, Bezombes C (2006) Syk-dependent mTOR activation in follicular lymphoma cells. Blood.108(13):4156–62
Levine AM, Whitsett JA, Gwozdz JA, Richardson TR, Fisher JH, Burhans MS, Korfhagen TR (2000) Distinct effects of surfactant protein A or D deficiency during bacterial infection on the lung. J Immunol 165:3934–3940
Levine AM, Whitsett JA, Hartshorn KL, Crouch EC, Korfhagen TR (2001) Surfactant protein D enhances clearance of influenza A virus from the lung in vivo. J Immunol 167:5868–5873
Lim J, Coates CJ, Seoane PI, Garelnabi M, Taylor-Smith LM, Monteith P, Macleod CL, Escaron CJ, Brown GD, Hall RA, May RC (2018) Characterizing the mechanisms of nonopsonic uptake of *Cryptococci* by macrophages. J Immunol 200:3539–3546
Lin J, He K, Zhao G, Li C, Hu L, Zhu G, Niu Y, Hao G (2017) Mincle inhibits neutrophils and macrophages apoptosis in A. fumigatus keratitis. Int Immunopharmacol 52:101–109
Lionakis MS, Iliev ID, Hohl TM (2017) Immunity against fungi. JCI Insight 2
Lobato-Pascual A, Saether PC, Fossum S, Dissen E, Daws MR (2013) Mincle, the receptor for mycobacterial cord factor, forms a functional receptor complex with MCL and FcepsilonRI-gamma. Eur J Immunol 43:3167–3174
Londrigan SL, Turville SG, Tate MD, Deng YM, Brooks AG, Reading PC (2011) N-linked glycosylation facilitates sialic acid-independent attachment and entry of influenza A viruses into cells expressing DC-SIGN or L-SIGN. J Virol 85:2990–3000
Loures FV, Rohm M, Lee CK, Santos E, Wang JP, Specht CA, Calich VL, Urban CF, Levitz SM (2015) Recognition of *Aspergillus fumigatus* hyphae by human plasmacytoid dendritic cells is mediated by dectin-2 and results in formation of extracellular traps. PLoS Pathog 11:e1004643
Lugo-Villarino G, Hudrisier D, Tanne A, Neyrolles O (2011) C-type lectins with a sweet spot for *Mycobacterium tuberculosis*. Eur J Microbiol Immunol (Bp) 1:25–40
Lugo-Villarino G, Troegeler A, Balboa L, Lastrucci C, Duval C, Mercier I, Benard A, Capilla F, Al Saati T, Poincloux R, Kondova I, Verreck FAW, Cougoule C, Maridonneau-Parini I, Sasiain MDC, Neyrolles O (2018) The C-type lectin receptor DC-SIGN has an anti-inflammatory role in human M(IL-4) macrophages in response to *Mycobacterium tuberculosis*. Front Immunol 9:1123
Madan T, Eggleton P, Kishore U, Strong P, Aggrawal SS, Sarma PU, Reid KB (1997a) Binding of pulmonary surfactant proteins A and D *to Aspergillus fumigatus* conidia enhances phagocytosis and killing by human neutrophils and alveolar macrophages. Infect Immun 65:3171–3179
Madan T, Kishore U, Shah A, Eggleton P, Strong P, Wang JY, Aggrawal SS, Sarma PU, Reid KB (1997b) Lung surfactant proteins A and D can inhibit specific IgE binding to the allergens of Aspergillus fumigatus and block allergen-induced histamine release from human basophils. Clin Exp Immunol 110:241–249
Madan T, Kishore U, Singh M, Strong P, Hussain EM, Reid KB, Sarma PU (2001) Protective role of lung surfactant protein D in a murine model of invasive pulmonary aspergillosis. Infect Immun 69:2728–2731
Madan T, Reid KB, Clark H, Singh M, Nayak A, Sarma PU, Hawgood S, Kishore U (2010) Susceptibility of mice genetically deficient in SP-A or SP-D gene to invasive pulmonary aspergillosis. Mol Immunol 47:1923–1930
Maeda N, Nigou J, Herrmann JL, Jackson M, Amara A, Lagrange PH, Puzo G, Gicquel B, Neyrolles O (2003) The cell surface receptor DC-SIGN discriminates between Mycobacterium species through selective recognition of the mannose caps on lipoarabinomannan. J Biol Chem 278:5513–5516
Malhotra R, Haurum JS, Thiel S, Sim RB (1994) Binding of human collectins (SP-A and MBP) to influenza virus. Biochem J 304(Pt 2):455–461
Mansour MK, Latz E, Levitz SM (2006) Cryptococcus neoformans glycoantigens are captured by multiple lectin receptors and presented by dendritic cells. J Immunol 176:3053–3061

Mansour MK, Schlesinger LS, Levitz SM (2002) Optimal T cell responses to *Cryptococcus neoformans* mannoprotein are dependent on recognition of conjugated carbohydrates by mannose receptors. J Immunol 168:2872–2879

Mansour MK, Tam JM, Khan NS, Seward M, Davids PJ, Puranam S, Sokolovska A, Sykes DB, Dagher Z, Becker C, Tanne A, Reedy JL, Stuart LM, Vyas JM (2013) Dectin-1 activation controls maturation of beta-1,3-glucan-containing phagosomes. J Biol Chem 288:16043–16054

Marakalala MJ, Guler R, Matika L, Murray G, Jacobs M, Brombacher F, Rothfuchs AG, Sher A, Brown GD (2011) The Syk/CARD9-coupled receptor Dectin-1 is not required for host resistance to *Mycobacterium tuberculosis* in mice. Microbes Infect 13:198–201

Matsumoto Y, Matsuda S, Tegoshi T (1989) Yeast glucan in the cyst wall of *Pneumocystis carinii*. J Protozool 36:21S–22S

Matsushita M, Fujita T (1992) Activation of the classical complement pathway by mannose-binding protein in association with a novel C1s-like serine protease. J Exp Med 176:1497–1502

Mayer-Barber KD, Sher A (2015) Cytokine and lipid mediator networks in tuberculosis. Immunol Rev 264:264–275

McCormack FX, Festa AL, Andrews RP, Linke M, Walzer PD (1997) The carbohydrate recognition domain of surfactant protein A mediates binding to the major surface glycoprotein of *Pneumocystis carinii*. Biochemistry 36:8092–8099

McGeachy MJ, Cua DJ, Gaffen SL (2019) The IL-17 family of cytokines in health and disease. Immunity 50:892–906

McGreal EP, Rosas M, Brown GD, Zamze S, Wong SY, Gordon S, Martinez-Pomares L, Taylor PR (2006) The carbohydrate-recognition domain of Dectin-2 is a C-type lectin with specificity for high mannose. Glycobiology 16:422–430

Means TK, Wang S, Lien E, Yoshimura A, Golenbock DT, Fenton MJ (1999) Human toll-like receptors mediate cellular activation by *Mycobacterium tuberculosis*. J Immunol 163:3920–3927

Mitchell DA, Fadden AJ, Drickamer K (2001) A novel mechanism of carbohydrate recognition by the C-type lectins DC-SIGN and DC-SIGNR. Subunit organization and binding to multivalent ligands. J Biol Chem 276:28939–28945

Miyake Y, Masatsugu OH, Yamasaki S (2015) C-type lectin receptor MCL facilitates mincle expression and signaling through complex formation. J Immunol 194:5366–5374

Miyake Y, Toyonaga K, Mori D, Kakuta S, Hoshino Y, Oyamada A, Yamada H, Ono K, Suyama M, Iwakura Y, Yoshikai Y, Yamasaki S (2013) C-type lectin MCL is an FcRgamma-coupled receptor that mediates the adjuvanticity of mycobacterial cord factor. Immunity 38:1050–1062

Miyasaka T, Akahori Y, Toyama M, Miyamura N, Ishii K, Saijo S, Iwakura Y, Kinjo Y, Miyazaki Y, Oishi K, Kawakami K (2013) Dectin-2-dependent NKT cell activation and serotype-specific antibody production in mice immunized with pneumococcal polysaccharide vaccine. PLoS ONE 8:e78611

Mukaremera L, Nielsen K (2017) Adaptive immunity to *Cryptococcus neoformans* infections. J Fungi (Basel) 3

Muntoni F, Mateddu A, Cau M, Congiu R, Puddu R, Cossu P, Cao A, Melis MA (1993) Diagnosis of DMD carrier status in a family with no known affected males. Dev Med Child Neurol 35:70–73

Nakamura K, Kinjo T, Saijo S, Miyazato A, Adachi Y, Ohno N, Fujita J, Kaku M, Iwakura Y, Kawakami K (2007) Dectin-1 is not required for the host defense to *Cryptococcus neoformans*. Microbiol Immunol 51:1115–1119

Nakamura Y, Sato K, Yamamoto H, Matsumura K, Matsumoto I, Nomura T, Miyasaka T, Ishii K, Kanno E, Tachi M, Yamasaki S, Saijo S, Iwakura Y, Kawakami K (2015) Dectin-2 deficiency promotes Th2 response and mucin production in the lungs after pulmonary infection with *Cryptococcus neoformans*. Infect Immun 83:671–681

Neth O, Jack DL, Dodds AW, Holzel H, Klein NJ, Turner MW (2000) Mannose-binding lectin binds to a range of clinically relevant microorganisms and promotes complement deposition. Infect Immun 68:688–693

Neumann K, Castineiras-Vilarino M, Hockendorf U, Hannesschlager N, Lemeer S, Kupka D, Meyermann S, Lech M, Anders HJ, Kuster B, Busch DH, Gewies A, Naumann R, Gross O, Ruland J (2014) Clec12a is an inhibitory receptor for uric acid crystals that regulates inflammation in response to cell death. Immunity 40:389–399

Ng WC, Liong S, Tate MD, Irimura T, Denda-Nagai K, Brooks AG, Londrigan SL, Reading PC (2014) The macrophage galactose-type lectin can function as an attachment and entry receptor for influenza virus. J Virol 88:1659–1672

Ng WC, Londrigan SL, Nasr N, Cunningham AL, Turville S, Brooks AG, Reading PC (2016) The C-type lectin langerin functions as a receptor for attachment and infectious entry of influenza a virus. J Virol 90:206–221

Nigou J, Zelle-Rieser C, Gilleron M, Thurnher M, Puzo G (2001) Mannosylated lipoarabinomannans inhibit IL-12 production by human dendritic cells: evidence for a negative signal delivered through the mannose receptor. J Immunol 166:7477–7485

Ofek I, Mesika A, Kalina M, Keisari Y, Podschun R, Sahly H, Chang D, McGregor D, Crouch E (2001) Surfactant protein D enhances phagocytosis and killing of unencapsulated phase variants of *Klebsiella pneumoniae*. Infect Immun 69:24–33

Paczosa MK, Mecsas J (2016) Klebsiella pneumoniae: going on the offense with a strong defense. Microbiol Mol Biol Rev 80:629–661

Palomino-Segura M, Perez L, Farsakoglu Y, Virgilio T, Latino I, D'Antuono R, Chatziandreou N, Pizzagalli DU, Wang G, Garcia-Sastre A, Sallusto F, Carroll MC, Neyrolles O, Gonzalez SF (2019) Protection against influenza infection requires early recognition by inflammatory dendritic cells through C-type lectin receptor SIGN-R1. Nat Microbiol 4:1930–1940

Park CG, Takahara K, Umemoto E, Yashima Y, Matsubara K, Matsuda Y, Clausen BE, Inaba K, Steinman RM (2001) Five mouse homologues of the human dendritic cell C-type lectin, DC-SIGN. Int Immunol 13:1283–1290

Phelps DS, Umstead TM, Rose RM, Fishman JA (1996) Surfactant protein-A levels increase during *Pneumocystis carinii* pneumonia in the rat. Eur Respir J 9:565–570

Philips JA, Ernst JD (2012) Tuberculosis pathogenesis and immunity. Annu Rev Pathol 7:353–384

Pitarque S, Herrmann JL, Duteyrat JL, Jackson M, Stewart GR, Lecointe F, Payre B, Schwartz O, Young DB, Marchal G, Lagrange PH, Puzo G, Gicquel B, Nigou J, Neyrolles O (2005) Deciphering the molecular bases of *Mycobacterium tuberculosis* binding to the lectin DC-SIGN reveals an underestimated complexity. Biochem J 392:615–624

Qu J, He L, Rong Z, Pan J, Chen X, Morrison DC, Li X (2001) Alteration of surfactant proteins A and D in bronchoalveolar lavage fluid of *Pneumocystis carinii* pneumonia. Chin Med J (Engl) 114:1143–1146

Rabes A, Zimmermann S, Reppe K, Lang R, Seeberger PH, Suttorp N, Witzenrath M, Lepenies B, Opitz B (2015) The C-type lectin receptor Mincle binds to *Streptococcus pneumoniae* but plays a limited role in the anti-pneumococcal innate immune response. PLoS ONE 10:e0117022

Rajaram MVS, Arnett E, Azad AK, Guirado E, Ni B, Gerberick AD, He LZ, Keler T, Thomas LJ, Lafuse WP, Schlesinger LS (2017) M. tuberculosis-initiated human mannose receptor signaling regulates macrophage recognition and vesicle trafficking by FcRgamma-Chain, Grb2, and SHP-1. Cell Rep 21:126–140

Rapaka RR, Goetzman ES, Zheng M, Vockley J, McKinley L, Kolls JK, Steele C (2007) Enhanced defense against *Pneumocystis carinii* mediated by a novel dectin-1 receptor Fc fusion protein. J Immunol 178:3702–3712

Reading PC, Miller JL, Anders EM (2000) Involvement of the mannose receptor in infection of macrophages by influenza virus. J Virol 74:5190–5197

Ricks DM, Chen K, Zheng M, Steele C, Kolls JK (2013) Dectin immunoadhesins and pneumocystis pneumonia. Infect Immun 81:3451–3462

Rivera A, Hohl TM, Collins N, Leiner I, Gallegos A, Saijo S, Coward JW, Iwakura Y, Pamer EG (2011) Dectin-1 diversifies *Aspergillus fumigatus*-specific T cell responses by inhibiting T helper type 1 CD4 T cell differentiation. J Exp Med 208:369–381

Rohatgi S, Pirofski LA (2015) Host immunity to *Cryptococcus neoformans*. Future Microbiol 10:565–581

Rothfuchs AG, Bafica A, Feng CG, Egen JG, Williams DL, Brown GD, Sher A (2007) Dectin-1 interaction with *Mycobacterium tuberculosis* leads to enhanced IL-12p40 production by splenic dendritic cells. J Immunol 179:3463–3471

Saba K, Denda-Nagai K, Irimura T (2009) A C-type lectin MGL1/CD301a plays an anti-inflammatory role in murine experimental colitis. Am J Pathol 174:144–152

Sahly H, Ofek I, Podschun R, Brade H, He Y, Ullmann U, Crouch E (2002) Surfactant protein D binds selectively to *Klebsiella pneumoniae* lipopolysaccharides containing mannose-rich O-antigens. J Immunol 169:3267–3274

Saijo S, Fujikado N, Furuta T, Chung SH, Kotaki H, Seki K, Sudo K, Akira S, Adachi Y, Ohno N, Kinjo T, Nakamura K, Kawakami K, Iwakura Y (2007) Dectin-1 is required for host defense against *Pneumocystis carinii* but not against *Candida albicans*. Nat Immunol 8:39–46

Sassi M, Kutty G, Ferreyra GA, Bishop LR, Liu Y, Qiu J, Huang DW, Kovacs JA (2018) The major surface glycoprotein of *Pneumocystis murina* does not activate dendritic cells. J Infect Dis 218:1631–1640

Schaefer M, Reiling N, Fessler C, Stephani J, Taniuchi I, Hatam F, Yildirim AO, Fehrenbach H, Walter K, Ruland J, Wagner H, Ehlers S, Sparwasser T (2008) Decreased pathology and prolonged survival of human DC-SIGN transgenic mice during mycobacterial infection. J Immunol 180:6836–6845

Schelenz S, Malhotra R, Sim RB, Holmskov U, Bancroft GJ (1995) Binding of host collectins to the pathogenic yeast *Cryptococcus neoformans*: human surfactant protein D acts as an agglutinin for acapsular yeast cells. Infect Immun 63:3360–3366

Schlesinger LS (1993) Macrophage phagocytosis of virulent but not attenuated strains of *Mycobacterium tuberculosis* is mediated by mannose receptors in addition to complement receptors. J Immunol 150:2920–2930

Schlesinger LS, Hull SR, Kaufman TM (1994) Binding of the terminal mannosyl units of lipoarabinomannan from a virulent strain of *Mycobacterium tuberculosis* to human macrophages. J Immunol 152:4070–4079

Schlesinger LS, Kaufman TM, Iyer S, Hull SR, Marchiando LK (1996) Differences in mannose receptor-mediated uptake of lipoarabinomannan from virulent and attenuated strains of *Mycobacterium tuberculosis* by human macrophages. J Immunol 157:4568–4675

Schoenen H, Bodendorfer B, Hitchens K, Manzanero S, Werninghaus K, Nimmerjahn F, Agger EM, Stenger S, Andersen P, Ruland J, Brown GD, Wells C, Lang R (2010) Cutting edge: mincle is essential for recognition and adjuvanticity of the mycobacterial cord factor and its synthetic analog trehalose-dibehenate. J Immunol 184:2756–2760

Serrano-Gomez D, Dominguez-Soto A, Ancochea J, Jimenez-Heffernan JA, Leal JA, Corbi AL (2004) Dendritic cell-specific intercellular adhesion molecule 3-grabbing nonintegrin mediates binding and internalization of *Aspergillus fumigatus* conidia by dendritic cells and macrophages. J Immunol 173:5635–5643

Serrano-Gomez D, Leal JA, Corbi AL (2005) DC-SIGN mediates the binding of *Aspergillus fumigatus* and keratinophylic fungi by human dendritic cells. Immunobiology 210:175–183

Sharma A, Simonson TJ, Jondle CN, Mishra BB, Sharma J (2017) Mincle-mediated neutrophil extracellular trap formation by regulation of autophagy. J Infect Dis 215:1040–1048

Sharma A, Steichen AL, Jondle CN, Mishra BB, Sharma J (2014) Protective role of Mincle in bacterial pneumonia by regulation of neutrophil mediated phagocytosis and extracellular trap formation. J Infect Dis 209:1837–1846

Sokulska M, Kicia M, Wesolowska M, Hendrich AB (2015) Pneumocystis jirovecii–from a commensal to pathogen: clinical and diagnostic review. Parasitol Res 114:3577–3585

Srivastava S, Ernst JD, Desvignes L (2014) Beyond macrophages: the diversity of mononuclear cells in tuberculosis. Immunol Rev 262:179–192

Stappers MHT, Clark AE, Aimanianda V, Bidula S, Reid DM, Asamaphan P, Hardison SE, Dambuza IM, Valsecchi I, Kerscher B, Plato A, Wallace CA, Yuecel R, Hebecker B, da Gloria Teixeira Sousa M, Cunha C, Liu Y, Feizi T, Brakhage AA, Kwon-Chung KJ, Gow NAR, Zanda M, Piras M, Zanato C, Jaeger M, Netea MG, Van de Veerdonk FL, Lacerda JF, Campos A, Carvalho A, Willment JA, Latge JP, Brown GD (2018) Recognition of DHN-melanin by a C-type lectin receptor is required for immunity to *Aspergillus*. Nature 555:382–386

Steele C, Marrero L, Swain S, Harmsen AG, Zheng M, Brown GD, Gordon S, Shellito JE, Kolls JK (2003) Alveolar macrophage-mediated killing *of Pneumocystis carinii* f. sp. muris involves molecular recognition by the Dectin-1 beta-glucan receptor. J Exp Med 198:1677–1688

Steele C, Rapaka RR, Metz A, Pop SM, Williams DL, Gordon S, Kolls JK, Brown GD (2005) The beta-glucan receptor dectin-1 recognizes specific morphologies of *Aspergillus fumigatus*. PLoS Pathog 1:e42

Steichen AL, Binstock BJ, Mishra BB, Sharma J (2013) C-type lectin receptor Clec4d plays a protective role in resolution of Gram-negative pneumonia. J Leukoc Biol 94:393–398

Stringer JR, Beard CB, Miller RF, Wakefield AE (2002) A new name (*Pneumocystis jiroveci*) for Pneumocystis from humans. Emerg Infect Dis 8:891–896

Sun H, Xu XY, Shao HT, Su X, Wu XD, Wang Q, Shi Y (2013) Dectin-2 is predominately macrophage restricted and exhibits conspicuous expression during *Aspergillus fumigatus* invasion in human lung. Cell Immunol 284:60–67

Sun H, Xu XY, Tian XL, Shao HT, Wu XD, Wang Q, Su X, Shi Y (2014) Activation of NF-kappaB and respiratory burst following *Aspergillus fumigatus* stimulation of macrophages. Immunobiology 219:25–36

Swain SD, Lee SJ, Nussenzweig MC, Harmsen AG (2003) Absence of the macrophage mannose receptor in mice does not increase susceptibility to *Pneumocystis carinii* infection in vivo. Infect Immun 71:6213–6221

Syme RM, Spurrell JC, Amankwah EK, Green FH, Mody CH (2002) Primary dendritic cells phagocytose *Cryptococcus neoformans* via mannose receptors and Fcgamma receptor II for presentation to T lymphocytes. Infect Immun 70:5972–5981

Tailleux L, Pham-Thi N, Bergeron-Lafaurie A, Herrmann JL, Charles P, Schwartz O, Scheinmann P, Lagrange PH, de Blic J, Tazi A, Gicquel B, Neyrolles O (2005) DC-SIGN induction in alveolar macrophages defines privileged target host cells for mycobacteria in patients with tuberculosis. PLoS Med 2:e381

Tailleux L, Schwartz O, Herrmann JL, Pivert E, Jackson M, Amara A, Legres L, Dreher D, Nicod LP, Gluckman JC, Lagrange PH, Gicquel B, Neyrolles O (2003) DC-SIGN is the major *Mycobacterium tuberculosis* receptor on human dendritic cells. J Exp Med 197:121–127

Tanne A, Ma B, Boudou F, Tailleux L, Botella H, Badell E, Levillain F, Taylor ME, Drickamer K, Nigou J, Dobos KM, Puzo G, Vestweber D, Wild MK, Marcinko M, Sobieszczuk P, Stewart L, Lebus D, Gicquel B, Neyrolles O (2009) A murine DC-SIGN homologue contributes to early host defense against *Mycobacterium tuberculosis*. J Exp Med 206:2205–2220

Tecle T, White MR, Sorensen G, Gantz D, Kacak N, Holmskov U, Smith K, Crouch EC, Hartshorn KL (2008) Critical role for cross-linking of trimeric lectin domains of surfactant protein D in antiviral activity against influenza A virus. Biochem J 412:323–329

Teng O, Chen ST, Hsu TL, Sia SF, Cole S, Valkenburg SA, Hsu TY, Zheng JT, Tu W, Bruzzone R, Peiris JSM, Hsieh SL, Yen HL (2017) CLEC5A-mediated enhancement of the inflammatory response in myeloid cells contributes to influenza virus pathogenicity in vivo. J Virol 91

Toyonaga K, Torigoe S, Motomura Y, Kamichi T, Hayashi JM, Morita YS, Noguchi N, Chuma Y, Kiyohara H, Matsuo K, Tanaka H, Nakagawa Y, Sakuma T, Ohmuraya M, Yamamoto T, Umemura M, Matsuzaki G, Yoshikai Y, Yano I, Miyamoto T, Yamasaki S (2016) C-type lectin receptor dcar recognizes mycobacterial phosphatidyl-inositol mannosides to promote a Th1 response during infection. Immunity 45:1245–1257

Troegeler A, Mercier I, Cougoule C, Pietretti D, Colom A, Duval C, Vu Manh TP, Capilla F, Poincloux R, Pingris K, Nigou J, Rademann J, Dalod M, Verreck FA, Al Saati T, Lugo-Villarino G, Lepenies B, Hudrisier D, Neyrolles O (2017) C-type lectin receptor DCIR modulates immunity to tuberculosis by sustaining type I interferon signaling in dendritic cells. Proc Natl Acad Sci USA 114:E540–E549

Underhill DM, Ozinsky A, Smith KD, Aderem A (1999) Toll-like receptor-2 mediates mycobacteria-induced proinflammatory signaling in macrophages. Proc Natl Acad Sci U S A 96:14459–14463

Upham JP, Pickett D, Irimura T, Anders EM, Reading PC (2010) Macrophage receptors for influenza A virus: role of the macrophage galactose-type lectin and mannose receptor in viral entry. J Virol 84:3730–3737

van der Poll T, Opal SM (2009) Pathogenesis, treatment, and prevention of pneumococcal pneumonia. Lancet 374:1543–1556

van der Vlist M, de Witte L, de Vries RD, Litjens M, de Jong MA, Fluitsma D, de Swart RL, Geijtenbeek TB (2011) Human Langerhans cells capture measles virus through Langerin and present viral antigens to CD4(+) T cells but are incapable of cross-presentation. Eur J Immunol 41:2619–2631

Vassallo R, Standing JE, Limper AH (2000) Isolated *Pneumocystis carinii* cell wall glucan provokes lower respiratory tract inflammatory responses. J Immunol 164:3755–3763

von Meyenn F, Schaefer M, Weighardt H, Bauer S, Kirschning CJ, Wagner H, Sparwasser T (2006) Toll-like receptor 9 contributes to recognition of *Mycobacterium* bovis Bacillus Calmette-Guerin by Flt3-ligand generated dendritic cells. Immunobiology 211:557–565

Vuk-Pavlovic Z, Standing JE, Crouch EC, Limper AH (2001) Carbohydrate recognition domain of surfactant protein D mediates interactions with *Pneumocystis carinii* glycoprotein A. Am J Respir Cell Mol Biol 24:475–484

Walenkamp AM, Verheul AF, Scharringa J, Hoepelman IM (1999) Pulmonary surfactant protein A binds to *Cryptococcus neoformans* without promoting phagocytosis. Eur J Clin Invest 29:83–92

Walsh NM, Wuthrich M, Wang H, Klein B, Hull CM (2017) Characterization of C-type lectins reveals an unexpectedly limited interaction between *Cryptococcus neoformans* spores and Dectin-1. PLoS ONE 12:e0173866

Wang SF, Huang JC, Lee YM, Liu SJ, Chan YJ, Chau YP, Chong P, Chen YM (2008) DC-SIGN mediates avian H5N1 influenza virus infection in cis and in trans. Biochem Biophys Res Commun 373:561–566

Warris A (2014) The biology of pulmonary aspergillus infections. J Infect 69(Suppl 1):S36–S41

Wartha F, Beiter K, Albiger B, Fernebro J, Zychlinsky A, Normark S, Henriques-Normark B (2007) Capsule and D-alanylated lipoteichoic acids protect *Streptococcus pneumoniae* against neutrophil extracellular traps. Cell Microbiol 9:1162–1171

Werner JL, Gessner MA, Lilly LM, Nelson MP, Metz AE, Horn D, Dunaway CW, Deshane J, Chaplin DD, Weaver CT, Brown GD, Steele C (2011) Neutrophils produce interleukin 17A (IL-17A) in a dectin-1- and IL-23-dependent manner during invasive fungal infection. Infect Immun 79:3966 3977

Werner JL, Metz AE, Horn D, Schoeb TR, Hewitt MM, Schwiebert LM, Faro-Trindade I, Brown GD, Steele C (2009) Requisite role for the dectin-1 beta-glucan receptor in pulmonary defense against *Aspergillus fumigatus*. J Immunol 182:4938–4946

Werninghaus K, Babiak A, Gross O, Holscher C, Dietrich H, Agger EM, Mages J, Mocsai A, Schoenen H, Finger K, Nimmerjahn F, Brown GD, Kirschning C, Heit A, Andersen P, Wagner H, Ruland J, Lang R (2009) Adjuvanticity of a synthetic cord factor analogue for subunit *Mycobacterium tuberculosis* vaccination requires FcRgamma-Syk-Card9-dependent innate immune activation. J Exp Med 206:89–97

Wiesner DL, Smith KD, Kotov DI, Nielsen JN, Bohjanen PR, Nielsen K (2016) Regulatory T cell induction and retention in the lungs drives suppression of detrimental Type 2 Th cells during pulmonary cryptococcal infection. J Immunol 196:365–374

Wilson GJ, Marakalala MJ, Hoving JC, van Laarhoven A, Drummond RA, Kerscher B, Keeton R, van de Vosse E, Ottenhoff TH, Plantinga TS, Alisjahbana B, Govender D, Besra GS, Netea MG, Reid DM, Willment JA, Jacobs M, Yamasaki S, van Crevel R, Brown GD (2015) The C-type lectin receptor CLECSF8/CLEC4D is a key component of anti-mycobacterial immunity. Cell Host Microbe 17:252–259

Wurzburg BA, Tarchevskaya SS, Jardetzky TS (2006) Structural changes in the lectin domain of CD23, the low-affinity IgE receptor, upon calcium binding. Structure 14:1049–1058

Yadav M, Schorey JS (2006) The beta-glucan receptor dectin-1 functions together with TLR2 to mediate macrophage activation by mycobacteria. Blood 108:3168–3175

Yamamoto H, Nakamura Y, Sato K, Takahashi Y, Nomura T, Miyasaka T, Ishii K, Hara H, Yamamoto N, Kanno E, Iwakura Y, Kawakami K (2014) Defect of CARD9 leads to impaired accumulation of gamma interferon-producing memory phenotype T cells in lungs and increased susceptibility to pulmonary infection with *Cryptococcus neoformans*. Infect Immun 82:1606–1615

Yonekawa A, Saijo S, Hoshino Y, Miyake Y, Ishikawa E, Suzukawa M, Inoue H, Tanaka M, Yoneyama M, Oh-Hora M, Akashi K, Yamasaki S (2014) Dectin-2 is a direct receptor for mannose-capped lipoarabinomannan of mycobacteria. Immunity 41:402–413

Zamze S, Martinez-Pomares L, Jones H, Taylor PR, Stillion RJ, Gordon S, Wong SY (2002) Recognition of bacterial capsular polysaccharides and lipopolysaccharides by the macrophage mannose receptor. J Biol Chem 277:41613–41623

Zenaro E, Donini M, Dusi S (2009) Induction of Th1/Th17 immune response by *Mycobacterium tuberculosis*: role of dectin-1, Mannose Receptor, and DC-SIGN. J Leukoc Biol 86:1393–1401

Zhang J, Tachado SD, Patel N, Zhu J, Imrich A, Manfruelli P, Cushion M, Kinane TB, Koziel H (2005) Negative regulatory role of mannose receptors on human alveolar macrophage proinflammatory cytokine release in vitro. J Leukoc Biol 78:665–674

Zhao G, Xu Q, Lin J, Chen W, Cui T, Hu L, Jiang N (2017) The role of Mincle in innate immune to fungal keratitis. J Infect Dev Ctries 11:89–97

Zhao XQ, Zhu LL, Chang Q, Jiang C, You Y, Luo T, Jia XM, Lin X (2014) C-type lectin receptor dectin-3 mediates trehalose 6,6'-dimycolate (TDM)-induced Mincle expression through CARD9/Bcl10/MALT1-dependent nuclear factor (NF)-kappaB activation. J Biol Chem 289:30052–30062

Zhu LL, Zhao XQ, Jiang C, You Y, Chen XP, Jiang YY, Jia XM, Lin X (2013) C-type lectin receptors Dectin-3 and Dectin-2 form a heterodimeric pattern-recognition receptor for host defense against fungal infection. Immunity 39:324–334

Signaling C-Type Lectin Receptors in Antifungal Immunity

Maxine A. Höft, J. Claire Hoving, and Gordon D. Brown

Contents

1 Introduction 64
2 The Big 4 Fungal Pathogens of Humans 65
3 Fungal Cell Wall Structure 67
4 Fungal Recognition Systems 69
5 C-Type Lectin Receptors 70
5.1 Signaling Pathways 72
5.2 Signaling C-Type Lectin Receptors in Antifungal Immunity 74
6 Conclusions 88
References 88

Abstract We are all exposed to fungal organisms daily, and although many of these organisms are not harmful, billions of people a year contract a fungal infection. Most of these infections are not fatal and can be cleared by the host immune response. However, due to an increase in high-risk populations, the global fungal burden has increased, with more than 1.5 million deaths per year caused by invasive fungal infections. The fungal cell wall is an important surface for interacting with the host immune system as it contains pathogen-associated molecular

M. A. Höft · J. C. Hoving
AFGrica Medical Mycology Research Unit, Institute of Infectious Disease and Molecular Medicine (IDM) at the University of Cape Town, Werner & Beit South Building, Anzio Road, Observatory, 7925 Cape Town, South Africa
e-mail: HFTMAX001@myuct.ac.za

J. C. Hoving
e-mail: jennifer.hoving@uct.ac.za

G. D. Brown (✉)
Medical Research Council Centre for Medical Mycology at the University of Exeter, Geoffrey Pope Building, Stocker Road, EX4 4QD Exeter, UK
e-mail: Gordon.Brown@exeter.ac.uk

Current Topics in Microbiology and Immunology (2020) 429: 63–101
https://doi.org/10.1007/82_2020_224

Published Online: 17 September 2020

patterns (PAMPs) which are detected as being foreign by the host pattern recognition receptors (PRRs). C-type lectin receptors are a group of PRRs that play a central role in the protection against invasive fungal infections. Following the recognition of fungal PAMPs, CLRs trigger various innate and adaptive immune responses. In this chapter, we specifically focus on C-type lectin receptors capable of activating downstream signaling pathways, resulting in protective antifungal immune responses. The current roles that these signaling CLRs play in protection against four of the most prevalent fungal infections affecting humans are reviewed. These include *Candida albicans, Aspergillus fumigatus, Cryptococcus neoformans* and *Pneumocystis jirovecii.*

1 Introduction

Fungi are multicellular eukaryotic organisms including yeasts, molds and mushrooms, all classified under the kingdom fungi. There are millions of species of fungi, which can grow and colonize a diverse range of habitats and living conditions. These include the ocean, freshwaters, plants, animals and are particularity abundant in organic substrates such as soil and plant debris. Humans are known to have commensal fungal organisms associated with the gastrointestinal tract, skin, mouth and other mucosal surfaces (e.g., *Candida, Saccharomyces* and *Malassezia* species), and they typically do not cause infection and disease in healthy individuals (Limon et al. 2017). However, fungi can also have harmful relationships with the host by commensal overgrowth termed as fungal dysbiosis. An opportunistic fungal pathogen can also cause infection if there is a break in the physical barriers of the innate immune system or if the host is immunocompromised (Iliev and Leonardi 2017; Brown 2011). An estimated 600 species of fungi are able to cause disease in humans. Increased incident and prevalence rates of invasive fungal infections are due to the increased number of at-risk populations (Bongomin et al. 2017) owing to the HIV/AIDS pandemic, modern medical practices and other diseases. These include tuberculosis, chronic obstructive pulmonary disease (COPD) and cancer; ultimately leading to more frequent use of chemotherapy, corticosteroids and various other immune suppressant drugs (Brown et al. 2012a; Armstrong-James et al. 2014).

The severity of fungal infections can range from superficial cutaneous infection, to life-threatening systemic infection. Depending on fungal virulence factors, pathology can result in inflammation of infected areas or extensive tissue damage. Approximately, 1.5 million people are killed annually by infections caused by a variety of invasive fungal pathogens. The majority of these deaths, approximately 90%, are related to infections caused by *Candida albicans, Aspergillus fumigatus, Cryptococcus neoformans* and *Pneumocystis jirovecii* (https://www.gaffi.org/why/fungal-disease-frequency/) (Brown et al. 2012a, b).

Fungi have evolved mechanisms to change their cell shape, form, and developmental strategies in response to signals in their immediate environment. The fungal cell contains various pathogen-associated molecular patterns (PAMPs) that are recognized by cells of our innate immune system, and the fungal cell wall is often the target of antifungal drugs. PAMP epitopes are recognized by pattern recognition receptors (PRRs) on the surface of innate immune cells and are responsible for initiating an immune response in the host. This chapter focuses on a type of PRR called C-type lectin receptors (CLRs). In particular, we will describe the CLRs that are able to induce intracellular activation pathways that activate and modulate important immune responses.

Below is a brief description of the four most lethal invasive fungal pathogens of humans, followed by a brief introduction to their fungal cell wall structure.

We will then discuss the main signaling CLRs responsible for recognizing fungal pathogens, with descriptions of the current findings and roles these signaling CLRs play in fungal immunity.

2 The Big 4 Fungal Pathogens of Humans

Candida albicans is a common commensal fungus making up part of the microflora of 30–70% of healthy adults. It is found on the skin, in the oral cavity, genitourinary tract and the epithelial and mucosal surface of the gastrointestinal (GI) tract (Rizzetto et al. 2015). *Candida* species as a whole have caused more than 750,000 invasive infections with a mortality rate of 40% (if treated) (https://www.gaffi.org/why/fungal-disease-frequency/) and is the fourth most common pathogen to cause nosocomial infections (Hidron et al. 2008; Alangaden 2011). *Candida albicans* is polymorphic, having three morphologies, with the ability to grow as a yeast, pseudohyphae, and as hyphae (Saville et al. 2003). The ability to switch morphologies has been shown to give the pathogen increased virulence (Wächtler et al. 2011). There are different types of *C. albicans* infection, classified by where the infection occurs: mucosal candidiasis (e.g., oral, GI tract, respiratory), cutaneous candidiasis (e.g., skin, nail bed) and systemic candidiasis (e.g., disseminated infection or organ infection). As a commensal, the fungus does not cause disease; however, if the anatomical physical barriers of the host's innate immune system are damaged (e.g., skin and mucous membranes), the fungus can cause mucosal and systemic invasive candidiasis. *C. albicans* can disseminate into the bloodstream and deep tissues after crossing the epithelial layer of the GI tract (Gow et al. 2012) causing serious and often lethal infection (30–40% mortality rate) (Perlroth et al. 2007). *C. albicans* can be an opportunistic pathogen in immune-compromised individuals. Chronic mucocutaneous candidiasis is the human infection caused by an overgrowth of the fungus on the skin, nail beds and vaginal and oral mucosal surfaces. This severe and chronic infection is caused by the dysfunction of T cells which are unable to effectively activate phagocytic cells.

Aspergillus fumigatus is a saprotrophic fungus and is the most common species within the genus *Aspergillus* (comprising of approximately 200 species) to cause infections in humans (Gugnani 2003). Unlike *Candida*, which is a commensal fungus found in the gut or mucosal surfaces, *A. fumigatus* is extremely abundant in the environment, with the ecological niche being the soil. The filamentous life form is found growing in the soil producing conidia which is dispersed into the air and inhaled by the host (Latgé and Chamilos 2019). *A. fumigatus* is not dimorphic (does not form yeast cells) yet has successfully evolved to survive in the human host causing devasting infections. Hypersensitivity to *A. fumigatus* allergens can cause pulmonary disease such as allergic bronchopulmonary aspergillosis (ABPA). ABPA has been reported in patients with cystic fibrosis (1–15%) and in individuals suffering from asthma (2%). *A. fumigatus* can also cause invasive aspergillosis (IA) in individuals with weakened immune or lung function. Over 300,000 cases of IA have been reported with mortality rates of between 30 and 50%. Conidia in the alveoli of the lung and can enter the lung epithelium, and following germination, hyphae can invade the alveolar epithelium resulting in IA (Latgé and Chamilos 2019). Patients at risk of IA include those with immunodeficiencies (e.g., HIV and chronic granulomatous disease), those receiving immunosuppressive therapy, hematopoietic stem cell transplant (HSCT) and solid-organ (especially lung) transplant recipients, patients suffering from chronic obstructive pulmonary disease (COPD) and those admitted to intensive care units (7% incidence rates) (Denning 1998; Kousha et al. 2011; Baddley et al. 2013). IA is a severe disease requiring long-term hospitalization and antifungal treatments with high mortality rates particularly in neutropenic patients and HSCT recipients (50–90%) (Kousha et al. 2011).

Cryptococcus neoformans is an encapsulated yeast, producing hyphae during its sexual phase which give rise to basidiospores and eventually spores (Zhao et al. 2019). It is a ubiquitous fungus particularly found in soil, trees and pigeon feces whereby spores are released into the air leading to exposure by inhalation. Cryptococcus has two separate species, *Cryptococcus neoformans* (serotypes A and D cause prevalent disease in immunocompromised patients) and *Cryptococcus gattii* (serotypes B and C are known to infect immunocompetent people) (Sloan and Parris 2014). *C. neoformans* is the cause of 15% of AIDS-related deaths worldwide, and in sub-Saharan Africa, it is the main cause of meningitis in HIV positive adults (approximately 223,000 reported cases in patients with advanced HIV disease) (Rajasingham et al. 2017). In healthy individuals, exposure is normally asymptomatic, and the organism is typically cleared or becomes dormant. If the patient is immunocompromised, yeast cells can disseminate from the lung into the bloodstream causing disseminated cryptococcosis (Zhao et al. 2019). The opportunistic fungus is able to infect various tissues (e.g., kidney and liver), particularly targeting the central nervous system causing cryptococcal meningitis and meningoencephalitis resulting in 100% mortality if not treated (Buchanan and Murphy 1998; Perfect 2005). If treated, mortality rates range from 15 to 20% in developed countries such as the USA, and at least 50% in developing countries. *C. neoformans* can also form enlarged encapsulated titan cells during infection (up to 100 μm cell diameter) (Zaragoza et al. 2009). Titan cells are large polyploid cells with thick cell walls and are a small subset of *Cryptococcus* cells found in infected lung tissue (10–20%). Due to their large size, phagocytosis does not occur and the thick cell wall poses increased resistance to antifungal drugs (Zaragoza et al. 2009).

Pneumocystis jirovecii is an ascomycetous fungus and the cause of potentially severe fungal pneumonia. After taxonomic and DNA analysis of the different strains of *Pneumocystis*, renaming and clarification of species were reported based on the host specificity. There are various strains for different hosts: *P. jirovecii* refers to the fungus isolated from humans (Stringer et al. 2002; Frenkel 1999), *P. carinii* refers to the fungus infecting rats and *P. murina,* the organism causing infection in mice. Person-to-person transmission of *Pneumocystis* is thought to occur via an airborne route, including from those who are asymptomatic carriers. *P. jirovecii* only grows in human lung tissue, and therefore, culturing the fungus in vitro has not been possible (Thomas and Limper 2004). Approximately, 20% of immunocompetent individuals acquire the fungus, which is typically controlled by the immune system without causing disease (Medrano et al. 2005). This exposure can happen in early childhood and has been detected in healthy infants (Vargas et al. 2001). However, *Pneumocystis* pneumonia is a serious opportunistic infection of the lungs predominantly occurring in immunocompromised individuals, of which many are HIV positive patients (over 400 000 cases in AIDS patients and over 100,000 cases in non-AIDS patients). Incidence rates are also rising in developed countries (e.g., UK and USA) due to an increase in immunosuppression drugs or therapies. *Pneumocystis* pneumonia has mortality rates of approximately 50% in non-AIDS patients, and in AIDS patients, it ranges from 15% if treated to 100% mortality if untreated (https://www.gaffi.org/why/fungal-disease-frequency/).

3 Fungal Cell Wall Structure

The fungal cell wall is comprised of approximately 90% carbohydrates and 10% proteins. This is generally made up of an outer layer of mannans meshed with α- and β-glucans and an inner layer of chitins. The composition of these components varies in different species and their different life forms (e.g., yeast, conidia, hyphae, pseudohyphae). The cell wall provides various functions such as maintaining the cell shape and rigidity, iron exchange, metabolism and of course is the surface that interacts with the host immune and defense mechanisms.

Candida albicans outer cell wall consists predominately of O-linked, N-linked and phosphorylated mannans which form glycoproteins with other protein associations (Erwig and Gow 2016) (Fig. 1d). The dense inner cell wall consists of chitin, galactomannan, and other proteins which are covalently bound to β-glucans. The transition from yeast to hyphae alters the cell wall protein composition, increasing chitin levels, and causes modification to closed β-glucan linked structures and glycosylation (Lowman et al. 2014; Chaffin et al. 1998; Heilmann et al. 2011). It has also been reported that *C. albicans* can alter its cell wall components when exposed to lactate, a metabolite produced by the host, as a mechanism of evading immune recognition (Ballou et al. 2017).

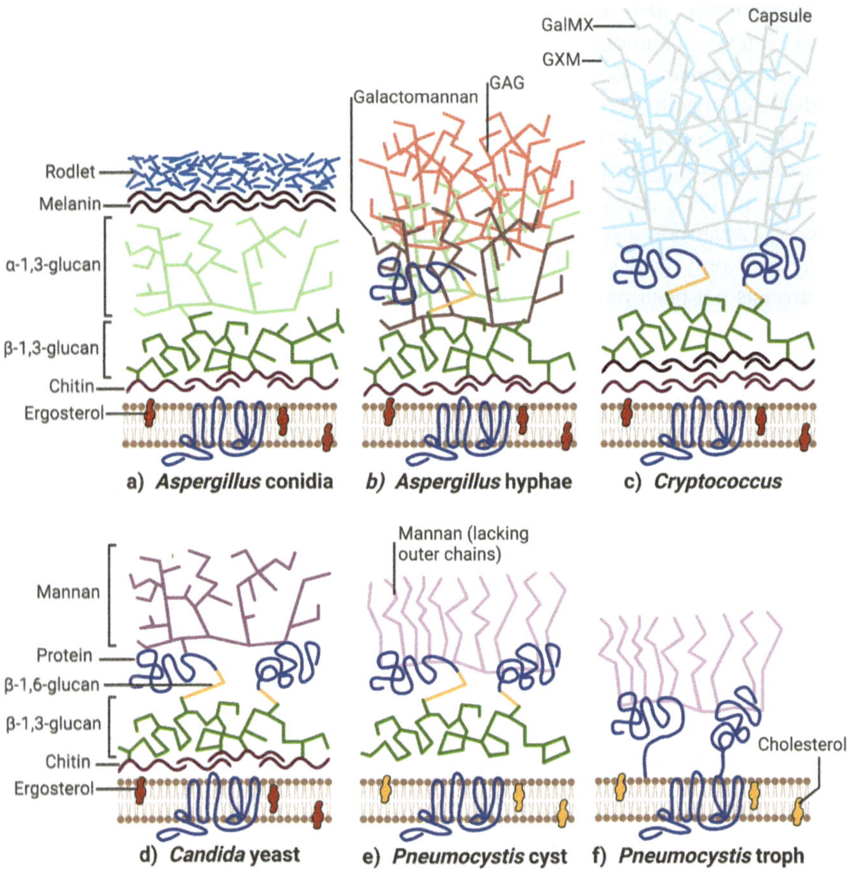

Fig. 1 Fungal cell wall composition of the big 4 fungal pathogens of humans. Adapted from Erwig and Gow (2016) with permission. A schematic depicting the cell wall components of the four most prevalent fungal pathogens is shown, presenting the various layers. **a–d** The inner fungal cell wall layer is made up of highly proinflammatory β-1,3-glucan, β-1,6-glucan and chitin, with ergosterol present in the cell membrane. **e**, **f** However, *Pneumocystis* lacks the enzymes required to synthesize chitin and outer N-mannan chains, and has cholesterol in the cell membrane which is derived from the host. **f** *Pneumocystis* troph also lacks β-1,3-glucan and β-1,6-glucan. **a** The outer cell wall of *Aspergillus spp* conidia contains an exterior hydrophobin rodlet layer, including melanin and α-1,3-glucan **b** while the outer layer of *Aspergillus fumigatus* hyphae contains galactomannan and galactosaminoglycan (GAG) and α-1,3-glucan. **c** The *Cryptococcus* cell wall is encapsulated by a thick layer of glucuronoxylomannan (GXM) and galactoxylomannan (GalXM) and contains melanin in the inner cell wall. **d** The outer cell wall of *Candida spp* is predominantly made up of proinflammatory mannan proteins (phosphorylated mannans, O-linked mannans and N-linked mannans). **e**, **f** *Pneumocystis* outer cell walls contain N-mannan and O-mannan proteins

Aspergillus fumigatus fungal cells have different cell wall compositions depending on its morphology, exposing different PAMPs to the host immune system. Conidia contain α-1,3-glucan, 1,8-dihydroxynaphthalene-(DHN) melanin and a rodlet layer comprised of hydrophobins that coat the outer cell wall (Fig. 1a) (van de Veerdonk et al. 2017). Conidia germinate to form hyphae, which lack the rodlet and melanin layers of the cell wall (Fig. 1b).

Hyphal cell walls contain α-1,3-glucans, galactomannan, galactosaminogalactan (GAG) and other proteins that make up the extracellular matrix (Fig. 1b). This matrix plays a role in enhancing virulence and disease progression.

Cryptococcus neoformans has a distinctly unique gelatinous outer coating that encapsulates the cell, contributing to its very thick cell wall structure. The capsule is comprised of glucuronoxylomannan (GXM) and galactoxylomannan (GalXM) polysaccharides that mask β-glucan exposure (Fig. 1c). This masking inhibits immune cell detection and activation of phagocytosis and proinflammatory responses induced by certain cell receptors (Buchanan and Murphy 1998). The inner cell wall of *C. neoformans* contains 3,4-dihydroxyphenylalanine (DOPA) melanin formed by laccase and accumulates in melanin granules that consist of fungal melanosomes (Fig. 1c) (Camacho et al. 2019). Melanin in *C. neoformans* cells have been reported to increase virulence of the fungal pathogen by enhancing resistance to host immune effector mechanisms (Casadevall et al. 2000). Titan cells have different cell wall and capsule compositions with altered levels of chitin and glucans which improves immune evasion (Mukaremera et al. 2018).

Pneumocystis jirovecii has three distinct life cycle stages while growing in the lung: trophic form, precyst form and mature cyst (or ascus) which differ in structure and size. The cyst has heavily mannosylated (*N*-mannan and O-mannan) glycoproteins (including glycoprotein A (gpA) or MSG) in the outer cell wall that mask the *β*-glucan exposure) (Fig. 1e). *P. jirovecii* lacks chains of N-mannans on the outer cell wall which are present in *C. albicans* and *A. fumigatus* (Ma et al. 2016). While both the trophic (Fig. 1f) and the precystic forms lack β-glucans (Chabé et al. 2011; Nollstadt et al. 1994). It has adapted to grow especially in the host lung and has lost the ability to grow independently due to a loss of specific genes and pathways that are not required for surviving in the host. In fact, it uses host derived cholesterol in its cell membrane rather than synthesizing ergosterol as do many other fungi (Fig. 1e, f). It is unculturable in vitro suggesting that it is an obligate biotroph, explaining the differences seen in the cell wall of this fungus. It is still unclear as to whether *P. jirovecii* has any chitin content within the inner cell wall layer (Skalski et al. 2015; Ma et al. 2016).

4 Fungal Recognition Systems

Despite the prevalence and high mortality rates of invasive fungal infections, they continue to be misdiagnosed or underdiagnosed and are largely understudied as compared to bacterial and viral infections. Due to an increase in prevalence of disease, the importance of understanding how our immune system interacts with these severe fungal pathogens has been recognized. A key part of our innate immune system is physical barriers such as our skin, mucosal surfaces and the antimicrobial peptides (AMPs), that are released by epithelial cells making up these barriers. These AMPs are effective in protecting us from infections caused by many microbes including fungi (Brown 2011; Schop 2007). If these protective barriers are compromised, the pathogen can enter tissues where cells of the innate immune system recognize non-self from self, via specific receptors which activate the cells and promote killing. It is well accepted that phagocytes such as macrophages, neutrophils, monocytes and dendritic cells play an important protective role against fungal infections. Indeed, the loss of these cells, or faults in their effector functions, increases the hosts susceptibility to disease.

These phagocytes can directly (non-opsonic) recognize PAMPs on cell wall components of fungi using membrane-bound PRRs, which are widely expressed across all cell types (Janeway 1992). There are four main families of PRRs: Toll-like receptors (TLRs), Nod-like receptors (NLRs), RIG-I-like receptors (RLRs) and C-type lectin receptors (CLRs). Indirect recognition can occur through membrane-bound opsonic receptors [e.g., complement receptor 3 (CR3)] which recognize soluble PRRs (secreted by various cells including non-phagocytic cells) that opsonize (coat) the pathogen to allow for easy detection (Brown 2011). Soluble PRRs include complement, surfactant proteins (e.g., SP-A and SP-D), collectins (e.g., CL-K1), mannan-binding lectin (MBL), etc. (Brown et al. 2018). Adaptive immune responses can also be activated, producing fungal-specific antibodies which also increases opsonic recognition of some fungal pathogens. These PRRs not only mediate the binding of pathogens but also induce intracellular signaling pathways which activate cellular immune responses involved in microbial take up, killing and other proinflammatory immune responses. PRRs that are directly involved in antifungal immunity include TLRs: TLR1, TLR2, TLR4 and TLR6; NLRs: NOD2 and NLRP3; CLRs: Dectin-1, Dectin-2, Mincle, MCL, DC-SIGN, MR and MelLec, CD23; and others including FcγR, CR3, CD36 and Galectin 3.

Although there are multiple mechanisms involved in fungal protection, CLRs have been reported to be vital in antifungal immunity and will be the focus of this chapter.

5 C-Type Lectin Receptors

C-type lectin receptors make up a large family of carbohydrate-binding proteins consisting of over 1000 soluble and membrane-bound proteins clustered into at least 17 different groups based on their structures and function (for a complete list of CLRs refer to: http://www.imperial.ac.uk/research/animallectins/ctld/mammals/humanvmousedata.html) (Zelensky and Gready 2005). CLRs all have at least one C-type lectin-like domain (CTLD). The binding specificity of the CTLD is determined by its EPN (Glu-Pro-Asn) and QPD (Gln-Pro-Asp) motifs which recognize mannose-like and galactose-like carbohydrate structures, respectively. These receptors and binding domains are highly conserved in vertebrates (Zelensky and Gready 2005). They were originally named after their ability to bind to carbohydrates in a calcium-dependent manner, but it has since been reported that some CLRs can bind ligand in a calcium-independent manner (e.g., β-glucan recognition by Dectin-1). CLRs can also bind non-carbohydrate ligands such as lipids, proteins and lipolipids (Zelensky and Gready 2005). For an extensive overview on the structure and function of the other lectin families, the reader is referred to a highly informative Web site http://www.imperial.ac.uk/research/animallectins/ctld/lectins.html. Some CLRs bind self-ligands such as dead or damaged cells (e.g., DNGR-1 and Lox-1), others can bind both microbial and self-ligands and may respond differently to each of them (e.g., Mincle and DC-SIGN). CLRs such as Dectin-1 and Dectin-2 predominantly recognize microbial organisms.

Of particular importance to antifungal immunity, are the membrane-bound CLRs (belonging to groups II, V and VI) that are predominantly expressed on myeloid cells (Fig. 2). Group II includes the calcium-dependent CLRs with single carbohydrate recognition domains (CRDs) (e.g., Dectin-2, Mincle, DC-SIGN, MCL),

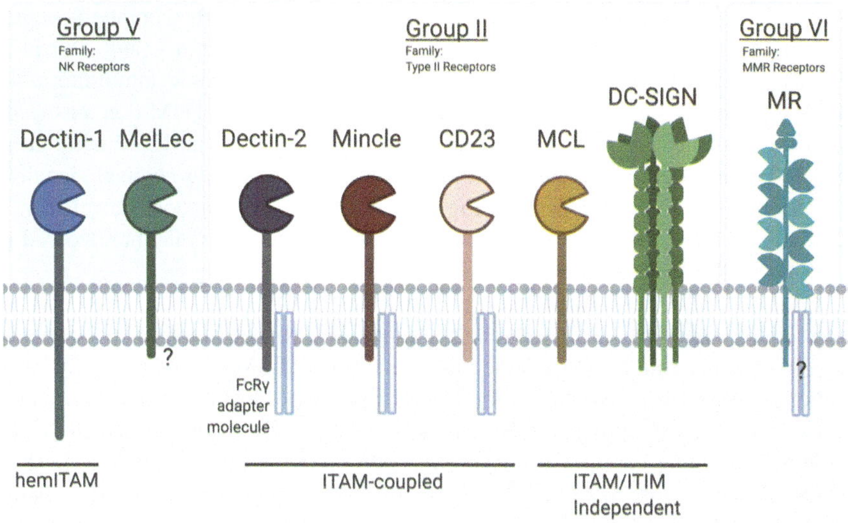

Fig. 2 Schematic representation of signaling C-type lectin receptors involved in antifungal immunity. Dectin-1 and MelLec belong to the family of Natural Killer (NK) receptors within Group V of CLRs. Dectin-1 contains a hemITAM motif however MelLec does not contain conserved motifs in the cytoplasmic tail region. Dectin-2, Mincle, CD23, MCL and DC-SIGN belong to the group II, type II of CLRs. Dectin-2, Mincle and CD23 have been shown to associate with FcRγ chain which contains ITAM tyrosine motifs. Upon ligand binding, the ITAMs are phosphorylated to initiate downstream signaling, however MCL and DC-SIGN have been shown to induce signaling independent of known ITAM containing adaptor molecules such as FcRγ chain. MCL has been reported to form a functional complex with Mincle, which may explain the downstream signaling capabilities of the receptor. MR (macrophage mannose receptor family) belongs to group VI of CLRs and it remains unclear what adaptor molecule MR associates with

group V includes calcium-independent CLRs with single CTLDs (e.g., Dectin-1 and MelLec) and group VI contains calcium-independent CLRs with multiple CRDs (e.g., MR) (Fig. 2) (Zelensky and Gready 2005; Sancho and Reis e Sousa 2012; Hoving et al. 2014). These receptors function as PRRs that recognize PAMPs of fungal pathogens and trigger immune responses such as phagocytosis, respiratory burst, inflammasome activation, etc. In some cases, these responses can occur independent of Syk (see below) and transcription factor activation (Drummond et al. 2011; Strasser et al. 2012).

In this chapter, we will be focusing on the major CLRs that activate downstream signaling pathways to deploy effector functions that play a central role in antifungal immunity (Fig. 2). These include Dectin-1, Dectin-2, Mincle, CD23, MCL, DC-SIGN, MR and MelLec.

5.1 Signaling Pathways

CLRs can be further subdivided into different types based on their signaling potential: (i) activation Syk-coupled CLRs possessing immunoreceptor tyrosine-based activation motif (ITAM) domains (e.g., Dectin-1 and Dectin-2; Fig. 2); (ii) inhibitory CLRs with immunoreceptor tyrosine-based inhibition motif (ITIM) domains (e.g., DCIR) or (iii) CLRs without ITAM or ITIM domains (e.g., MR, MCL and DC-SIGN; Fig. 2) (Sancho and Reis e Sousa 2012; Hoving et al. 2014). The intracellular signaling mechanisms of CLRs lacking atypical signaling domains (ITAM/ITIM independent Fig. 2) are not fully understood; however, it is thought that these CLRs do not function in isolation and some may be involved in modulating the signaling outcomes of other receptors.

CARD9 Syk-dependent signaling: Activation CLRs (e.g., Dectin-1) trigger intracellular signaling directly through a single ITAM motif (consensus sequence YxxL/I) found in their cytoplasmic tail (termed hemITAM) or indirectly by association with an ITAM-bearing FcRγ adaptor molecule. ITAMs are phosphorylated and subsequently recruit spleen tyrosine kinase (Syk) (Kerrigan and Brown 2011). CLR-Syk coupling can also occur through other adaptor proteins such as DAP12 which also bear Syk-recruiting ITAM motifs. Activating Syk results in a phosphorylation cascade that recruits caspase recruitment domain-containing protein 9 (CARD9), B-cell lymphoma/leukemia 10 (BCL-10) and mucosa-associated lymphoid tissue 1 (MALT1) (Gringhuis et al. 2011). A complex is formed, and its activation induces the expression of transcription factors such as NF-κB which stimulates the production of cytokines and chemokines (Fig. 3).

Raf-1 Syk-independent signaling: Expression of NF-κB can also be induced by CLR signaling pathways that are Syk-independent (e.g., Dectin-1 and DC-SIGN). Activated CLRs can recruit GTPase Ras proteins which subsequently activate serine/threonine-protein kinase Raf-1. This results in Raf-1 mediated phosphorylation of the p65 subunit of NF-κB (Geijtenbeek and Gringhuis 2009) (Fig. 3).

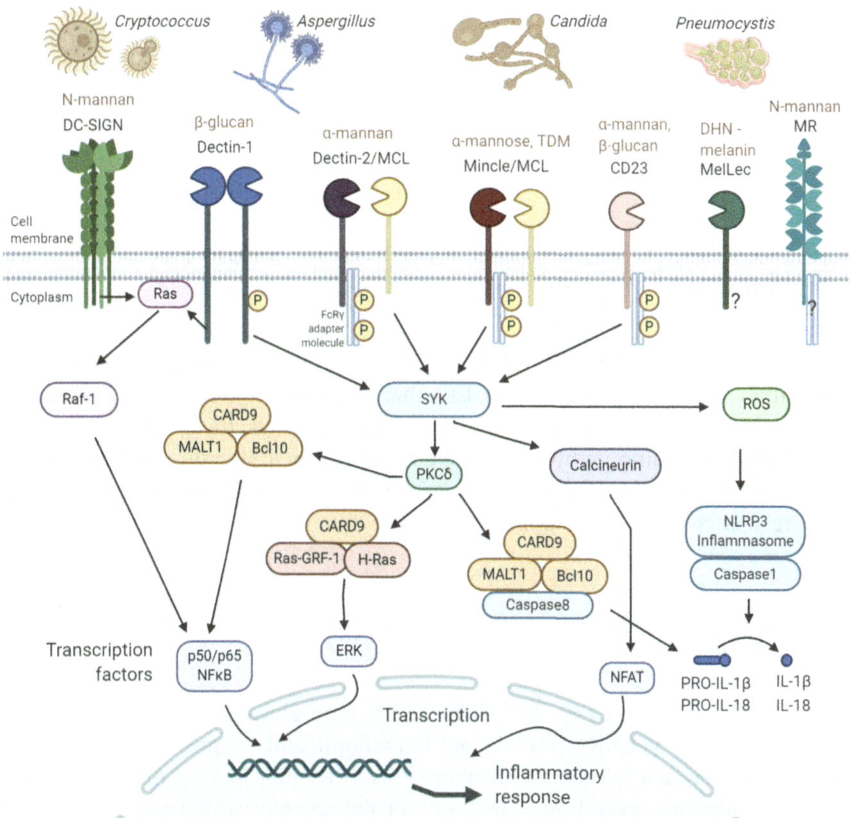

Fig. 3 CLR downstream signaling pathways occurring during innate immune recognition of fungi. CLRs recognize and are activated by various fungal pathogens inducing intracellular signaling cascades that result in multiple immune responses. Dectin-1, Dectin-2, Mincle and CD23 trigger intracellular signaling through a single ITAM motif (consensus sequence YxxL/I). The motif is either found in the cytoplasmic tail (termed hemITAM for Dectin-1) or by association with an ITAM-bearing FcRγ adaptor molecule (Dectin-2, Mincle and CD23). Upon ligand binding, ITAMs are phosphorylated and subsequently recruit spleen tyrosine kinase (Syk) resulting in a phosphorylation cascade that recruits caspase recruitment domain-containing protein 9 (CARD9), B-cell lymphoma/leukemia 10 (BCL-10) and mucosa-associated lymphoid tissue 1 (MALT1). A complex is formed, and its activation induces the expression of transcription factors such as NF-κB which stimulates the production of inflammatory cytokines and chemokines. DC-SIGN and Dectin-1 can also recruit GTPase Ras proteins which subsequently activate serine/threonine-protein kinase Raf-1. This results in Raf-1 mediated phosphorylation of the p65 subunit of NF-κB. The intracellular signaling mechanism of MCL and MR and MelLec is not fully understood, however, it is likely that MCL associates with other receptors such as Mincle and Dectin-2

Some CLRs can trigger multiple pathways, such as Dectin-1 which activates both CARD9 Syk-dependent and Raf-1 Syk-independent signaling, both resulting in the activation of transcription factors and subsequent gene expression (Fig. 3).

Inhibitory CLRs trigger intracellular signaling through ITIM motifs (consensus sequence I/V/L/SxYxxI/L/V) located in their cytoplasmic tails. Upon ligand activation, tyrosine and inositol phosphatases such as SHIP, SHP-1 and SHP-2 are recruited causing the inhibition of certain activation pathways, resulting in suppression of immune cell responses.

5.2 Signaling C-Type Lectin Receptors in Antifungal Immunity

CLRs are the most important PRR family responsible for the detection and recognition of fungi. The signaling CLRs that play a key role in tailoring protective antifungal immune responses are described below. Here we focus on CLRs capable of mediating signaling pathways and we highlight their structure and function as well as emphasizing what is currently known about the specific roles that each of these receptors plays in response to the big four fungal pathogens of humans mentioned above.

5.2.1 Dectin-1

Dectin-1 (gene name: CLEC7A; CLECSF12, Beta-glucan receptor BGR) forms part of the group V CLRs (Fig. 2) and is predominantly expressed as two major isoforms, either as a full-length receptor or as a receptor lacking the stalk region; however, there are several other isoforms of the receptor which are generated by alternative splicing (Brown 2006; Heinsbroek et al. 2006). The isoforms are expressed at different levels depending on the cell type or mouse strain.

Dectin-1 is primarily expressed on the surface of myeloid cells including macrophages, dendritic cells, monocytes, neutrophils as well as certain subsets of T lymphocyte cells and B cells in humans (Brown et al. 2018; Brown and Crocker 2016). Dectin-1 is also expressed in epithelial cells, with differential expression in different tissue types (Brown et al. 2018; Sun et al. 2012). High expression levels have been reported in the lung and intestine. This is thought to be due to the higher risk of pathogen entry at these tissue sites.

Dectin-1 recognizes fungal cell wall β-glucans (Brown and Crocker 2016). Tropomyosin, an arthropod-derived molecule, binds Dectin-1 (Gour et al. 2018), as well as unidentified ligands in *Mycobacteria* (Wagener et al. 2018) and *Leishmania* (Lima-Junior et al. 2017). Additionally, endogenous ligands such as vimentin (Thiagarajan et al. 2013) and galectin-9 (Daley et al. 2017) have been reported to be recognized by Dectin-1.

Dectin-1 β-glucan recognition and binding occurs in a calcium-independent manner. Subsequent to ligand binding, the hem-ITAM region of the receptors cytoplasmic tail, triggers activation of several pathways such as the Syk/CARD9 and Raf-1 signaling pathways described above (Fig. 3). Although Dectin-1 can bind to soluble β-glucans, it will only activate signaling pathways once bound to β-glucan-containing particles such as yeast, as this binding induces the formation of phagocytic synapses (Goodridge et al. 2011). These synapses form by Dectin-1 clustering and the isolation from regulatory tyrosine phosphatases CD45 and CD148, allowing for successful downstream signaling (Goodridge et al. 2011). This mechanism allows innate immune cells to differentiate between direct binding to a microbial cell and the identification of components shed from microbial cell surfaces which would suggest the presence of a nearby infection. SHIP-1 is also a regulatory protein that has been shown to bind to the hemITAM cytoplasmic tail region of Dectin-1 and modulate reactive oxygen species (ROS) production in dendritic cells (Blanco-Menéndez et al. 2015). Other regulators causing the termination of signaling have been described, including the internalization of Dectin-1 during phagocytosis of ligand baring particles (Hernanz-Falcón et al. 2009).

Dectin-1 signaling can therefore lead to the induction or regulation of cellular and immunological responses such as phagocytosis, respiratory burst (resulting in production of ROS), neutrophil extracellular trap formation (NETosis), dendritic cell maturation and antigen presentation, as well as the production of cytokines, chemokines, eicosanoids and other inflammatory mediators (Brown and Crocker 2016). In addition, this receptor can induce inflammasome activation including the NLRP3 and non-canonical caspase-8 inflammasomes, aiding in IL-1β production (Fig. 3) (Brown and Crocker 2016). Other pathogen recognition receptors induce cellular responses which can be modulated by Dectin-1, influencing innate immune memory formation (Brown et al. 2018). Moreover, Dectin-1 plays a role in regulating T-helper type (Th)-1, Th17 and Th2 adaptive immunity and can induce the development of CD4 and CD8 T cells as well as B-cell responses (Brown et al. 2018; Kaisar et al. 2018; Zhao et al. 2016).

Dectin-1 involvement in fungal immunity

Dectin-1 is the most characterized CLR in antifungal immunity, and our understanding has been primarily derived using mouse models. Dectin-1 is required to initiate protective immune responses to various fungal pathogens including specific strains of *Aspergillus, Candida* and *Pneumocystis*. Based on the various cell wall structures shown above (Fig. 1), Dectin-1 interacts with multiple fungal pathogens in different ways, through its ability to recognize the β-1,3-linked glucans. Importantly, a polymorphism of Dectin-1 was found in humans and was linked to an increased susceptibility to fungal infections and disease (Brown and Crocker 2016).

Candida albicans: *C. albicans* has been studied extensively, particularly looking at the role Dectin-1 plays in infection. In vitro experiments show that Dectin-1 is able to bind and recognize *C. albicans* and induce phagocytosis of the pathogen, stimulating the production of ROS, nitric oxide (NO) and cytokines such as proinflammatory TNF-α and IL-6 (Drummond and Brown 2011; Pinke et al. 2016). The protective role for Dectin-1 in the GI tract during systemic *C. albicans* infection has been highlighted in vivo. Dectin-1 deficient dendritic cells were shown to be defective in CD4+ T cell activation (with less IL-2 production), but did not affect CD8+ T cells (Drummond et al. 2016). Dectin-1 involvement in clearing systemic *Candida* infection in mice has been shown to be both protective and redundant depending on the strain of *C. albicans* (Marakalala et al. 2013; Netea et al. 2010; Taylor et al. 2007; Saijo et al. 2007). The route of infection and the strain of mouse used also plays a role in Dectin-1 infection phenotypes. Different mouse strains have different dominant isoforms of Dectin-1, such a full-length Dectin-1A and stalkless Dectin-1B, which induce different immune responses to soluble and particulate β-glucans (Heinsbroek et al. 2006). The role of Dectin-1 in mucosal candidiasis has also been investigated extensively in vivo, many results showing an important protective role for Dectin-1; however, this is dependent on strains of *C. albicans* and the site of infection. Dectin-1 played a key role in controlling gastrointestinal and vaginal candidiasis in C57BL/6 but not BALB/c mice (Carvalho et al. 2012).

Dectin-1 has been shown to play a protective role in controlling *Candida* infection during colitis in mice (Iliev et al. 2012). The importance of Dectin-1 in controlling fungal infections in the GI tract has been shown, and mice deficient for Dectin-1 had impaired macrophage responses to *C. albicans* infection, with reduced phagocytosis of *Candida* yeast cells and a decrease in ROS production (Galès et al. 2010). Although Dectin-1 plays a protective role *in C. albicans* GI tract infection, following disruption of the mucosal barrier, by inducing killing mechanisms and cytokine production, it has a redundant role in the control of GI colonization in healthy animals (Vautier et al. 2012).

Using a mouse skin infection model, it was shown that different morphologies of *C. albicans* (yeast or filamentous forms) determine specific T helper cell responses, and that these responses are tissue specific (Kashem et al. 2015). In this model, yeast was detected by Langerhans cells, and though Dectin-1 signaling, IL-6 is produced driving Th17 cell responses that are protective against cutaneous infection. In contrast, Dectin-1 was unable to bind to filamentous forms of *C. albicans* and did not induce Th17 responses. Instead Th1 responses were induced, and these T helper cells provided protection against systemic infection (Kashem et al. 2015).

Osteomyelitis is an infection caused by the entry of *C. albicans* into skeletal tissue causing serious inflammation and bone destruction. Recently, Dectin-1 was reported to stimulate Nav1.8-positive-nociceptors which induce calcitonin gene-related peptide resulting in decreased osteoinflammation during osteomyelitis and osteoporosis (Maruyama et al. 2017).

The contribution of mycobiota and Dectin-1 has been linked to alcoholic liver disease and cirrhosis. Long-term administration of alcohol to mice caused an increase in commensal fungal growth resulting in translocation and systemic circulation of fungal β-glucans. Dectin-1 induced liver inflammation and IL-1β in response to circulating β-glucans causing hepatic cell and tissue damage (Yang et al. 2017). Upon investigation of alcohol-related cirrhosis in humans, *Candida* overgrowth was observed with increased fungal immune responses and high mortality rates (Yang et al. 2017).

Research done in humans also shows that Dectin-1 plays a role in protection against mucosal and systemic *C. albicans* infection. A single nucleotide polymorphism in the Dectin-1 gene (CLEC7A) has been associated with higher susceptibility to chronic mucocutaneous candidiasis as well other diseases, which is caused by a reduction in Th17 responses (Ferwerda et al. 2009). A link was shown between patients with CLEC7A polymorphism (rs2078178) and severe ulcerative colitis (Iliev et al. 2012). Humans can also completely lack Dectin-1 cell surface expression, due to the Tyr238X Dectin-1 polymorphism, which increases the chances of *Candida* colonization and systemic infection (Plantinga et al. 2009). This is due to a decrease in cytokine production from immune cells containing this polymorphism (Plantinga et al. 2009).

Aspergillus fumigatus: Dectin-1 is involved in innate recognition and induces immune responses toward *A. fumigatus*. In particular, Dectin-1 was reported to play a key role in human dendritic cells in vitro. Inhibition of Dectin-1 in immature dendritic cells and myeloid dendritic cells produced less proinflammatory cytokines in response to *A. fumigatus* (conidia, swollen conidia and geminated spores) (Mezger et al. 2008; Hefter et al. 2017). Human plasmacytoid dendritic cells (pDCs) were shown to express Dectin-1 and induced phagocytosis and killing of *A. fumigatus* conidia (Maldonado et al. 2015). Contradicting this, it was demonstrated that the *A. fumigatus* hyphae are recognized by human pDCs through Dectin-2 rather than Dectin-1 in vitro (Loures et al. 2015). Additionally, the importance of Dectin-1 was demonstrated in *Aspergillus* keratitis (fungal disease of the cornea) and was shown to be responsible for mediating *A. fumigatus* killing by neutrophil recruitment and inducing cytokine and chemokine production (e.g., IL-1β, CXCL1) (Yuan et al. 2017; Leal et al. 2010). In addition to myeloid cells, Dectin-1 expression was shown in non-phagocytic human bronchial epithelial cells and induced innate immune responses against *A. fumigatus* in conjunction with TLR2 (Sun et al. 2012).

A role for Dectin-1 in adaptive immunity has also been shown. Dectin-1 knockout mice were more susceptible during survival studies, with increased fungal burden in the lung due to decreased neutrophil recruitment and IL-17 production (Werner et al. 2009). Dectin-1 was required for the induction of Th1 and Th17 immune responses toward pulmonary *A. fumigatus* infection in vivo and shapes CD4 T cell differentiation (Rivera et al. 2011).

A human Dectin-1 TyrY238X polymorphism was discovered in invasive aspergillosis patients. Polymorphonuclear cells (PMCs) containing the polymorphism produced less cytokines (e.g., IFN-γ) (Chai et al. 2011), and individuals were reported to have increased susceptibility to invasive aspergillosis especially in transplant patients (Fisher et al. 2017).

Cryptococcus neoformans: There are contradicting results regarding the role of Dectin-1 in *Cryptococcus* protection. It was first reported that Dectin-1 interacts with exposed β-glucans on *C. neoformans* conidia cells and induces phagocytosis, but was later shown to be redundant (Walsh et al. 2017). Another report suggested that it was the collaboration of Dectin-1 and Dectin-2 that enabled the phagocytosis of *C. neoformans* (Lim et al. 2018). In vivo experiments show that Dectin-1 deficient mice are not susceptible to *C. neoformans* infection (Nakamura et al. 2007).

Pneumocystis jirovecii: Dectin-1 on alveolar macrophages has been reported to recognize *P. jirovecii* cysts in vitro and enhances phagocytosis and fungal clearance. Supporting this, Dectin-1 overexpressing macrophage cell lines have been shown to increase efficiency of *P. jirovecii* binding (Steele et al. 2003). Mice deficient for Dectin-1 proved to have increased fungal burden following *Pneumocystis* lung infection with a notable decrease in ROS production, and no differences in cytokine production (Saijo et al. 2007).

5.2.2 Dectin-2

Dectin-2 (gene name: CLEC4N; CLEC6A, CLECSF10) forms part of the group II of CLRs (Fig. 2) and is encoded by the natural killer (NK)-gene cluster. The structure of Dectin-2 is similar to that of Dectin-1 (single extracellular C-type lectin-like domain and a transmembrane domain), apart from the fact that it only has a short cytoplasmic tail which lacks detectable signaling domains (Graham and Brown 2009). Characteristic of other CLRs in group II, Dectin-2 dimerises with the ITAM-containing FcRγ chain for mediating surface expression and intracellular signaling (Sato et al. 2006). This association between ITAM-FcRγ adapter molecule is facilitated by a membrane-proximal region found on the intracellular tail of Dectin-2, rather than through transmembrane arginine residues like other similar receptors such as DCAR (Sato et al. 2006). On the surface of certain myeloid cells, Dectin-2 forms a heterodimer with MCL (later discussed) which increases the ligand binding efficiency (Zhu et al. 2013). As mentioned, Dectin-2 is expressed predominantly on myeloid cells such as tissue macrophages, Kupffer cells, dendritic cell subsets, Langerhans cells and neutrophils (Gringhuis et al. 2011; Ariizumi et al. 2000; Loures et al. 2015). Activation of inflammatory responses can also cause upregulation of Dectin-2 on monocytes (Taylor et al. 2005).

Dectin-2 recognizes mannose-based structures, in a calcium-dependent manner (McGreal et al. 2006). This leads to the recognition of multiple pathogens including α-mannans on various fungal species, mannose-capped lipoarabinomannan of mycobacteria, pathogen-derived molecules such as house dust mite (HDM) allergens and schistosome egg antigen, nematodes and other bacteria (Parsons et al. 2014; Ritter et al. 2010; Yonekawa et al. 2014; Shiokawa et al. 2017; Sato et al. 2006). Upon ligand binding through the CTLD [which possess a conserved mannose-binding motif (EPN motif) (McGreal et al. 2006)], signaling from the ITAM region occurs through the Syk/CARD9 pathway (Fig. 3), mentioned above, and mediated by phospholipase Cγ2 (PLCγ2) (Gorjestani et al. 2011). Comparable to Dectin-1, Dectin-2 is involved in induction or regulation of cellular and immunological responses such as phagocytosis, respiratory burst, plasmacytoid dendritic cell extracellular trap formation (Loures et al. 2015), Nlrp3 inflammasome activation, and the production of cytokines, chemokines and ROS (Brown and Crocker 2016). Dectin-2 also plays a role in the production of cysteinyl leukotrienes caused by house dust mite and induces airway inflammation and the development of Th2 responses (Barrett et al. 2009). Complementary findings show a reduction in

HDM-derived airway inflammation in Dectin-2 deficient mice, due to dampened Th17 and Th2 responses and cell differentiation (Plantinga et al. 2013; Norimoto et al. 2014). Moreover, Dectin-2 has also been shown to drive the development of Th1 and Th17 immunity which provides protection responses toward fungi (Wang et al. 2017).

Dectin-2 involvement in fungal immunity

Candida albicans: Dectin-2 recognizes both yeast and filamentous forms of *C. albicans* and is involved in protective immune responses toward infection. Dectin-2 has been well-characterized in mouse models. Dectin-2 recognizes mannose-rich structures such as α-mannans on the surface of *C. albicans* hyphae and induces inflammatory responses by signaling through the Syk pathway and NF-κB expression (Fig. 3) (Sato et al. 2006; Bi et al. 2010; Saijo et al. 2010). Dectin-2 deficient mice were also shown to have an increased susceptibility to *Candida glabrata*, with decreased Th1 and Th17 cytokines and defective phagocytosis in Dectin-2 deficient macrophages and neutrophils (Ifrim et al. 2014). As a result, ROS production from neutrophils, which is an effective mechanism for killing *Candida,* was also reduced in response to *C. albicans.*

Dectin-2 readily forms a heterodimeric complex with MCL which increases binding and recognition efficacy of α-mannans on *C. albicans* hyphae. In vivo studies showed that mice were highly susceptible to *C. albicans* when Dectin-2 activity was blocked using antibodies. The Dectin-2 and MCL heterodimer provided better protection and increased inflammatory responses against *C. albicans* than their respective homodimers (Zhu et al. 2013).

Aspergillus fumigatus: Dectin-2 recognizes *A. fumigatus* and has been shown to play an important role in triggering protective inflammatory responses during infection (Loures et al. 2015). The RodA hydrophobin (a cell wall component of *A. fumigatus* conidia) was reported to mask Dectin-2 mediated immune responses resulting in increased fungal burden in mice (Carrion et al. 2013). *A. fumigatus* produces proteases (neutrophil elastase) that can cleave CLRs including Dectin-2 resulting in abnormal immune responses in mice. This may also be linked to disease progression in cystic fibrosis patients (Griffiths et al. 2018). Dectin-2 together with MR expression has been reported to be expressed by alveolar macrophages found in the lungs of invasive aspergillosis and allergic bronchopulmonary aspergillosis patients suggesting a role for these CLRs in pulmonary aspergillosis defenses (Sun et al. 2013). Dectin-2 has also been shown to induce IL-17RC expression in both mouse and human bone marrow neutrophils in response to *Aspergillus* keratitis (Taylor et al. 2014).

Cryptococcus neoformans: Dectin-2 was demonstrated to play a role in inflammatory responses toward *C. neoformans* in vitro. Dectin-2 deficient dendritic cells and alveolar macrophages produced less IL-12p40 and TNF-α. This is contradicted with in vivo data stating that Dectin-2 deficient mice did not show an increase in susceptibility to *C. neoformans* and that Th1 and Th17 immune responses were not altered. However, mucin production and Th2 immune responses were enhanced (Nakamura et al. 2015). There has been one report of a human Dectin-2 polymorphism leading to pulmonary cryptococcosis (Hu et al. 2015).

Pneumocystis jirovecii: Dectin-2 has been shown to bind and recognize *Pneumocystis* and play a role in innate immune signaling in response to infection. Alveolar macrophages isolated from Dectin-2 deficient mice showed a decrease in Syk kinase activation and a decreased IL-6 and TNF-α cytokine production in response to *Pneumocystis*. However, it was reported that Dectin-2 plays a redundant role in controlling *Pneumocystis* infection in vivo (Kottom et al. 2018).

5.2.3 Mincle

Macrophage-inducible C-type lectin or Mincle (gene name: CLEC4E; CLECSF9) forms part of the group II of CLRs, which is the same group as Dectin-2, hence the similar structure of the receptors (Fig. 2). Mincle has a single extracellular C-type lectin-like domain, with a short cytoplasmic tail, and a transmembrane domain which associates with the ITAM-containing FcRγ chain after ligand recognition via a positively charged arginine residue (Fig. 2) (Yamasaki et al. 2008). Mincle and FcRγ form a heterotrimeric complex with MCL (see below) which allows for cell surface expression and enhanced phagocytosis and downstream signaling (Fig. 3) (Kerscher et al. 2016). Mincle is expressed mainly on myeloid cells such as macrophages, neutrophils, dendritic cells, monocytes, other leukocytes and certain B-cell subsets (Williams 2017). Expression levels of Mincle have been shown to increase during exposure to inflammatory stimuli, as well as through TLR signaling triggered by fungal and other microbial ligands (Patin et al. 2017; Kerscher et al. 2016).

Mincle recognizes mannosyloxystearyl mannitol, galectin and b-gentiobiosyl diglycerides on fungal cells, and mycobacterial ligands such as cord factor (Ishikawa et al. 2009, 2013; Hattori et al. 2014). Additionally, the CRD of Mincle can bind to endogenous ligands such as spliceosome-associated protein (SAP) 130, sterols, and b-glucosylceramides. Like Dectin-2, the CTLD of Mincle also possess an EPN (Glu-Pro-Asn) tripeptide motif, which allows potential binding of mannose, N-acetylglucosamine L-fucose and glucose. However, glycolipid recognition seems more prevalent (Furukawa et al. 2013). Recognition and binding of ligand occur in a calcium-dependent manner.

Following the binding of ligands, the heterotricomplex induces signaling through the FcRγ adaptor molecule via the Syk/CARD9 pathway, as well as the mitogen-activated protein kinase pathways (Fig. 3) (Strasser et al. 2012; Yamasaki et al. 2008; Zhao et al. 2014). Mincle activation leads to the induction and regulation of multiple cellular and immunological responses including phagocytosis, endocytosis, respiratory burst, NET formation, Nlrp3 inflammasome activation and the production of inflammatory cytokines and chemokines (Yamasaki et al. 2008; Lobato-Pascual et al. 2013; Schweneker et al. 2013). Mincle has also been shown to promote Th1, Th17 and Th2 (Wevers et al. 2014) immune development (Brown and Crocker 2016).

Mincle involvement in fungal immunity

Candida albicans: Mincle is able to recognize *C. albicans* yeast. Macrophages lacking Mincle produced less TNF-α both in vitro and in vivo. Mincle knockout mice demonstrated an increase in *C. albicans* fungal burden during systemic candidiasis, suggesting an important role for Mincle in inflammatory responses toward the infection (Wells et al. 2008).

Aspergillus fumigatus: Mincle expression is initially increased upon *A. fumigatus* infection as well as cytokines such as TNF-α and IL-1β in rat and human cornea. This suggests that Mincle may play a role in innate immune recognition and responses of corneal resistance to infection (Zhao et al. 2017). There is, however, a lack of studies exploring the role of this receptor in vivo.

Cryptococcus neoformans: Phagocytic cells lacking Mincle did not show an impairment of *Cryptococcal* binding or uptake, and the receptor was reported to play a redundant role in innate antifungal immunity toward *C. neoformans* (Walsh et al. 2017). Nothing is known about the role of this receptor in vivo during infection with this pathogen.

Pneumocystis jirovecii: *P. murina* was recently reported to be recognized by the CLR Mincle, inducing FcRγ-mediated intracellular signaling. A significant role for Mincle was described during *Pneumocystis* infection (Kottom et al. 2017). Murine macrophages overexpressing Mincle had increased binding to *Pneumocystis* followed by an increase in activation of downstream signaling pathways. Mincle deficient mice infected with *P. murina* did not show an increase in susceptibility in terms of survival, but did show a significant increase in fungal burden. However, mice deficient for both Mincle and CD4 T cells were highly susceptible to infection with defects in fungal clearance and cytokine production (Kottom et al. 2017).

5.2.4 MCL

Originally named, murine macrophage C-type lectin as it was first identified in mice, MCL (gene name: CLEC4D; Dectin-3, CLECSF8, CLEC-6, MPCL) forms part of the group II of CLRs (Fig. 2) and is the least characterized among the receptors in its family. As other CLRs in group II, MCL consists of one extracellular CTLD, a short cytoplasmic tail and a transmembrane domain (Fig. 2). MCL has a similar structure and function as Mincle as they function as a complex (see above). The association of MCL with Mincle does not occur by charged amino acid residues in the transmembrane or cytoplasmic domain, but rather by a different mechanism within the stalk region (Miyake et al. 2013). Intracellular retention of MCL is not mediated by the transmembrane region but by the C-type lectin domain (Kerscher et al. 2016). Unlike Dectin-2 and Mincle, MCL does not associate with the typical adaptor molecules, such as FcRγ chain, DAP12 or DAP10, which are known to be involved in C-type lectin signaling. However, it is still able to trigger cellular activation utilizing the Syk kinase pathway. The mechanism responsible for mediating intracellular retention and activation is not fully understood. MCL has been reported to associate with a similar adaptor protein, FcεRIγ (Lobato-Pascual et al. 2013) and has been shown to be linked to FcRγ (Miyake et al. 2013) but this is likely due to the functional complexes formed with Mincle and Dectin-2. Evidently, MCL has been shown to be required for induction of Mincle surface expression (Lobato-Pascual et al. 2013; Miyake et al. 2013).

Similar to Mincle, MCL is expressed mainly on myeloid cells such as macrophages, neutrophils, dendritic cells, monocytes and other leukocyte and B-cell subsets (Graham et al. 2012; Brown and Crocker 2016). Additionally, the conserved amino acids found on CTLD of other group II CLRs, are absent in the MCL CTLD and therefore are unable to detect unconjugated sugars (Graham et al. 2012). Instead, MCL contains a surface hydrophobic domain that allows recognition of glycolipids and other ligands such as mycobacterial cord factor (Miyake et al. 2013; Furukawa et al. 2013). MCL and Mincle have been shown to be coordinately expressed under naïve and inflammatory responses (Kerscher et al. 2016; Miyake et al. 2015) and is able to trigger phagocytosis, endocytosis, respiratory burst, NET formation, Nlrp3 inflammasome activation and the production of inflammatory cytokines and chemokines (Strasser et al. 2012; Zhu et al. 2013; Yamasaki et al. 2008; Lobato-Pascual et al. 2013; Miyake et al. 2013; Graham et al. 2012; Arce et al. 2004).

MCL involvement in fungal immunity

Candida albicans: Similarly to Dectin-2, it has been demonstrated in mice that MCL recognizes mannose-rich structures such as α-mannans on the surface of *C. albicans* hyphae and is able to induce inflammatory responses through NF-κB-mediated Syk signaling (Zhu et al. 2013). MCL was reported to be unnecessary for protection against high doses of *C. albicans* in a study comparing wild type and MCL deficient mice (Graham et al. 2012). These results were contradicted by a low does infection of *C. albicans* in wild type and MCL deficient mice, demonstrating that MCL can detect low levels of fungal ligands (Zhu et al. 2013). This study also showed that MCL deficient mice had defective activation and subsequent recruitment of innate cells (e.g., neutrophils) to infected organs due to reduced proinflammatory cytokines (Zhu et al. 2013). This suggests a critical role for MCL in the recognition and protection against *C. albicans*.

Aspergillus fumigatus: The role MCL plays in *Aspergillus* infection is unclear and requires further research.

Cryptococcus neoformans: There have been variable reports regarding the role that MCL plays in protection against *C. neoformans* infection, which may be as a result of differences in virulence between the different fungal serotypes. Pulmonary macrophages and dendritic cells did not require MCL for *C. neoformans* uptake in vitro (Campuzano et al. 2017). However, studies using bone marrow derived macrophages showed that MCL recognizes the GXM component of *C. neoformans* cell wall, inducing enhanced proinflammatory cytokine production which is lost upon MCL or CARD9 deficiency (Huang et al. 2018). Mice lacking MCL did not show an increase in susceptibility toward *C. neoformans* (serotype A and serotype D) with unaltered survivals rates, leukocyte recruitment and cytokine production

(Campuzano et al. 2017). In contrast, MCL-deficient mice were highly susceptible to *C. neoformans* (serotype AD), due to defective alveolar macrophage accumulation and killing function (Huang et al. 2018).

Pneumocystis jirovecii: MCL has been shown to bind to *Pneumocystis* with high affinity (Kottom et al. 2019). The role this receptor plays in vivo has not yet been demonstrated for *Pneumocystis*.

5.2.5 DC-SIGN

Dendritic cell-specific ICAM-3-grabbing non-integrin or DC-SIGN (gene name: CD209; CLEC4L) also belongs to the group II of CLRs (Fig. 2). It consists of two orthologues in humans, namely DC-SIGN and liver/lymph node-specific ICAM-3-grabbing integrin (L-SIGN) (Soilleux et al. 2000; Bashirova et al. 2001) and eight orthologues in mice which vary in expression profiles and structure: mDC-SIGN, L-SIGN, DC-SIGNR1,2-4 and DC-SIGNR6-8 (Park et al. 2001; Koppel et al. 2004; Powlesland et al. 2006). DC-SIGN is predominantly present on the cell surfaces of certain macrophage subsets, immature dendritic cells found in peripheral tissues and mature dendritic cells in lymphoid tissue (Koppel et al. 2004; Krutzik et al. 2005). DC-SIGN signals through the Raf-1 pathway, independent of Syk (Fig. 3) (Gringhuis et al. 2007). DC-SIGN recognizes mannose containing structures, for example galactomannans, and binds ligand in a calcium-dependent manner. The variety of recognized structures range from endogenous ligands, to various fungal pathogens such as bacteria including *Mycobacterium tuberculosis*, viruses such as HIV, Ebola and Hepatitis, as well as *Leishmania* (Garcia-Vallejo and Van Kooyk 2013). In fact, it has been shown that numerous pathogens target DC-SIGN in order to promote their survival by modulating dendritic cell-mediated immune responses. Most studies done investigating the role of DC-SIGN have been in vitro utilizing human cell lines, with little work done investigating the immune responses implicated by DC-SIGN orthologues in mouse models. In vitro studies show an association of DC-SIGN with allergic responses, as human dendritic cells lacking DC-SIGN polarize toward Th2 differentiation in autologous dendritic cell-T cell cocultures (Emara et al. 2012). DC-SIGN has also been shown to recognize allergens from Bermuda grass pollen, peanut, dog and HDM. The protease Derp1, from HDM, has been shown to contribute to driving allergic responses through the cleavage of DC-SIGN (Furmonaviciene et al. 2007).

DC-SIGN involvement in fungal immunity

Candida albicans: DC-SIGN is able to recognize *C. albicans* and has been shown to mediate mannan-binding and phagocytosis of *C. albicans* leading to the production of proinflammatory cytokines and adaptive immune responses (Cambi et al. 2008; Vendele et al. 2020).

Aspergillus fumigatus: DC-SIGN recognizes live *A. fumigatus* (not heat-killed conidia) and has been shown to mediate binding and uptake of conidia in IL-4 stimulated monocyte-derived macrophages and dendritic cells (Serrano-Gómez et al. 2004, 2005). Contributing to this, DC-SIGN overexpression in dendritic cells enhanced *A. fumigatus* uptake and resulted in increased Th1 cytokine production (Li et al. 2018). There is, however, a lack of studies exploring the role of this receptor in vivo. Multiple DC-SIGN polymorphisms (rs4804800, rs11465384, rs7248637, rs7252229) have been identified and may be linked to an increased risk in invasive aspergillosis (Sainz et al. 2012).

Cryptococcus neoformans: In humans, DC-SIGN has been shown to bind mannoproteins in vitro, however additional studies are required to better understand the role of this receptor during *Cryptococcus* infection in vivo (Mansour et al. 2006).

Pneumocystis jirovecii: DC-SIGN has been shown to recognize *Pneumocystis;* however, its exact role needs further investigation (Kottom et al. 2019).

5.2.6 Mannose Receptor (MR)

The MR (gene name: MRC1; CD206, CLEC13D) is a type I transmembrane protein belonging to group VI of CLRs (Fig. 2). MR has a soluble form that arises after proteolytic cleavage, which has been shown to occur subsequent to Dectin-1 signaling (Gazi et al. 2011). MR contains three extracellular structures: a cysteine-rich domain, a fibronectin type II domain and eight CTLDs located closest to the plasma membrane (Fig. 2) (Martinez-Pomares 2012; Taylor et al. 1990). MR also has a transmembrane domain and a short cytoplasmic tail that lacks detectable signaling motifs. The surface expression of MR on macrophages has, however, been shown to be mediated by FcRγ chain and soluble MR has been shown to associate with adaptor molecules such as Grb2 and SHP-1 which mediate immune responses (Rajaram et al. 2017).

MR is mainly expressed intracellularly during the endocytic pathway, in tissue macrophages, certain subtypes of dendritic cells, and nonvascular endothelial cells (Martinez-Pomares 2012; McKenzie et al. 2007). Upregulation of MR can be caused by increased IL-4 production as well as several other cytokines (Stein et al. 1992). The three extracellular domains of MR each recognize different ligands. Multiple CTLDs of MR recognize and bind to N-linked mannans, fructose-based structures and N-acetylglucosamine in a calcium-dependent manner (Martinez-Pomares 2012). The cysteine-rich domain recognizes sulfated carbohydrates in lymphoid and kidney tissues, and the fibronectin domain binds collagen I–IV (Martinez-Pomares 2012). Due to the wide range of ligands MR can bind, it is not surprising that the receptor can recognize a wide range of pathogens. These include various fungi; bacteria such as *Mycobacterium tuberculosis, Streptococcus pneumoniae* and *Klebsiella pneumoniae*; viruses such as HIV, Dengue virus and hepatitis B; parasites such as helminths and *Trichuris muris*; as well as trypanosome protozoa (Martinez-Pomares 2012; Lefèvre et al. 2013).

The MR has been shown to induce endocytosis, phagocytosis, phagosomal trafficking, antigen cross-presentation and the production of cytokines (Martinez-Pomares 2012). Additionally, MR has been shown to drive adaptive immune responses such as the development of antifungal Th17 responses (Van De Veerdonk et al. 2009; Romani 2011). MR has also been implicated in allergic responses and asthma as it drives allergen-induced (e.g., HDM) Th2 polarization and has been shown to be upregulated on dendritic cells in allergy patients (Deslée et al. 2002; Royer et al. 2010).

MR involvement in fungal immunity

MR recognizes α-1,3-glucans, chitin (Cambi et al. 2008; Romani 2011) and glycoprotein A (O'Riordan et al. 1995) in various fungal pathogens such as *P. jirovecii* and *C. neoformans* (Martinez-Pomares 2012). Currently, no human MR polymorphisms have been linked to susceptibility to fungal infections, but have been reported for tuberculosis, asthma and sarcoidosis (Brown and Crocker 2016).

Candida albicans: Studies describing the roles of MR in immune responses have been controversial; however, these effects may derive from associations with other receptors. Initially, MR was described as a opsonin-independent phagocytic receptor with reported abilities to bind and phagocytose zymosan particles (Harris et al. 1992) as well as *C. albicans* yeast cells (Porcaro et al. 2003). These findings were challenged by studies showing that MR was unable to mediate phagocytosis (Le Cabec et al. 2005).

Due to the ability of Dectin-2 and Mincle to recognize many of the same ligands as MR, the exact role it plays in inducing these responses is unclear. MR involvement in association with Dectin-1, TLRs and CR3 has been shown to play a role in the initial macrophage recognition, and phagosome maturation in response to *C. albicans*, with subsequent production of IL-17 and Th17 cells (Galès et al. 2010; Van De Veerdonk et al. 2009; Netea 2006; Heinsbroek et al. 2008). Ultimately, in vivo studies have shown that mice deficient for MR did not change the susceptibility to *C. albicans* infection (Lee et al. 2003).

Aspergillus fumigatus: Human corneal epithelial cells are known to express MR, and following infection with *A. fumigatus*, MR expression was increased, similarly to the expression of Dectin-1, suggesting an interaction. Blocking MR with antibody resulted in a decrease in inflammatory cytokine production in response to *A. fumigatus* infection in vitro (Wang et al. 2016). The role for MR in vivo has been shown to be redundant as no increase in susceptibility to *A. fumigatus* was seen in knockout mice.

Cryptococcus neoformans: MR has been shown to recognize *C. neoformans* and was reported necessary for the binding and initiation of phagocytosis by dendritic cells (Syme et al. 2002). Mice deficient for MR were reported to have defective development of CD4 T cells with a slight increase in susceptibility to *C. neoformans* infection (Dan et al. 2008).

Pneumocystis jirovecii: MR is known to recognize *P. jirovecii* and bind gpA in the cell wall, inducing phagocytosis (Ezekowitz et al. 1991; O'Riordan et al. 1995). Interestingly, *Pneumocystis* is able to shed gpA from its cell wall, which then competitively binds to MR on the surface of alveolar macrophages, blocking detection of live *Pneumocystis* cells and preventing phagocytosis (Lasbury et al. 2004). This is a clever immune evasion mechanism, as alveolar macrophages are the first innate cells that function to recognize and kill *P. jirovecii* in the lung. Alveolar macrophages isolated from HIV patients had downregulated levels of MR expression and were reported to have a decrease in binding and killing function when infected with *P. murina* in vitro (Koziel et al. 1998). In fact, *Pneumocystis* has also been shown to promote the shedding of MR from alveolar macrophages, releasing soluble MR that then coats the fungal cells, causing a delay in recognition and non-opsonic phagocytosis of the organism (Fraser et al. 2000). However, there is a report that soluble MR can significantly enhance phagocytosis of *Pneumocystis* and may have a valuable therapeutic role (Stehle et al. 2000). In any event, although mice deficient for MR had increased phagocytic cell infiltration to infected lung tissue, the lack of MR did not change the susceptibility to *P. murina* infection in mice (Swain et al. 2003).

5.2.7 CD23

CD23 (gene name: FCER2; FcεRII, Clec4J, IgE Fc receptor) is a CLR belonging to group II (Fig. 2); however, it is also known as an Fc receptor as it is a low-affinity receptor for IgE. It is different to other known Fc receptors as it consists of a single extracellular C-type lectin-like domain, a transmembrane domain and a cytoplasmic tail containing an ITAM-like motif (Kijimoto-Ochiai 2002). CD23 is expressed in activated macrophages, follicular dendritic cells, eosinophils, monocytes, mature B cells and also keratinocytes (Guo et al. 2018). CD23 has also been shown to be a B-cell activation marker (Bonnefoy et al. 1997).

CD23 can directly recognize α-mannans and β-glucans on fungi and the Fc portion of IgE antibodies (Guo et al. 2018). CD23 has been shown to mediate macrophage responses to fungi, *Mycobacteria* (*M. tuberculosis* and *M. avium*) (Mossalayi et al. 2009) and parasites (*Leishmania*) (Vouldoukis et al. 1995) consequently facilitating in clearance. Upon ligand binding and downstream Syk signaling through association with FcRγ, CD23 has been shown to initiate antifungal immune responses including nitric oxide production by activated macrophages (Guo et al. 2018).

CD23 involvement in fungal immunity

It has recently been shown that CD23 acts as a typical CLR in recognizing certain fungi and modulating clearance primarily through macrophage responses (Guo et al. 2018). Human CD23 polymorphisms have been found and linked to susceptibility to allergy and asthma; however, further research is required to determine the links to antifungal immunity.

Candida albicans: CD23 binds to both α-mannan and β-glucan in the cell wall of *Candida* and can induce downstream signaling pathways leading to the production of NF-κB. Mice deficient in CD23 were highly susceptible to *C. albicans* and showed a decrease in proinflammatory cytokines and nitric oxide production as well as an increase in fungal burden. Macrophages were reported to be responsible for the defective antifungal immune responses in mice lacking CD23 (Guo et al. 2018).

Aspergillus fumigatus: CD23 recognizes the α-mannan and β-glucan components of *A. fumigatus* cell wall and through association with FcRγ adaptor molecule, activates intracellular signaling pathways. A significant increase in susceptibility was reported in mice lacking CD23 with an increase in fungal burden (Guo et al. 2018).

Cryptococcus neoformans: CD23 was unable to detect or bind to glucuronoxylomannan found in the cell wall of *C. neoformans*. CD23-deficient mice did not show an increase in susceptibility to *C. neoformans*, with unaffected cytokine production and fungal clearance (Guo et al. 2018).

Pneumocystis jirovecii: No research has been done investigating the role of CD23 in *Pneumocystis* infection.

5.2.8 MelLec

Mellec (gene name: CLEC1A; CLEC-1) forms part of the group V CLRs (Fig. 2) and is structurally similar to Dectin-1. In humans, MelLec is expressed on both myeloid and endothelial cells (Sattler et al. 2012; Sobanov et al. 2001; Colonna et al. 2000).

However, in mice MelLec is expressed on CD31+ endothelial cells and not on myeloid cells, unlike other fungal detecting CLRs mentioned above. This includes a CD31+ subpopulation of epithelial cells co-expressing epithelial adhesion molecule EpCAM, found only in the lung and liver of mice (Sattler et al. 2012). MelLec recognizes a specific form of melanin, 1,8-dihydroxynaphthalene (DHN)-melanin which is synthesized by numerous fungi such as *A. fumigatus* and *Fonsecaea pedrosoi*. MelLec does not contain conserved signaling motifs in the cytoplasmic tail region, and although MelLec has been shown to recognize certain fungi, the intracellular signaling mechanisms are unknown.

MelLec involvement in fungal immunity

MelLec recognizes fungi containing DHN-melanin in their fungal cell wall and plays an important role in protection against certain melanized fungi in mice and humans. *Pneumocystis* does not produce melanin in its cell wall, and although *Candida* and *Cryptococcus* do produce L-3,4-dihydroxyphenylalanine (L-DOPA) melanin, MelLec does not recognize melanin synthesized by other biosynthetic pathways and is specific for DHN-melanin (Stappers et al. 2018). Therefore, MelLec does not play a role in recognizing these pathogens.

Aspergillus fumigatus: MelLec recognizes *A. fumigatus* conidia, and mice lacking the receptor are more susceptible to infection, suggesting a protective role. This was linked to a decrease in cytokine production, including KC, CCL2 and IL-1β, and an increase in disseminated fungal burden (Stappers et al. 2018). A single nucleotide polymorphism of human MelLec (rs2306894) was identified, affecting the cytoplasmic tail region of the receptor, and is linked to an increased risk of aspergillosis. Cells carrying the variant were reported to have impaired inflammatory responses in vitro, with decreased IL-1β and IL-8 cytokine production (Stappers et al. 2018).

6 Conclusions

Fungi surround us during our everyday life, from the air we breathe, to the commensal fungi in our gut and mucosal surfaces. As more is revealed about the role CLRs play in fungal immunity, exciting new potential methods for diagnosis, treatment, immunotherapies and vaccines begin to arise. Improving our understanding of the mechanisms involved in CLR-induced inflammatory responses may lead to novel therapies, especially for patients with co-morbidities or high-risk individuals (e.g., Dectin-1 mutations, severe forms of ulcerative colitis and advanced HIV disease). Identifying polymorphisms in human CLRs that are linked to fungal immunity is an important step toward identifying high-risk patients prone to invasive fungal infections. These patients could benefit from antifungal prophylaxis treatment, especially during immunosuppressing treatments. Understanding how CLRs interact with colonizing and infectious fungal pathogens may provide insightful information for therapies involving the regulation of fungal immune responses. Another area for future development is understanding the role of CLRs in shaping the mycobiome. We also need to know how different fungal morphologies are recognized by CLRs of innate cells, and how this influences resultant T-cell responses. Indeed, such information could impact the development of vaccines. What we have learnt so far, however, is that CLRs are essential components of the arsenal required to combat fungal infections.

Acknowledgements MAH is supported by the National Research Foundation of South Africa and the Oppenheimer Memorial Trust; JCH is supported by the Wellcome Trust (209293) and Carnegie Corporation; GDB is supported by the Wellcome Trust and the Medical Research Council (UK) Centre for Medical Mycology.

References

Alangaden GJ (2011) Nosocomial fungal infections: epidemiology, infection control, and prevention. Infect Dis Clin North Am 25(1):201–225. https://doi.org/10.1016/j.idc.2010.11.003

Arce I, Martínez-Muñoz L, Roda-Navarro P, Fernández-Ruiz E (2004) The human C-type lectin CLECSF8 is a novel monocyte/macrophage endocytic receptor. Eur J Immunol 34(1):210–220. https://doi.org/10.1002/eji.200324230

Ariizumi K, Shen G-L, Shikano S, Ritter R, Zukas P, Edelbaum D, Morita A, Takashima A (2000) Cloning of a second dendritic cell-associated C-type lectin (Dectin-2) and its alternatively spliced isoforms. J Biol Chem 275(16):11957–11963. https://doi.org/10.1074/jbc.275.16.11957

Armstrong-James D, Meintjes G, Brown GD (2014) A neglected epidemic: fungal infections in HIV/AIDS. Trends Microbiol 22(3):120–127. https://doi.org/10.1016/j.tim.2014.01.001

Baddley JW, Stephens JM, Ji X, Gao X, Schlamm HT, Tarallo M (2013) Aspergillosis in Intensive Care Unit (ICU) patients: epidemiology and economic outcomes. BMC Infect Dis 13(1):29. https://doi.org/10.1186/1471-2334-13-29

Ballou ER, Avelar GM, Childers DS, Mackie J, Bain JM, Wagener J, Kastora SL, Panea MD, Hardison SE, Walker LA, Erwig LP, Munro CA, Gow NAR, Brown GD, Maccallum DM, Brown AJP (2017) Lactate signalling regulates fungal β-glucan masking and immune evasion. Nat Microbiol 2(2):16238. https://doi.org/10.1038/nmicrobiol.2016.238

Barrett NA, Maekawa A, Rahman OM, Austen KF, Kanaoka Y (2009) Dectin-2 recognition of house dust mite triggers cysteinyl leukotriene generation by dendritic cells. J Immunol 182 (2):1119–1128. https://doi.org/10.4049/jimmunol.182.2.1119

Bashirova AA, Geijtenbeek TBH, Van Duijnhoven GCF, Van Vliet SJ, Eilering JBG, Martin MP, Wu L, Martin TD, Viebig N, Knolle PA, Kewalramani VN, Van Kooyk Y, Carrington M (2001) A dendritic cell–specific intercellular adhesion molecule 3–grabbing nonintegrin (Dc-Sign)–related protein is highly expressed on human liver sinusoidal endothelial cells and promotes HIV-1 infection 193 (6):671–678. https://doi.org/10.1084/jem.193.6.671

Bi L, Gojestani S, Wu W, Hsu Y-MS, Zhu J, Ariizumi K, Lin X (2010) CARD9 mediates dectin-2-induced IκBα kinase ubiquitination leading to activation of NF-κB in response to stimulation by the hyphal form of *Candida albicans*. J Biol Chem 285(34):25969–25977. https://doi.org/10.1074/jbc.m110.131300

Blanco-Menéndez N, Del Fresno C, Fernandes S, Calvo E, Conde-Garrosa R, Kerr WG, Sancho D (2015) SHIP-1 couples to the dectin-1 hemITAM and selectively modulates reactive oxygen species production in dendritic cells in response to *Candida albicans*. J Immunol 195(9):4466–4478. https://doi.org/10.4049/jimmunol.1402874

Bongomin F, Gago S, Oladele R, Denning D (2017) Global and multi-national prevalence of fungal diseases—estimate precision. J Fungi 3(4):57. https://doi.org/10.3390/jof3040057

Bonnefoy JY, Lecoanet-Henchoz S, Gauchat JF, Graber P, Aubry JP, Jeannin P, Plater-Zyberk C (1997) Structure and functions of CD23. Int Rev Immunol 16(1–2):113–128. https://doi.org/10.3109/08830189709045705

Brown GD (2006) Dectin-1: a signalling non-TLR pattern-recognition receptor. Nat Rev Immunol 6(1):33–44. https://doi.org/10.1038/nri1745

Brown GD (2011) Innate antifungal immunity: the key role of phagocytes. Annu Rev Immunol 29 (1):1–21. https://doi.org/10.1146/annurev-immunol-030409-101229

Brown GD, Crocker PR (2016) Lectin receptors expressed on myeloid cells. Microbiol Spectr 4 (5). https://doi.org/10.1128/microbiolspec.mchd-0036-2016

Brown GD, Denning DW, Gow NAR, Levitz SM, Netea MG, White TC (2012a) Hidden killers: human fungal infections. Sci Transl Med 4(165). https://doi.org/10.1126/scitranslmed.3004404

Brown GD, Denning DW, Levitz SM (2012b) Tackling human fungal infections. Science 336 (6082):647–647. https://doi.org/10.1126/science.1222236

Brown GD, Willment JA, Whitehead L (2018) C-type lectins in immunity and homeostasis. Nat Rev Immunol 18(6):374–389. https://doi.org/10.1038/s41577-018-0004-8

Buchanan KL, Murphy JW (1998) What makes cryptococcus neoformans a pathogen? Emerg Infect Dis 4(1):71–83. https://doi.org/10.3201/eid0401.980109

Camacho E, Vij R, Chrissian C, Prados-Rosales R, Gil D, O'Meally RN, Cordero RJB, Cole RN, McCaffery JM, Stark RE, Casadevall A (2019) The structural unit of melanin in the cell wall of the fungal pathogen *Cryptococcus neoformans*. J Biol Chem 294(27):10471–10489. https://doi.org/10.1074/jbc.ra119.008684

Cambi A, Netea MG, Mora-Montes HM, Gow NAR, Hato SV, Lowman DW, Kullberg B-J, Torensma R, Williams DL, Figdor CG (2008) Dendritic cell interaction with *Candida albicans* critically depends on N-linked mannan. J Biol Chem 283(29):20590–20599. https://doi.org/10.1074/jbc.m709334200

Campuzano A, Castro-Lopez N, Wozniak KL, Leopold Wager CM, Wormley FL (2017) Dectin-3 is not required for protection against *Cryptococcus neoformans* infection. PLoS ONE 12(1): e0169347. https://doi.org/10.1371/journal.pone.0169347

Carrion SDJ, Leal SM, Ghannoum MA, Aimanianda V, Latgé J-P, Pearlman E (2013) The RodA hydrophobin on *Aspergillus fumigatus* spores masks dectin-1– and dectin-2–dependent responses and enhances fungal survival in vivo. J Immunol 191(5):2581–2588. https://doi.org/10.4049/jimmunol.1300748

Carvalho A, Giovannini G, De Luca A, D'Angelo C, Casagrande A, Iannitti RG, Ricci G, Cunha C, Romani L (2012) Dectin-1 isoforms contribute to distinct Th1/Th17 cell activation in *Mucosal candidiasis*. Cell Mol Immunol 9(3):276–286. https://doi.org/10.1038/cmi.2012.1

Casadevall A, Rosas AL, Nosanchuk JD (2000) Melanin and virulence in *Cryptococcus neoformans*. Curr Opin Microbiol 3(4):354–358. https://doi.org/10.1016/s1369-5274(00)00103-x

Chabé M, Aliouat-Denis C-M, Delhaes L, Aliouat EM, Viscogliosi E, Dei-Cas E (2011) Pneumocystis: from a doubtful unique entity to a group of highly diversified fungal species. FEMS Yeast Res 11(1):2–17. https://doi.org/10.1111/j.1567-1364.2010.00698.x

Chaffin WL, López-Ribot JL, Casanova M, Gozalbo D, Martínez JP (1998) Cell wall and secreted proteins of *Candida albicans*: identification, function, and expression. Microbiol Mol Biol Rev MMBR 62(1)

Chai LYA, De Boer MGJ, Van Der Velden WJFM, Plantinga TS, Van Spriel AB, Jacobs C, Halkes CJM, Vonk AG, Blijlevens NM, Van Dissel JT, Donnelly PJ, Kullberg B-J, Maertens J, Netea MG (2011) The Y238X stop codon polymorphism in the human β-glucan receptor dectin-1 and susceptibility to invasive Aspergillosis. J Infect Dis 203(5):736–743. https://doi.org/10.1093/infdis/jiq102

Colonna M, Samaridis J, Angman L (2000) Molecular characterization of two novel C-type lectin-like receptors, one of which is selectively expressed in human dendritic cells. Eur J Immunol 30(2):697–704. https://doi.org/10.1002/1521-4141(200002)30:2%3c697:aid-immu697%3e3.0.co;2-m

Daley D, Mani VR, Mohan N, Akkad N, Ochi A, Heindel DW, Lee KB, Zambirinis CP, Pandian GSB, Savadkar S, Torres-Hernandez A, Nayak S, Wang D, Hundeyin M, Diskin B, Aykut B, Werba G, Barilla RM, Rodriguez R, Chang S, Gardner L, Mahal LK, Ueberheide B, Miller G (2017) Dectin 1 activation on macrophages by galectin 9 promotes pancreatic carcinoma and peritumoral immune tolerance. Nat Med 23(5):556–567. https://doi.org/10.1038/nm.4314

Dan JM, Kelly RM, Lee CK, Levitz SM (2008) Role of the mannose receptor in a murine model of *Cryptococcus neoformans* infection. Infect Immun 76(6):2362–2367. https://doi.org/10.1128/iai.00095-08

Denning DW (1998) Invasive Aspergillosis. Clin Infect Dis 26(4):781–803. https://doi.org/10.1086/513943

Deslée GT, Charbonnier A-S, Hammad H, Angyalosi G, Tillie-Leblond I, Mantovani A, Tonnel AÉ-B, Pestel J (2002) Involvement of the mannose receptor in the uptake of der p 1, a major mite allergen, by human dendritic cells. J Allergy Clin Immunol 110(5):763–770. https://doi.org/10.1067/mai.2002.129121

Drummond RA, Brown GD (2011) The role of Dectin-1 in the host defence against fungal infections. Curr Opin Microbiol 14(4):392–399. https://doi.org/10.1016/j.mib.2011.07.001

Drummond RA, Dambuza IM, Vautier S, Taylor JA, Reid DM, Bain CC, Underhill DM, Masopust D, Kaplan DH, Brown GD (2016) CD4+ T-cell survival in the GI tract requires dectin-1 during fungal infection. Mucosal Immunol 9(2):492–502. https://doi.org/10.1038/mi.2015.79

Drummond RA, Saijo S, Iwakura Y, Brown GD (2011) The role of Syk/CARD9 coupled C-type lectins in antifungal immunity. Eur J Immunol 41(2):276–281. https://doi.org/10.1002/eji.201041252

Emara M, Royer P-J, Mahdavi J, Shakib F, Ghaemmaghami AM (2012) Retagging Identifies dendritic cell-specific intercellular adhesion molecule-3 (ICAM3)-grabbing non-integrin (DC-SIGN) protein as a novel receptor for a major allergen from house dust mite 287 (8):5756–5763. https://doi.org/10.1074/jbc.m111.312520

Erwig LP, Gow NAR (2016) Interactions of fungal pathogens with phagocytes. Nat Rev Microbiol 14(3):163–176. https://doi.org/10.1038/nrmicro.2015.21

Ezekowitz RA, Williams DJ, Koziel H, Armstrong MY, Warner A, Richards FF, Rose RM (1991) Uptake of *Pneumocystis carinii* mediated by the macrophage mannose receptor. Nature 351 (6322):155–158. https://doi.org/10.1038/351155a0

Ferwerda B, Ferwerda G, Plantinga TS, Willment JA, Van Spriel AB, Venselaar H, Elbers CC, Johnson MD, Cambi A, Huysamen C, Jacobs L, Jansen T, Verheijen K, Masthoff L, Morré SA, Vriend G, Williams DL, Perfect JR, Joosten LAB, Wijmenga C, Van Der Meer JWM, Adema GJ, Kullberg BJ, Brown GD, Netea MG (2009) Human dectin-1 deficiency and mucocutaneous fungal infections. N Engl J Med 361(18):1760–1767. https://doi.org/10.1056/nejmoa0901053

Fisher CE, Hohl TM, Fan W, Storer BE, Levine DM, Zhao LP, Martin PJ, Warren EH, Boeckh M, Hansen JA (2017) Validation of single nucleotide polymorphisms in invasive aspergillosis following hematopoietic cell transplantation. Blood 129(19):2693–2701. https://doi.org/10.1182/blood-2016-10-743294

Fraser IP, Takahashi K, Koziel H, Fardin B, Harmsen A, Ezekowitz RAB (2000) *Pneumocystis carinii* enhancessoluble mannose receptor production by macrophages. Microbes Infect 2 (11):1305–1310. https://doi.org/10.1016/s1286-4579(00)01283-1

Frenkel JK (1999) Pneumocystis pneumonia, an immunodeficiency-dependent disease (IDD): a critical historical overview. J Eukaryot Microbiol 46(5):89s–92s

Furmonaviciene R, Ghaemmaghami AM, Boyd SE, Jones NS, Bailey K, Willis AC, Sewell HF, Mitchell DA, Shakib F (2007) The protease allergen Der p 1 cleaves cell surface DC-SIGN and DC-SIGNR: experimental analysis of in silico substrate identification and implications in allergic responses. Clin Exp Allergy: J Br Soc Allergy Clin Immunol 37(2):231–242

Furukawa A, Kamishikiryo J, Mori D, Toyonaga K, Okabe Y, Toji A, Kanda R, Miyake Y, Ose T, Yamasaki S, Maenaka K (2013) Structural analysis for glycolipid recognition by the C-type lectins Mincle and MCL. Proc Natl Acad Sci 110(43):17438–17443. https://doi.org/10.1073/pnas.1312649110

Galès A, Conduché A, Bernad J, Lefevre L, Olagnier D, Béraud M, Martin-Blondel G, Linas M-D, Auwerx J, Coste A, Pipy B (2010) PPARγ controls dectin-1 expression required for host antifungal defense against *Candida albicans* 6(1):e1000714. https://doi.org/10.1371/journal.ppat.1000714

Garcia-Vallejo JJ, Van Kooyk Y (2013) The physiological role of DC-SIGN: a tale of mice and men. Trends Immunol 34(10):482–486. https://doi.org/10.1016/j.it.2013.03.001

Gazi U, Rosas M, Singh S, Heinsbroek S, Haq I, Johnson S, Brown GD, Williams DL, Taylor PR, Martinez-Pomares L (2011) Fungal recognition enhances mannose receptor shedding through dectin-1 engagement. J Biol Chem 286(10):7822–7829. https://doi.org/10.1074/jbc.m110.185025

Geijtenbeek TBH, Gringhuis SI (2009) Signalling through C-type lectin receptors: shaping immune responses. Nat Rev Immunol 9(7):465–479. https://doi.org/10.1038/nri2569

Goodridge HS, Reyes CN, Becker CA, Katsumoto TR, Ma J, Wolf AJ, Bose N, Chan ASH, Magee AS, Danielson ME, Weiss A, Vasilakos JP, Underhill DM (2011) Activation of the innate immune receptor dectin-1 upon formation of a 'phagocytic synapse'. Nature 472 (7344):471–475. https://doi.org/10.1038/nature10071

Gorjestani S, Yu M, Tang B, Zhang D, Wang D, Lin X (2011) Phospholipase Cγ2 (PLCγ2) is key component in dectin-2 signaling pathway, mediating anti-fungal innate immune responses. J Biol Chem 286(51):43651–43659. https://doi.org/10.1074/jbc.m111.307389

Gour N, Lajoie S, Smole U, White M, Hu D, Goddard P, Huntsman S, Eng C, Mak A, Oh S, Kim J-H, Sharma A, Plante S, Salem IH, Resch Y, Xiao X, Yao N, Singh A, Vrtala S, Chakir J, Burchard EG, Lane AP, Wills-Karp M (2018) Dysregulated invertebrate tropomyosin–dectin-1 interaction confers susceptibility to allergic diseases. Sci Immunol 3 (20):eaam9841. https://doi.org/10.1126/sciimmunol.aam9841

Gow NAR, Van De Veerdonk FL, Brown AJP, Netea MG (2012) *Candida albicans* morphogenesis and host defence: discriminating invasion from colonization. Nat Rev Microbiol 10(2):112–122. https://doi.org/10.1038/nrmicro2711

Graham LM, Brown GD (2009) The dectin-2 family of C-type lectins in immunity and homeostasis. Cytokine 48(1–2):148–155. https://doi.org/10.1016/j.cyto.2009.07.010

Graham LM, Gupta V, Schafer G, Reid DM, Kimberg M, Dennehy KM, Hornsell WG, Guler R, Campanero-Rhodes MA, Palma AS, Feizi T, Kim SK, Sobieszczuk P, Willment JA, Brown GD (2012) The C-type lectin receptor CLECSF8 (CLEC4D) is expressed by myeloid cells and triggers cellular activation through Syk kinase. J Biol Chem 287(31):25964–25974. https://doi.org/10.1074/jbc.m112.384164

Griffiths JS, Thompson A, Stott M, Benny A, Lewis NA, Taylor PR, Forton J, Herrick S, Orr SJ, McGreal EP (2018) Differential susceptibility of dectin-1 isoforms to functional inactivation by neutrophil and fungal proteases. FASEB J 32(6):3385–3397. https://doi.org/10.1096/fj.201701145R

Gringhuis SI, den Dunnen J, Litjens M, van het Hof B, van Kooyk Y, Geijtenbeek Teunis BH (2007) C-type lectin DC-SIGN modulates toll-like receptor signaling via raf-1 kinase-dependent acetylation of transcription factor NF-κB. Immunity 26(5):605–616. https://doi.org/10.1016/j.immuni.2007.03.012

Gringhuis SI, Wevers BA, Kaptein TM, Van Capel TMM, Theelen B, Boekhout T, De Jong EC, Geijtenbeek TBH (2011) Selective C-rel activation via Malt1 controls anti-fungal TH-17 immunity by dectin-1 and dectin-2. PLoS Pathog 7(1):e1001259. https://doi.org/10.1371/journal.ppat.1001259

Gugnani HC (2003) Ecology and taxonomy of pathogenic aspergilli. Front Biosci 8(6):s346–357. https://doi.org/10.2741/1002

Guo Y, Chang Q, Cheng L, Xiong S, Jia X, Lin X, Zhao X (2018) C-type lectin receptor CD23 Is required for host defense against *Candida albicans* and *Aspergillus fumigatus* infection. J Immunol 201(8):2427–2440. https://doi.org/10.4049/jimmunol.1800620

Harris N, Super M, Rits M, Chang G, Ezekowitz RA (1992) Characterization of the murine macrophage mannose receptor: demonstration that the downregulation of receptor expression mediated by interferon-gamma occurs at the level of transcription. Blood 80(9):2363–2373

Hattori Y, Morita D, Fujiwara N, Mori D, Nakamura T, Harashima H, Yamasaki S, Sugita M (2014) Glycerol monomycolate is a novel ligand for the human, but not mouse macrophage inducible C-type lectin, Mincle. J Biol Chem 289(22):15405–15412. https://doi.org/10.1074/jbc.m114.566489

Hefter M, Lother J, Weiß E, Schmitt AL, Fliesser M, Einsele H, Loeffler J (2017) Human primary myeloid dendritic cells interact with the opportunistic fungal pathogen *Aspergillus fumigatusvia* the C-type lectin receptor Dectin-1. Medical Mycology:myw105. https://doi.org/10.1093/mmy/myw105

Heilmann CJ, Sorgo AG, Siliakus AR, Dekker HL, Brul S, De Koster CG, De Koning LJ, Klis FM (2011) Hyphal induction in the human fungal pathogen *Candida albicans* reveals a characteristic wall protein profile. Microbiology 157(8):2297–2307. https://doi.org/10.1099/mic.0.049395-0

Heinsbroek SEM, Taylor PR, Martinez FO, Martinez-Pomares L, Brown GD, Gordon S (2008) Stage-specific sampling by pattern recognition receptors during *Candida albicans* phagocytosis. PLoS Pathog 4(11):e1000218. https://doi.org/10.1371/journal.ppat.1000218

Heinsbroek SEM, Taylor PR, Rosas M, Willment JA, Williams DL, Gordon S, Brown GD (2006) Expression of functionally different dectin-1 isoforms by Murine macrophages. J Immunol 176 (9):5513–5518. https://doi.org/10.4049/jimmunol.176.9.5513

Hernanz-Falcón P, Joffre O, Williams DL, Reis e Sousa C (2009) Internalization of dectin-1 terminates induction of inflammatory responses. Eur J Immunol 39 (2):507–513. https://doi.org/10.1002/eji.200838687

Hidron AI, Edwards JR, Patel J, Horan TC, Sievert DM, Pollock DA, Fridkin SK (2008) Antimicrobial-resistant pathogens associated with healthcare-associated infections: annual summary of data reported to the national healthcare safety network at the centers for disease control and prevention, 2006–2007. Infect Control Hosp Epidemiol 29(11):996–1011. https://doi.org/10.1086/591861

Hoving JC, Wilson GJ, Brown GD (2014) Signalling C-type lectin receptors, microbial recognition and immunity. Cell Microbiol 16(2):185–194. https://doi.org/10.1111/cmi.12249

Hu X-P, Wang R-Y, Wang X, Cao Y-H, Chen Y-Q, Zhao H-Z, Wu J-Q, Weng X-H, Gao X-H, Sun R-H, Zhu L-P (2015) Dectin-2 polymorphism associated with pulmonary cryptococcosis in HIV-uninfected Chinese patients. Med Mycol 53(8):810–816. https://doi.org/10.1093/mmy/myv043

Huang H-R, Li F, Han H, Xu X, Li N, Wang S, Xu J-F, Jia X-M (2018) Dectin-3 recognizes glucuronoxylomannan of *Cryptococcus neoformans* serotype AD and *Cryptococcus gattii* serotype B to initiate host defense against cryptococcosis. Front Immunol 9:1781

Ifrim DC, Bain JM, Reid DM, Oosting M, Verschueren I, Gow NAR, Van Krieken JH, Brown GD, Kullberg BJ, Joosten LAB, Van Der Meer JWM, Koentgen F, Erwig LP, Quintin J, Netea MG (2014) Role of dectin-2 for host defense against systemic infection with *Candida glabrata* 82(3):1064–1073. https://doi.org/10.1128/iai.01189-13

Iliev ID, Funari VA, Taylor KD, Nguyen Q, Reyes CN, Strom SP, Brown J, Becker CA, Fleshner PR, Dubinsky M, Rotter JI, Wang HL, McGovern DPB, Brown GD, Underhill DM (2012) Interactions between commensal fungi and the C-type lectin receptor dectin-1 influence colitis. Science 336(6086):1314–1317. https://doi.org/10.1126/science.1221789

Iliev ID, Leonardi I (2017) Fungal dysbiosis: immunity and interactions at mucosal barriers. Nat Rev Immunol 17(10):635–646. https://doi.org/10.1038/nri.2017.55

Ishikawa E, Ishikawa T, Morita YS, Toyonaga K, Yamada H, Takeuchi O, Kinoshita T, Akira S, Yoshikai Y, Yamasaki S (2009) Direct recognition of the mycobacterial glycolipid, trehalose dimycolate, by C-type lectin Mincle. J Exp Med 206(13):2879–2888. https://doi.org/10.1084/jem.20091750

Ishikawa T, Itoh F, Yoshida S, Saijo S, Matsuzawa T, Gonoi T, Saito T, Okawa Y, Shibata N, Miyamoto T, Yamasaki S (2013) Identification of distinct ligands for the C-type lectin receptors mincle and dectin-2 in the pathogenic fungus Malassezia. Cell Host Microbe 13 (4):477–488. https://doi.org/10.1016/j.chom.2013.03.008

Janeway CA (1992) The immune system evolved to discriminate infectious nonself from noninfectious self. Immunol Today 13(1):11–16. https://doi.org/10.1016/0167-5699(92)90198-G

Kaisar MMM, Ritter M, Del Fresno C, Jónasdóttir HS, Van Der Ham AJ, Pelgrom LR, Schramm G, Layland LE, Sancho D, Prazeres Da Costa C, Giera M, Yazdanbakhsh M, Everts B (2018) Dectin-1/2–induced autocrine PGE2 signaling licenses dendritic cells to prime Th2 responses. PLoS Biol 16(4):e2005504. https://doi.org/10.1371/journal.pbio.2005504

Kashem SW, Igyártó BZ, Gerami-Nejad M, Kumamoto Y, Mohammed J, Jarrett E, Drummond RA, Zurawski SM, Zurawski G, Berman J, Iwasaki A, Brown GD, Kaplan DH (2015) *Candida albicans* morphology and dendritic cell subsets determine T helper cell differentiation. Immunity 42(2):356–366. https://doi.org/10.1016/j.immuni.2015.01.008

Kerrigan AM, Brown GD (2011) Syk-coupled C-type lectins in immunity. Trends Immunol 32 (4):151–156. https://doi.org/10.1016/j.it.2011.01.002

Kerscher B, Wilson GJ, Reid DM, Mori D, Taylor JA, Besra GS, Yamasaki S, Willment JA, Brown GD (2016) Mycobacterial receptor, Clec4d (CLECSF8, MCL), is coregulated with Mincle and upregulated on mouse myeloid cells following microbial challenge. Eur J Immunol 46(2):381–389. https://doi.org/10.1002/eji.201545858

Kijimoto-Ochiai S (2002) CD23 (the low-affinity IgE receptor) as a C-type lectin: a multidomain and multifunctional molecule. Cell Mol Life Sci (CMLS) 59(4):648–664. https://doi.org/10.1007/s00018-002-8455-1

Koppel EA, Van Gisbergen KPJM, Geijtenbeek TBH, Van Kooyk Y (2004) Distinct functions of DC-SIGN and its homologues L-SIGN (DC-SIGNR) and mSIGNR1 in pathogen recognition and immune regulation 7 (2):157–165. https://doi.org/10.1111/j.1462-5822.2004.00480.x

Kottom TJ, Hebrink DM, Jenson PE, Marsolek PL, Wüthrich M, Wang H, Klein B, Yamasaki S, Limper AH (2018) Dectin-2 is a C-type lectin receptor that recognizes pneumocystis and participates in innate immune responses. Am J Respir Cell Mol Biol 58(2):232–240. https://doi.org/10.1165/rcmb.2016-0335oc

Kottom TJ, Hebrink DM, Jenson PE, Nandakumar V, Wüthrich M, Wang H, Klein B, Yamasaki S, Lepenies B, Limper AH (2017) The interaction of pneumocystis with the C-type lectin receptor mincle exerts a significant role in host defense against infection.1600744. https://doi.org/10.4049/jimmunol.1600744

Kottom TJ, Hebrink DM, Monteiro JT, Lepenies B, Carmona EM, Wuethrich M, Santo Dias LD, Limper AH (2019) Myeloid C-type lectin receptors that recognize fungal mannans interact with Pneumocystis organisms and major surface glycoprotein. J Med Microbiol 68(11):1649–1654. https://doi.org/10.1099/jmm.0.001062

Kousha M, Tadi R, Soubani AO (2011) Pulmonary aspergillosis: a clinical review. Eur Resp Rev 20(121):156–174. https://doi.org/10.1183/09059180.00001011

Koziel H, Eichbaum Q, Kruskal BA, Pinkston P, Rogers RA, Armstrong MY, Richards FF, Rose RM, Ezekowitz RA (1998) Reduced binding and phagocytosis of *Pneumocystis carinii* by alveolar macrophages from persons infected with HIV-1 correlates with mannose receptor downregulation 102 (7):1332–1344. https://doi.org/10.1172/jci560

Krutzik SR, Tan B, Li H, Ochoa MT, Liu PT, Sharfstein SE, Graeber TG, Sieling PA, Liu Y-J, Rea TH, Bloom BR, Modlin RL (2005) TLR activation triggers the rapid differentiation of monocytes into macrophages and dendritic cells. Nat Med 11(6):653–660. https://doi.org/10.1038/nm1246

Lasbury ME, Lin P, Tschang D, Durant PJ, Lee CH (2004) Effect of bronchoalveolar lavage fluid from *Pneumocystis carinii*—infected hosts on phagocytic activity of alveolar macrophages. Infect Immunity 72(4):2140–2147. https://doi.org/10.1128/iai.72.4.2140-2147.2004

Latgé J-P, Chamilos G (2019) *Aspergillus fumigatus* and *Aspergillosis* in 2019. Clin Microbiol Rev 33(1):e00140–00118. https://doi.org/10.1128/CMR.00140-18

Le Cabec V, Emorine LJ, Toesca I, Cougoule C, Maridonneau-Parini I (2005) The human macrophage mannose receptor is not a professional phagocytic receptor. J Leukoc Biol 77 (6):934–943. https://doi.org/10.1189/jlb.1204705

Leal SM, Cowden S, Hsia Y-C, Ghannoum MA, Momany M, Pearlman E (2010) Distinct roles for dectin-1 and TLR4 in the pathogenesis of *Aspergillus fumigatus* keratitis. PLoS Pathog 6(7): e1000976. https://doi.org/10.1371/journal.ppat.1000976

Lee SJ, Zheng NY, Clavijo M, Nussenzweig MC (2003) Normal host defense during systemic *Candidiasis* in mannose receptor-deficient mice. Infect Immun 71(1):437–445. https://doi.org/10.1128/iai.71.1.437-445.2003

Lefèvre L, Lugo-Villarino G, Meunier E, Valentin A, Olagnier D, Authier H, Duval C, Dardenne C, Bernad J, Jean Auwerx J, Neyrolles O, Pipy B, Coste A (2013) The C-type lectin receptors dectin-1, MR, and SIGNR3 contribute both positively and negatively to the macrophage response to *Leishmania infantum*. Immunity 38(5):1038–1049. https://doi.org/10.1016/j.immuni.2013.04.010

Li L-Y, Zhang H-R, Jiang Z-L, Chang Y-Z, Shao C-Z (2018) Overexpression of dendritic cell-specific intercellular adhesion molecule-3-grabbing nonintegrin in dendritic cells protecting against aspergillosis. Chin Med J 131(21)

Lim J, Coates CJ, Seoane PI, Garelnabi M, Taylor-Smith LM, Monteith P, Macleod CL, Escaron CJ, Brown GD, Hall RA, May RC (2018) Characterizing the mechanisms of nonopsonic uptake of cryptococci by macrophages. J Immunol 200(10):3539–3546. https://doi.org/10.4049/jimmunol.1700790

Lima-Junior DS, Mineo TWP, Calich VLG, Zamboni DS (2017) Dectin-1 activation during leishmania amazonensis phagocytosis prompts Syk-dependent reactive oxygen species production to trigger inflammasome assembly and restriction of parasite replication. J Immunol 199(6):2055–2068. https://doi.org/10.4049/jimmunol.1700258

Limon JJ, Skalski JH, Underhill DM (2017) Commensal fungi in health and disease. Cell Host Microbe 22(2):156–165. https://doi.org/10.1016/j.chom.2017.07.002

Lobato-Pascual A, Saether PC, Fossum S, Dissen E, Daws MR (2013) Mincle, the receptor for mycobacterial cord factor, forms a functional receptor complex with MCL and FcεRI-γ. Eur J Immunol 43(12):3167–3174. https://doi.org/10.1002/eji.201343752

Loures FV, Röhm M, Lee CK, Santos E, Wang JP, Specht CA, Calich VLG, Urban CF, Levitz SM (2015) Recognition of *Aspergillus fumigatus* hyphae by human plasmacytoid dendritic cells is mediated by dectin-2 and results in formation of extracellular traps 11(2):e1004643. https://doi.org/10.1371/journal.ppat.1004643

Lowman DW, Greene RR, Bearden DW, Kruppa MD, Pottier M, Monteiro MA, Soldatov DV, Ensley HE, Cheng S-C, Netea MG, Williams DL (2014) Novel structural features in *Candida albicans* hyphal glucan provide a basis for differential innate immune recognition of hyphae versus yeast. J Biol Chem 289(6):3432–3443. https://doi.org/10.1074/jbc.m113.529131

Ma L, Chen Z, Huang DW, Kutty G, Ishihara M, Wang H, Abouelleil A, Bishop L, Davey E, Deng R, Deng X, Fan L, Fantoni G, Fitzgerald M, Gogineni E, Goldberg JM, Handley G, Hu X, Huber C, Jiao X, Jones K, Levin JZ, Liu Y, Macdonald P, Melnikov A, Raley C, Sassi M, Sherman BT, Song X, Sykes S, Tran B, Walsh L, Xia Y, Yang J, Young S, Zeng Q, Zheng X, Stephens R, Nusbaum C, Birren BW, Azadi P, Lempicki RA, Cuomo CA, Kovacs JA (2016) Genome analysis of three Pneumocystis species reveals adaptation

mechanisms to life exclusively in mammalian hosts. Nat Commun 7(1):10740. https://doi.org/10.1038/ncomms10740
Maldonado SG, Dai J, Singh S, Mwangi D, Rivera A, Fitzgerald-Bocarsly P (2015) Human pDCs express the C-type lectin receptor dectin-1 and uptake and kill *Aspergillus fumigatus* spores in vitro (MPF4P.734). J Immunol 194
Mansour MK, Latz E, Levitz SM (2006) *Cryptococcus neoformans* glycoantigens are captured by multiple lectin receptors and presented by dendritic cells. J Immunol 176(5):3053–3061. https://doi.org/10.4049/jimmunol.176.5.3053
Marakalala MJ, Vautier S, Potrykus J, Walker LA, Shepardson KM, Hopke A, Mora-Montes HM, Kerrigan A, Netea MG, Murray GI, Maccallum DM, Wheeler R, Munro CA, Gow NAR, Cramer RA, Brown AJP, Brown GD (2013) Differential adaptation of *Candida albicans* in vivo modulates immune recognition by dectin-1 9(4):e1003315. https://doi.org/10.1371/journal.ppat.1003315
Martinez-Pomares L (2012) The mannose receptor. J Leukoc Biol 92(6):1177–1186. https://doi.org/10.1189/jlb.0512231
Maruyama K, Takayama Y, Kondo T, Ishibashi K-I, Sahoo BR, Kanemaru H, Kumagai Y, Martino MM, Tanaka H, Ohno N, Iwakura Y, Takemura N, Tominaga M, Akira S (2017) Nociceptors boost the resolution of fungal osteoinflammation via the TRP channel-CGRP-Jdp2 axis. Cell Rep 19(13):2730–2742. https://doi.org/10.1016/j.celrep.2017.06.002
McGreal EP, Rosas M, Brown GD, Zamze S, Wong SYC, Gordon S, Martinez-Pomares L, Taylor PR (2006) The carbohydrate-recognition domain of Dectin-2 is a C-type lectin with specificity for high mannose. Glycobiology 16(5):422–430. https://doi.org/10.1093/glycob/cwj077
McKenzie EJ, Taylor PR, Stillion RJ, Lucas AD, Harris J, Gordon S, Martinez-Pomares L (2007) Mannose receptor expression and function define a new population of murine dendritic cells. J Immunol 178(8):4975–4983. https://doi.org/10.4049/jimmunol.178.8.4975
Medrano FJ, Montes-Cano M, Conde M, Conde, de la Horra C, Respaldiza N, Gasch A, Perez-Lozano M, Varela JM, Calderon EJ (2005) *Pneumocystis jirovecii* in general population. Emerg Infect Dis J 11(2):245. https://doi.org/10.3201/eid1102.040487
Mezger M, Kneitz S, Wozniok I, Kurzai O, Einsele H, Loeffler J (2008) Proinflammatory response of immature human dendritic cells is mediated by dectin-1 after exposure to *Aspergillus fumigatus* germ tubes. J Infect Dis 197(6):924–931. https://doi.org/10.1086/528694
Miyake Y, Oh-Hora M, Yamasaki S (2015) C-type lectin receptor MCL facilitates mincle expression and signaling through complex formation. J Immunol 194(11):5366–5374. https://doi.org/10.4049/jimmunol.1402429
Miyake Y, Toyonaga K, Mori D, Kakuta S, Hoshino Y, Oyamada A, Yamada H, Ono K-I, Suyama M, Iwakura Y, Yoshikai Y, Yamasaki S (2013) C-type lectin MCL Is an FcRγ-coupled receptor that mediates the adjuvanticity of mycobacterial cord factor. Immunity 38(5):1050–1062. https://doi.org/10.1016/j.immuni.2013.03.010
Mossalayi MD, Vouldoukis I, Mamani-Matsuda M, Kauss T, Guillon J, Maugein J, Moynet D, Rambert JRM, Desplat V, Mazier D, Vincendeau P, Malvy D (2009) CD23 mediates antimycobacterial activity of human macrophages. Infect Immun 77(12):5537–5542. https://doi.org/10.1128/iai.01457-08
Mukaremera L, Lee KK, Wagener J, Wiesner DL, Gow NAR, Nielsen K (2018) Titan cell production in *Cryptococcus neoformans* reshapes the cell wall and capsule composition during infection. The Cell Surf. https://doi.org/10.1016/j.tcsw.2017.12.001
Nakamura K, Kinjo T, Saijo S, Miyazato A, Adachi Y, Ohno N, Fujita J, Kaku M, Iwakura Y, Kawakami K (2007) Dectin-1 is not required for the host defense to *Cryptococcus neoformans*. Microbiol Immunol 51(11):1115–1119. https://doi.org/10.1111/j.1348-0421.2007.tb04007.x
Nakamura Y, Sato K, Yamamoto H, Matsumura K, Matsumoto I, Nomura T, Miyasaka T, Ishii K, Kanno E, Tachi M, Yamasaki S, Saijo S, Iwakura Y, Kawakami K (2015) Dectin-2 deficiency promotes Th2 response and mucin production in the lungs after pulmonary infection with *Cryptococcus neoformans*. Infect Immun 83(2):671–681. https://doi.org/10.1128/iai.02835-14

Netea MG (2006) Immune sensing of *Candida albicans* requires cooperative recognition of mannans and glucans by lectin and toll-like receptors. J Clin Invest 116(6):1642–1650. https://doi.org/10.1172/jci27114

Netea MG, Gow NAR, Joosten LAB, Verschueren I, Van Der Meer JWM, Kullberg BJ (2010) Variable recognition of *Candida albicansstrains* by TLR4 and lectin recognition receptors. Med Mycol 48(7):897–903. https://doi.org/10.3109/13693781003621575

Nollstadt KH, Powles MA, Fujioka H, Aikawa M, Schmatz DM (1994) Use of beta-1,3-glucan-specific antibody to study the cyst wall of *Pneumocystis carinii* and effects of pneumocandin B0 analog L-733,560 38(10):2258–2265. https://doi.org/10.1128/aac.38.10.2258

Norimoto A, Hirose K, Iwata A, Tamachi T, Yokota M, Takahashi K, Saijo S, Iwakura Y, Nakajima H (2014) Dectin-2 promotes house dust mite-induced Th2 and Th17 cell differentiation and allergic airway inflammation in mice. Am J Resp Cell Mol Biol 140303075636004. https://doi.org/10.1165/rcmb.2013-0522oc

O'Riordan DM, Standing JE, Limper AH (1995) Pneumocystis carinii glycoprotein A binds macrophage mannose receptors. Infect Immun 63(3):779–784

Park CG, Takahara K, Umemoto E, Yashima Y, Matsubara K, Matsuda Y, Clausen BE, Inaba K, Steinman RM (2001) Five mouse homologues of the human dendritic cell C-type lectin, DC-SIGN. Int Immunol 13(10):1283–1290. https://doi.org/10.1093/intimm/13.10.1283

Parsons MW, Li L, Wallace AM, Lee MJ, Katz HR, Fernandez JM, Saijo S, Iwakura Y, Austen KF, Kanaoka Y, Barrett NA (2014) Dectin-2 regulates the effector phase of house dust mite-elicited pulmonary inflammation independently from its role in sensitization. J Immunol 192(4):1361–1371. https://doi.org/10.4049/jimmunol.1301809

Patin EC, Orr SJ, Schaible UE (2017) Macrophage inducible C-type lectin as a multifunctional player in immunity. Front Immunol 8:861

Perfect JR (2005) Cryptococcus neoformans: a sugar-coated killer with designer genes. FEMS Immunol Med Microbiol 45(3):395–404. https://doi.org/10.1016/j.femsim.2005.06.005

Perlroth J, Choi B, Spellberg B (2007) Nosocomial fungal infections: epidemiology, diagnosis, and treatment. Med Mycol 45(4):321–346. https://doi.org/10.1080/13693780701218689

Pinke KH, Lima HGD, Cunha FQ, Lara VS (2016) Mast cells phagocyte *Candida albicans* and produce nitric oxide by mechanisms involving TLR2 and dectin-1 221(2):220–227. https://doi.org/10.1016/j.imbio.2015.09.004

Plantinga M, Guilliams M, Vanheerswynghels M, Deswarte K, Branco-Madeira F, Toussaint W, Vanhoutte L, Neyt K, Killeen N, Malissen B, Hammad H, Bart (2013) Conventional and monocyte-derived CD11b + dendritic cells initiate and maintain T helper 2 cell-mediated immunity to house dust mite allergen. Immunity 38(2):322–335. https://doi.org/10.1016/j.immuni.2012.10.016

Plantinga TS, van der Velden WJFM, Ferwerda B, van Spriel AB, Adema G, Feuth T, Donnelly JP, Brown GD, Kullberg B-J, Blijlevens NMA, Netea MG (2009) Early stop polymorphism in human DECTIN-1 is associated with increased *Candida* colonization in hematopoietic stem cell transplant recipients. Clin Infect Dis 49(5):724–732. https://doi.org/10.1086/604714

Porcaro I, Vidal M, Jouvert S, Stahl PD, Giaimis J (2003) Mannose receptor contribution to Candida albicansphagocytosis by murine E-clone J774 macrophages. J Leukoc Biol 74 (2):206–215. https://doi.org/10.1189/jlb.1202608

Powlesland AS, Ward EM, Sadhu SK, Guo Y, Taylor ME, Drickamer K (2006) Widely divergent biochemical properties of the complete set of mouse DC-SIGN-related proteins. J Biol Chem 281(29):20440–20449. https://doi.org/10.1074/jbc.m601925200

Rajaram MVS, Arnett E, Azad AK, Guirado E, Ni B, Gerberick AD, He L-Z, Keler T, Thomas LJ, Lafuse WP, Schlesinger LS (2017) *M.tuberculosis*—initiated human mannose receptor signaling regulates macrophage recognition and vesicle trafficking by FcRγ-chain, Grb2, and SHP-1. Cell Rep 21 (1):126–140. https://doi.org/10.1016/j.celrep.2017.09.034

Rajasingham R, Smith RM, Park BJ, Jarvis JN, Govender NP, Chiller TM, Denning DW, Loyse A, Boulware DR (2017) Global burden of disease of HIV-associated cryptococcal meningitis: an updated analysis. Lancet Infect Dis 17(8):873–881. https://doi.org/10.1016/s1473-3099(17)30243-8

Ritter M, Gross O, Kays S, Ruland J, Nimmerjahn F, Saijo S, Tschopp J, Layland LE, Prazeres Da Costa C (2010) *Schistosoma mansoni* triggers Dectin-2, which activates the Nlrp3 inflammasome and alters adaptive immune responses. Proc Natl Acad Sci 107(47):20459–20464. https://doi.org/10.1073/pnas.1010337107

Rivera A, Hohl TM, Collins N, Leiner I, Gallegos A, Saijo S, Coward JW, Iwakura Y, Pamer EG (2011) Dectin-1 diversifies *Aspergillus fumigatus*–specific T cell responses by inhibiting T helper type 1 CD4 T cell differentiation. J Exp Med 208(2):369–381. https://doi.org/10.1084/jem.20100906

Rizzetto L, Weil T, Cavalieri D (2015) Systems level dissection of Candida recognition by dectins: a matter of fungal morphology and site of infection. Pathogens 4(3):639–661. https://doi.org/10.3390/pathogens4030639

Romani L (2011) Immunity to fungal infections. Nat Rev Immunol 11(4):275–288. https://doi.org/10.1038/nri2939

Royer PJ, Emara M, Yang C, Al-Ghouleh A, Tighe P, Jones N, Sewell HF, Shakib F, Martinez-Pomares L, Ghaemmaghami AM (2010) The mannose receptor mediates the uptake of diverse native allergens by dendritic cells and determines allergen-induced T cell polarization through modulation of IDO activity 185(3):1522–1531. https://doi.org/10.4049/jimmunol.1000774

Saijo S, Fujikado N, Furuta T, Chung S-h, Kotaki H, Seki K, Sudo K, Akira S, Adachi Y, Ohno N, Kinjo T, Nakamura K, Kawakami K, Iwakura Y (2007) Dectin-1 is required for host defense against Pneumocystis carinii but not against *Candida albicans*. Nat Immunol 8(1):39–46. https://doi.org/10.1038/ni1425

Saijo S, Ikeda S, Yamabe K, Kakuta S, Ishigame H, Akitsu A, Fujikado N, Kusaka T, Kubo S, Chung S-H, Komatsu R, Miura N, Adachi Y, Ohno N, Shibuya K, Yamamoto N, Kawakami K, Yamasaki S, Saito T, Akira S, Iwakura Y (2010) Dectin-2 recognition of α-mannans and induction of Th17 cell differentiation is essential for host defense against *Candida albicans*. Immunity 32(5):681–691. https://doi.org/10.1016/j.immuni.2010.05.001

Sainz J, Lupiáñez CB, Segura-Catena J, Vazquez L, Ríos R, Oyonarte S, Hemminki K, Försti A, Jurado M (2012) Dectin-1 and DC-SIGN polymorphisms associated with invasive pulmonary Aspergillosis infection. PLoS ONE 7(2):e32273. https://doi.org/10.1371/journal.pone.0032273

Sancho D, Reis e Sousa C (2012) Signaling by Myeloid C-type lectin receptors in immunity and homeostasis. Ann Rev Immunol 30 (1):491–529. https://doi.org/10.1146/annurev-immunol-031210-101352

Sato K, Yang X-L, Yudate T, Chung J-S, Wu J, Luby-Phelps K, Kimberly RP, Underhill D, Cruz PD, Ariizumi K (2006) Dectin-2 is a pattern recognition receptor for fungi that couples with the Fc receptor γ chain to induce innate immune responses. J Biol Chem 281(50):38854–38866. https://doi.org/10.1074/jbc.m606542200

Sattler S, Reiche D, Sturtzel C, Karas I, Richter S, Kalb ML, Gregor W, Hofer E (2012) The human C-type lectin-like receptor CLEC-1 is upregulated by TGF-β and primarily localized in the endoplasmic membrane compartment. Scand J Immunol 75(3):282–292. https://doi.org/10.1111/j.1365-3083.2011.02665.x

Saville SP, Lazzell AL, Monteagudo C, Lopez-Ribot JL (2003) Engineered control of cell morphology in vivo reveals distinct roles for yeast and filamentous forms of *Candida albicans* during infection. Eukaryot Cell 2(5):1053–1060. https://doi.org/10.1128/ec.2.5.1053-1060.2003

Schop J (2007) Protective immunity against cryptococcus neoformans infection. Mcgill J Med 10 (1):35–43

Schweneker K, Gorka O, Schweneker M, Poeck H, Tschopp J, Peschel C, Ruland J, Gross O (2013) The mycobacterial cord factor adjuvant analogue trehalose-6,6'-dibehenate (TDB) activates the Nlrp3 inflammasome. Immunobiology 218(4):664–673. https://doi.org/10.1016/j.imbio.2012.07.029

Serrano-Gómez D, Antonio Leal J, Corbí AL (2005) DC-SIGN mediates the binding of *Aspergillus fumigatus* and keratinophylic fungi by human dendritic cells. Immunobiology 210 (2–4):175–183. https://doi.org/10.1016/j.imbio.2005.05.011

Serrano-Gómez D, Domínguez-Soto A, Ancochea J, Jimenez-Heffernan JA, Leal JA, Corbí AL (2004) Dendritic cell-specific intercellular adhesion molecule 3-grabbing nonintegrin mediates binding and internalization of *Aspergillus fumigatus* conidia by dendritic cells and macrophages. J Immunol 173(9):5635–5643. https://doi.org/10.4049/jimmunol.173.9.5635

Shiokawa M, Yamasaki S, Saijo S (2017) C-type lectin receptors in anti-fungal immunity. Curr Opin Microbiol 40:123–130. https://doi.org/10.1016/j.mib.2017.11.004

Skalski JH, Kottom TJ, Limper AH (2015) Pathobiology of *Pneumocystis pneumonia*: life cycle, cell wall and cell signal transduction. FEMS Yeast Res 15(6). https://doi.org/10.1093/femsyr/fov046

Sloan D, Parris V (2014) Cryptococcal meningitis: epidemiology and therapeutic options. Clin Epidemiol 169. https://doi.org/10.2147/clep.s38850

Sobanov Y, Bernreiter A, Derdak S, Mechtcheriakova D, Schweighofer B, Düchler M, Kalthoff F, Hofer E (2001) A novel cluster of lectin-like receptor genes expressed in monocytic, dendritic and endothelial cells maps close to the NK receptor genes in the human NK gene complex. Eur J Immunol 31(12):3493–3503. https://doi.org/10.1002/1521-4141(200112)31:12%3c3493:aid-immu3493%3e3.0.co;2-9

Soilleux EJ, Barten R, Trowsdale J (2000) Cutting edge: DC-SIGN; a related gene, DC-SIGNR; and CD23 form a cluster on 19p13 165(6):2937–2942. https://doi.org/10.4049/jimmunol.165.6.2937

Stappers MHT, Clark AE, Aimanianda V, Bidula S, Reid DM, Asamaphan P, Hardison SE, Dambuza IM, Valsecchi I, Kerscher B, Plato A, Wallace CA, Yuecel R, Hebecker B, Da Glória Teixeira Sousa M, Cunha C, Liu Y, Feizi T, Brakhage AA, Kwon-Chung KJ, Gow NAR, Zanda M, Piras M, Zanato C, Jaeger M, Netea MG, Van De Veerdonk FL, Lacerda JF, Campos A, Carvalho A, Willment JA, Latgé J-P, Brown GD (2018) Recognition of DHN-melanin by a C-type lectin receptor is required for immunity to Aspergillus. Nature 555 (7696):382–386. https://doi.org/10.1038/nature25974

Steele C, Marrero L, Swain S, Harmsen AG, Zheng M, Brown GD, Gordon S, Shellito JE, Kolls JK (2003) Alveolar macrophage–mediated killing of *Pneumocystis carinii* f. sp. muris involves molecular recognition by the dectin-1 β-glucan receptor 198(11):1677–1688. https://doi.org/10.1084/jem.20030932

Stehle SE, Rogers RA, Harmsen AG, Ezekowitz RAB (2000) A soluble mannose receptor immunoadhesin enhances phagocytosis of *Pneumocystis carinii* by human polymorphonuclear leukocytes *in vitro*. Scand JImmunol 52:131–137

Stein M, Keshav S, Harris N, Gordon S (1992) Interleukin 4 potently enhances murine macrophage mannose receptor activity: a marker of alternative immunologic macrophage activation. J Exp Med 176(1):287–292. https://doi.org/10.1084/jem.176.1.287

Strasser D, Neumann K, Bergmann H, Mohlopheni Guler R, Rojowska A, Hopfner K-P, Brombacher F, Urlaub H, Baier G, Gordon Leitges M, Ruland J (2012) Syk kinase-coupled C-type lectin receptors engage protein kinase C-δ to elicit Card9 adaptor-mediated innate immunity. Immunity 36(1):32–42. https://doi.org/10.1016/j.immuni.2011.11.015

Stringer JR, Beard CB, Miller RF, Wakefield AE (2002) A new name for Pneumocystis from humans and new perspectives on the host-pathogen relationship. Emerg Infect Dis 8(9):891–896. https://doi.org/10.3201/eid0809.020096

Sun H, Xu X-Y, Shao H-T, Su X, Wu X-D, Wang Q, Shi Y (2013) Dectin-2 is predominately macrophage restricted and exhibits conspicuous expression during *Aspergillus fumigatus* invasion in human lung. Cell Immunol 284(1–2):60–67. https://doi.org/10.1016/j.cellimm.2013.06.013

Sun WK, Lu X, Li X, Sun QY, Su X, Song Y, Sun HM, Shi Y (2012) Dectin-1 is inducible and plays a crucial role in Aspergillus-induced innate immune responses in human bronchial epithelial cells. Eur J Clin Microbiol Infect Dis 31(10):2755–2764. https://doi.org/10.1007/s10096-012-1624-8

Swain SD, Lee SJ, Nussenzweig MC, Harmsen AG (2003) Absence of the macrophage mannose receptor in mice does not increase susceptibility to *Pneumocystis carinii* infection in vivo. Infect Immun 71(11):6213–6221. https://doi.org/10.1128/iai.71.11.6213-6221.2003

Syme RM, Spurrell JCL, Amankwah EK, Green FHY, Mody CH (2002) Primary dendritic cells phagocytose *Cryptococcus neoformans* via mannose receptors and Fc receptor II for presentation to T lymphocytes. Infect Immun 70(11):5972–5981. https://doi.org/10.1128/iai.70.11.5972-5981.2002

Taylor ME, Conary JT, Lennartz MR, Stahl PD, Drickamer K (1990) Primary structure of the mannose receptor contains multiple motifs resembling carbohydrate-recognition domains. J Biol Chem 265(21):12156–12162

Taylor PR, Reid DM, Heinsbroek SEM, Brown GD, Gordon S, Wong SYC (2005) Dectin-2 is predominantly myeloid restricted and exhibits unique activation-dependent expression on maturing inflammatory monocytes elicitedin vivo. Eur J Immunol 35(7):2163–2174. https://doi.org/10.1002/eji.200425785

Taylor PR, Roy S, Leal SM, Sun Y, Howell SJ, Cobb BA, Li X, Pearlman E (2014) Activation of neutrophils by autocrine IL-17A–IL-17RC interactions during fungal infection is regulated by IL-6, IL-23, RORγt and dectin-2. Nat Immunol 15(2):143–151. https://doi.org/10.1038/ni.2797

Taylor PR, Tsoni SV, Willment JA, Dennehy KM, Rosas M, Findon H, Haynes K, Steele C, Botto M, Gordon S, Brown GD (2007) Dectin-1 is required for β-glucan recognition and control of fungal infection. Nat Immunol 8(1):31–38. https://doi.org/10.1038/ni1408

Thiagarajan PS, Yakubenko VP, Elsori DH, Yadav SP, Willard B, Tan CD, Rene Rodriguez E, Febbraio M, Cathcart MK (2013) Vimentin is an endogenous ligand for the pattern recognition receptor Dectin-1 99(3):494–504. https://doi.org/10.1093/cvr/cvt117

Thomas CF, Limper AH (2004) Pneumocystis pneumonia. N Engl J Med 350(24):2487–2498. https://doi.org/10.1056/nejmra032588

van de Veerdonk FL, Gresnigt MS, Romani L, Netea MG, Latgé J-P (2017) *Aspergillus fumigatus* morphology and dynamic host interactions. Nat Rev Microbiol 15(11):661–674. https://doi.org/10.1038/nrmicro.2017.90

Van De Veerdonk FL, Marijnissen RJ, Kullberg BJ, Koenen HJPM, Cheng S-C, Joosten I, Van Den Berg WB, Williams DL, Van Der Meer JWM, Joosten LAB, Netea MG (2009) The macrophage mannose receptor induces IL-17 in response to *Candida albicans*. Cell Host Microbe 5(4):329–340. https://doi.org/10.1016/j.chom.2009.02.006

Vargas SL, Hughes WT, Santolaya ME, Ulloa AV, Ponce CA, Cabrera CE, Cumsille F, Gigliotti F (2001) Search for primary infection by *Pneumocystis carinii* in a cohort of normal, healthy infants. Clin Infect Dis 32(6):855–861. https://doi.org/10.1086/319340

Vautier S, Drummond RA, Redelinghuys P, Murray GI, Maccallum DM, Brown GD (2012) Dectin-1 is not required for controlling *Candida albicans* colonization of the gastrointestinal tract. Infect Immun 80(12):4216–4222. https://doi.org/10.1128/iai.00559-12

Vendele I, Willment JA, Silva LM, Palma AS, Chai W, Liu Y, Feizi T, Spyrou M, Stappers MHT, Brown GD, Gow NAR (2020) Mannan detecting C-type lectin receptor probes recognise immune epitopes with diverse chemical, spatial and phylogenetic heterogeneity in fungal cell walls. PLoS Pathog 16(1):e1007927. https://doi.org/10.1371/journal.ppat.1007927

Vouldoukis I, Riveros-Moreno V, Dugas B, Ouaaz F, Becherel P, Debre P, Moncada S, Mossalayi MD (1995) The killing of Leishmania major by human macrophages is mediated by nitric oxide induced after ligation of the Fc epsilon RII/CD23 surface antigen. Proc Natl Acad Sci 92(17):7804–7808. https://doi.org/10.1073/pnas.92.17.7804

Wächtler B, Wilson D, Haedicke K, Dalle F, Hube B (2011) From attachment to damage: defined genes of *Candida albicans* mediate adhesion, invasion and damage during interaction with oral epithelial cells. PLoS ONE 6(2):e17046. https://doi.org/10.1371/journal.pone.0017046

Wagener M, Hoving JC, Ndlovu H, Marakalala MJ (2018) Dectin-1-Syk-CARD9 signaling pathway in TB immunity. Front Immunol 9:225

Walsh NM, Wuthrich M, Wang H, Klein B, Hull CM (2017) Characterization of C-type lectins reveals an unexpectedly limited interaction between *Cryptococcus neoformans* spores and dectin-1. PLoS ONE 12(3):e0173866. https://doi.org/10.1371/journal.pone.0173866

Wang H, Lee T-J, Fites SJ, Merkhofer R, Zarnowski R, Brandhorst T, Galles K, Klein B, Wüthrich M (2017) Ligation of dectin-2 with a novel microbial ligand promotes adjuvant activity for vaccination. PLoS Pathog 13(8):e1006568. https://doi.org/10.1371/journal.ppat.1006568

Wang Q, Zhao G, Lin J, Li C, Jiang N, Xu Q, Wang Q, Zhang J (2016) Role of the mannose receptor during *Aspergillus fumigatus* infection and interaction with dectin-1 in corneal epithelial cells 35(2):267–273. https://doi.org/10.1097/ico.0000000000000710

Wells CA, Salvage-Jones JA, Li X, Hitchens K, Butcher S, Murray RZ, Beckhouse AG, Lo Y-L-S, Manzanero S, Cobbold C, Schroder K, Ma B, Orr S, Stewart L, Lebus D, Sobieszczuk P, Hume DA, Stow J, Blanchard H, Ashman RB (2008) The macrophage-inducible C-type lectin, mincle, is an essential component of the innate immune response to *Candida albicans*. J Immunol 180(11):7404–7413. https://doi.org/10.4049/jimmunol.180.11.7404

Werner JL, Metz AE, Horn D, Schoeb TR, Hewitt MM, Schwiebert LM, Faro-Trindade I, Brown GD, Steele C (2009) Requisite role for the dectin-1 β-glucan receptor in pulmonary defense against *Aspergillus fumigatus*. J Immunol 182(8):4938–4946. https://doi.org/10.4049/jimmunol.0804250

Wevers BA, Kaptein TM, Zijlstra-Willems EM, Theelen B, Boekhout T, Geijtenbeek TB, Gringhuis SI (2014) Fungal engagement of the C-type lectin mincle suppresses dectin-1-induced antifungal immunity. Cell Host Microbe 15(4):494–505. https://doi.org/10.1016/j.chom.2014.03.008

Williams SJ (2017) Sensing lipids with mincle: structure and function. Front Immunol 8:1662

Yamasaki S, Ishikawa E, Sakuma M, Hara H, Ogata K, Saito T (2008) Mincle is an ITAM-coupled activating receptor that senses damaged cells. Nat Immunol 9(10):1179–1188. https://doi.org/10.1038/ni.1651

Yang A-M, Inamine T, Hochrath K, Chen P, Wang L, Llorente C, Bluemel S, Hartmann P, Xu J, Koyama Y, Kisseleva T, Torralba MG, Moncera K, Beeri K, Chen C-S, Freese K, Hellerbrand C, Lee SML, Hoffman HM, Mehal WZ, Garcia-Tsao G, Mutlu EA, Keshavarzian A, Brown GD, Ho SB, Bataller R, Stärkel P, Fouts DE, Schnabl B (2017) Intestinal fungi contribute to development of alcoholic liver disease. J Clin Invest 127(7):2829–2841. https://doi.org/10.1172/jci90562

Yonekawa A, Saijo S, Hoshino Y, Miyake Y, Ishikawa E, Suzukawa M, Inoue H, Tanaka M, Yoneyama M, Oh-Hora M, Akashi K, Yamasaki S (2014) Dectin-2 is a direct receptor for mannose-capped lipoarabinomannan of mycobacteria. Immunity 41(3):402–413. https://doi.org/10.1016/j.immuni.2014.08.005

Yuan K, Zhao G, Che C, Li C, Lin J, Zhu G, He K (2017) Dectin-1 is essential for IL-1β production through JNK activation and apoptosis in *Aspergillus fumigatus* keratitis. Int Immunopharmacol 52:168–175. https://doi.org/10.1016/j.intimp.2017.09.008

Zaragoza O, Rodrigues ML, De Jesus M, Frases S, Dadachova E, Casadevall A (2009) Chapter 4 The capsule of the fungal pathogen *Cryptococcus neoformans*. Elsevier, pp 133–216. https://doi.org/10.1016/s0065-2164(09)01204-0

Zelensky AN, Gready JE (2005) The C-type lectin-like domain superfamily. FEBS J 272 (24):6179–6217. https://doi.org/10.1111/j.1742-4658.2005.05031.x

Zhao G, Xu Q, Lin J, Chen W, Cui T, Hu L, Jiang N (2017) The role of Mincle in innate immune to fungal keratitis. J Infect Develop Countries 11(01):89–97. https://doi.org/10.3855/jidc.7570

Zhao X-Q, Zhu L-L, Chang Q, Jiang C, You Y, Luo T, Jia X-M, Lin X (2014) C-type lectin receptor dectin-3 mediates trehalose 6,6′-dimycolate (TDM)-induced mincle expression through CARD9/Bcl10/MALT1-dependent nuclear factor (NF)-κB activation. J Biol Chem 289(43):30052–30062. https://doi.org/10.1074/jbc.m114.588574

Zhao Y, Chu X, Chen J, Wang Y, Gao S, Jiang Y, Zhu X, Tan G, Zhao W, Yi H, Xu H, Ma X, Lu Y, Yi Q, Wang S (2016) Dectin-1-activated dendritic cells trigger potent antitumour immunity through the induction of Th9 cells. Nat Commun 7(1):12368. https://doi.org/10.1038/ncomms12368

Zhao Y, Lin J, Fan Y, Lin X (2019) Life cycle of *Cryptococcus neoformans*. Annu Rev Microbiol 73(1):17–42. https://doi.org/10.1146/annurev-micro-020518-120210

Zhu L-L, Zhao X-Q, Jiang C, You Y, Chen X-P, Jiang Y-Y, Jia X-M, Lin X (2013) C-type lectin receptors dectin-3 and dectin-2 form a heterodimeric pattern-recognition receptor for host defense against fungal infection 39 (2):324–334. https://doi.org/10.1016/j.immuni.2013.05.017

Recognition of Mycobacteria by Dendritic Cell Immunoactivating Receptor

Kenji Toyonaga and Sho Yamasaki

Contents

1 Introduction 104
2 Identification of DCAR 106
3 Cellular Distribution of DCAR 106
4 DCAR Recognizes Acylated Phosphatidyl-Inositol Mannosides (PIMs) in Mycobacteria 107
5 PIMs Induce Chemokine Production via DCAR–FcRγ 109
6 DCAR Contributes to the Th1 Response Against Mycobacterial Infection 110
7 Conclusions 111
References 113

Abstract Mycobacteria have unique lipids on their cell walls, and the structures and physiological activities of these lipid components have been the subject of many studies. Although the host receptors for mycobacterial lipid have long been elusive, in recent years C-type lectin receptors (CLRs) have been reported to recognize these components. The dendritic cell immunoactivating receptor (DCAR), a CLR member, is encoded by *Clec4b1*. DCAR, which was identified in 2003, is reported to be associated with the immunoreceptor tyrosine-based activation motif (ITAM)-containing adaptor protein, the Fc receptor γ chain (FcRγ). However, its physiological ligand and biological function were unknown. We recently identified DCAR as an activating receptor for mycobacteria. DCAR recognizes acylated phosphatidyl-inositol mannosides (PIMs) in mycobacteria to promote Th1

K. Toyonaga (✉) · S. Yamasaki
Department of Molecular Immunology, Research Institute for Microbial Diseases, Osaka University, Suita, Osaka 565-0871, Japan
e-mail: toyonaga@biken.osaka-u.ac.jp

S. Yamasaki
e-mail: yamasaki@biken.osaka-u.ac.jp

Current Topics in Microbiology and Immunology (2020) 429: 103–115
https://doi.org/10.1007/82_2020_203

Published Online: 17 April 2020

responses during mycobacterial infection. This review summarizes recent discoveries about the ligands and immunological roles of DCAR.

1 Introduction

Macrophages and dendritic cells (DCs) are important for the detection of invading pathogens and the initiation of immune responses. They express innate immune receptors such as toll-like receptors (TLRs), NOD-like receptors (NLRs), and RIG-I-like receptors (RLRs), which detect pathogens or the damage induced by them (Takeuchi and Akira 2010). These receptors, also called pattern recognition receptors (PRRs), recognize simple molecules and the regular patterns in molecular structures (called pathogen-associated molecular patterns; PAMPs) derived from many microorganisms but not from host cells (Kawai and Akira 2010).

In recent years, the C-type lectin receptor (CLR) has come to light as a new innate immune receptor family (Dambuza and Brown 2015). C-type lectin is one of a family of proteins exhibiting calcium ion-dependent sugar-binding activity and is conserved in a wide variety of organisms (Zelensky and Gready 2005; Drickamer and Taylor 2015). Many CLRs expressed in macrophages and DCs are single-pass transmembrane proteins containing a carbohydrate recognition domain (CRD) in their extracellular regions (Sancho and Reis e Sousa 2012; Brown et al. 2018). These myeloid CLRs fall into the following three main groups based on their intracellular signaling motifs: the immunoreceptor tyrosine-based activation motif (ITAM)-bearing activating CLRs, the immunoreceptor tyrosine-based inhibitory motif (ITIM)-bearing inhibitory CLRs, and the ITAM/ITIM-independent CLRs (Fig. 1) (Sancho and Reis e Sousa 2012). Although some activating CLRs can transduce signals directly through the hemITAM motif in their cytoplasmic tails, most activating CLRs transduce signals indirectly by associating with ITAM-containing adaptor proteins such as the Fc receptor γ chain (FcRγ) and Dap12. ITAM-bearing CLRs were initially identified for their importance in antifungal immunity (Hardison and Brown 2012; Shiokawa et al. 2017). However, later studies revealed that they are also involved in anti-mycobacterial immunity (Ishikawa et al. 2017; Marakalala and Ndlovu 2017).

Tuberculosis caused by *Mycobacterium tuberculosis* (*M. tuberculosis*) is a major infectious disease along with acquired immunodeficiency syndrome (AIDS) and malaria. About one-third of the world's population is infected with this pathogen, and more than 1.5 million people die from it each year. The mycobacterial cell wall contains various lipid components that characterize the properties of mycobacteria (Fig. 2), such as acid resistance and hydrophobicity, and these lipid components have various effects on the host immune response. These lipids have been studied for many years as compounds involved in inflammation and pathogenicity (Jackson 2014; Minnikin et al. 2015). However, little was known about their host receptors. With the discovery of PRRs such as TLRs, it was thought that these host receptors were also responsible for the recognition of mycobacterial lipid components; but in

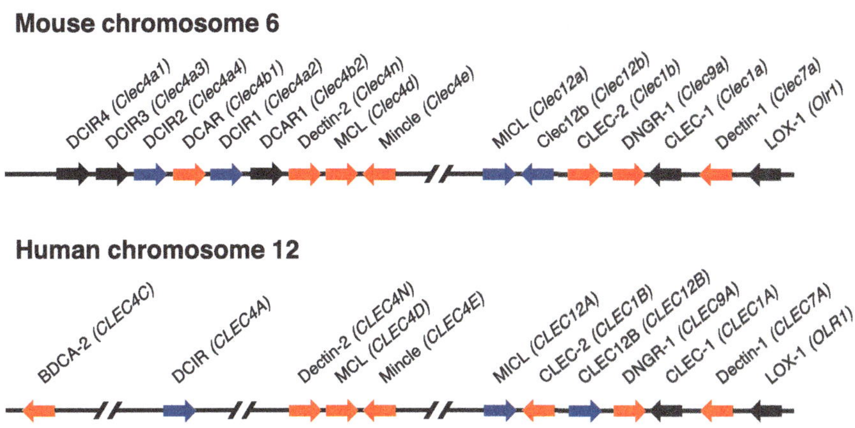

Fig. 1 Organization of CLR gene clusters on mouse chromosome 6 and human chromosome 12. ITAM- or hemITAM-bearing activating CLRs are shown in red; ITIM-bearing inhibitory CLRs are shown in blue; ITAM- or ITIM-independent CLRs are shown in black. Arrow heads indicate the transcripts' orientations

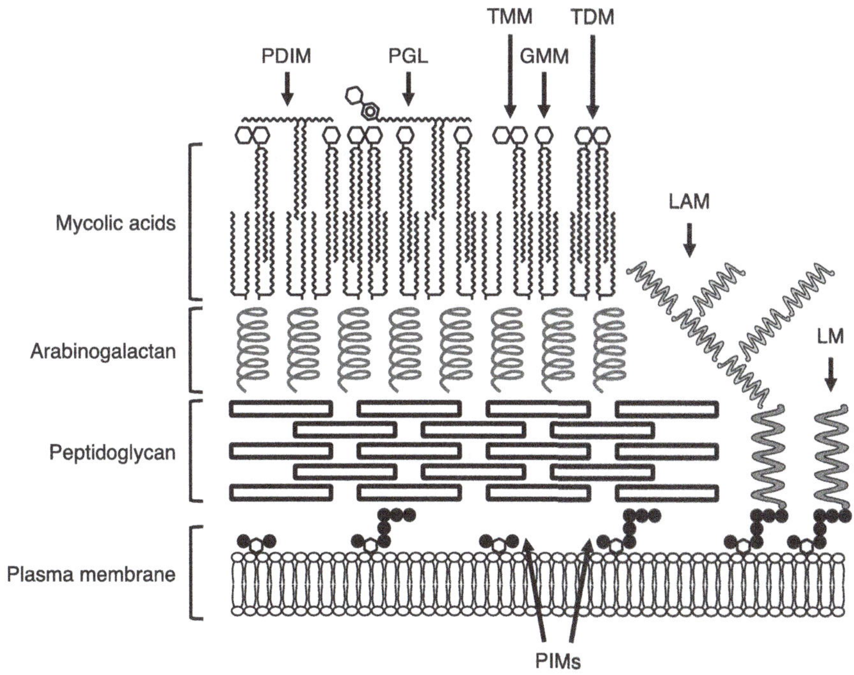

Fig. 2 Cell wall structure of mycobacteria. Phthiocerol dimycocerosate (PDIM), phenolic glycolipid (PGL), trehalose monomycolate (TMM), glucose monomycolate (GMM), trehalose dimycolate (TDM), lipoarabinomannan (LAM), lipomannan (LM), phosphatidy-linositol mannosides (PIMs)

recent years, it has been found that CLR family molecules are the direct receptors for the lipid components (Ishikawa et al. 2017; Marakalala and Ndlovu 2017). In this review, we summarize the recent discoveries made for the dendritic cell immunoactivating receptor (DCAR), which is an FcRγ-coupled activating receptor for mycobacteria.

2 Identification of DCAR

DCAR (gene symbol *Clec4b1*), also known as DCAR2, was initially identified by Kanazawa et al. in 2003 by database searches for cDNA and amino acid sequences homologous to the murine dendritic cell immunoreceptor (DCIR1, gene symbol *Clec4a2*), also known as DCIR (Kanazawa et al. 2002, 2003; Bates et al. 1999). DCAR and DCIR1 are type II transmembrane proteins with a single CRD in the extracellular region. The CRD in DCAR shares high homology with DCIR (91% amino acid identity), but their intracellular regions differ (Kanazawa et al. 2004; Kanazawa 2007). While DCIR contains ITIM in the cytoplasmic region, DCAR lacks a clear intracellular signaling motif and tyrosine residue in its short cytoplasmic region. However, like other CLRs, such as Mincle and Dectin-2, DCAR associates with the ITAM-containing adaptor FcRγ (Kanazawa et al. 2003; Kerscher et al. 2013). This association with FcRγ is required for efficient surface expression of DCAR, because when transfected into HEK293T cells, DCAR by itself is expressed weakly on the cell surface, whereas co-transfection with FcRγ enhances its surface expression. Although DCAR possesses a charged arginine residue in its transmembrane region, mutation analyses suggest that this charged residue only partially contributes to the interaction with FcRγ. In terms of signal transduction, the use of cell lines overexpressing a chimeric receptor containing cytoplasmic DCAR in the presence or absence of FcRγ revealed that Ca^{2+} influx and tyrosine phosphorylation of cellular proteins were dependent on FcRγ (Kanazawa et al. 2003).

3 Cellular Distribution of DCAR

DCAR mRNA has been reported to be highly expressed in several tissues such as lungs and spleen, as well as in the bone marrow-derived DCs induced by granulocyte-macrophage colony-stimulating factor (GM-CSF) (Kanazawa et al. 2003). However, because of the highly similar extracellular regions of DCAR and DCIR1, detailed analysis of DCAR protein expression has not been reported. As currently available monoclonal antibodies (mAbs) cross-react with DCIR1, we established specific mAbs against DCAR by immunizing DCAR-deficient mice with DCAR transfectants. We successfully obtained an anti-DCAR mAb that reacts with DCAR but not with other related CLRs including DCIR1. This mAb was used to analyze DCAR-expressing cells in various mouse tissues by flow cytometry. In peritoneal cavity, two resident macrophage subsets, small peritoneal macrophages (SPMs) and large peritoneal macrophages (LPMs), have been detected (Ghosn et al. 2010; T'Jonck et al. 2018). DCAR expression was only detected in the SPMs (Toyonaga et al. 2016) originating from adult-circulating monocytes (Cain et al. 2013; Kim et al. 2016; Bain et al. 2016). Recent gene expression analyses have also revealed higher DCAR expression in this peritoneal cell population (Gautier et al. 2013; Liao et al. 2016). LPMs derived from yolk-sac progenitors do not express DCAR, and DCAR expression in SPMs was not detectable in DCAR- and FcRγ-deficient mice, providing the first evidence of a role for FcRγ in the surface expression of endogenous DCAR. In addition to SPMs, we also found that DCAR is selectively expressed in the monocyte-derived cells present in spleen and lung tissues. Importantly, no abnormal development of any myeloid population in DCAR-deficient mice was observed, suggesting that loss of DCAR does not affect the generation of these populations in steady state.

Kishimoto et al. also established DCAR-specific mAbs by immunizing rats with DCAR transfectants and reported that DCAR is highly expressed on a subpopulation of conventional DCs ($CD11c^{hi}MHC\ II^{hi}$ cells), in bone marrow and skin-draining lymph nodes (Kishimoto et al. 2015). The authors performed FITC-painting experiments and showed that DCAR was expressed on all of the $FITC^{+}$ conventional DCs that had migrated from the skin after FITC painting, suggesting that DCAR expression can be used as a novel marker for migratory conventional DCs in skin-draining lymph nodes. That our experiments failed to confirm DCAR expression in these cells may be explained by methodological differences in preparing the primary cells. Kishimoto et al. also established a DCIR1-specific mAb using a similar strategy and reported that DCIR1 is widely expressed on myeloid cells including conventional DCs, macrophages, neutrophils, and eosinophils, in various organs (Kishimoto et al. 2015). The observed ubiquitous expression of DCIR in a variety of tissues may possibly contribute to the maintenance of immunological homeostasis.

4 DCAR Recognizes Acylated Phosphatidyl-Inositol Mannosides (PIMs) in Mycobacteria

DCAR is associated with FcRγ and is expressed in myeloid cells, suggesting that DCAR may act as a PRR like other CLRs, but its ligands remain elusive. To identify DCAR ligands, we established DCAR-expressing reporter cells for monitoring ligand recognition using green fluorescent protein (GFP). Because strong GFP expression was observed in the reporter cells upon stimulation with DCAR cross-linking, the reporter cells were considered useful for ligand screening. Consequently, these reporter cells were co-cultured with various pathogenic fungi and bacteria, and strong GFP expression was induced when they were co-cultured

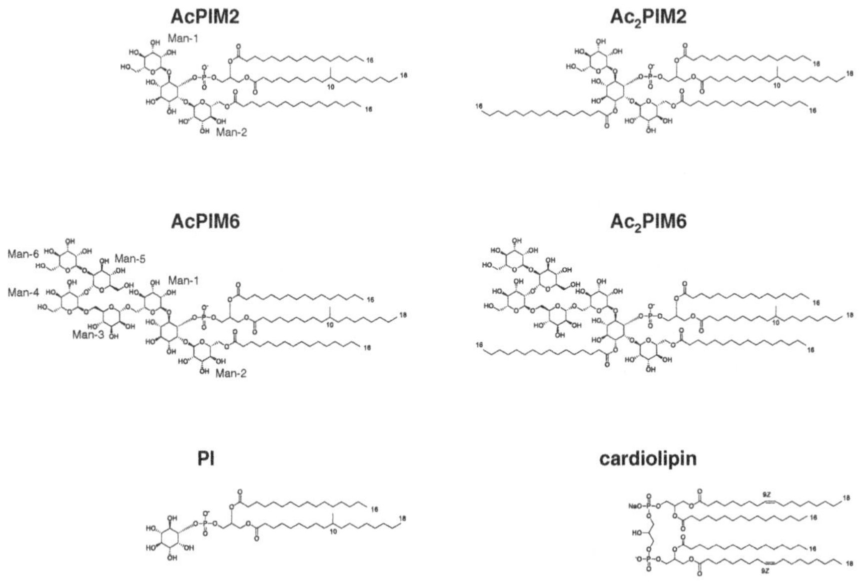

Fig. 3 Structures of DCAR ligands in *M. bovis* BCG. Chemical structures of AcPIM2, Ac_2PIM2, AcPIM6, Ac_2PIM6, PI, and cardiolipin from *M. bovis* BCG

with *Mycobacterium bovis* bacillus Calmette-Guérin (*M. bovis* BCG) (Toyonaga et al. 2016). Strong GFP expression was observed in *M. bovis* BCG and other strains such as *M. tuberculosis* H37Rv, *M. tuberculosis* H37Ra, and *M. smegmatis*, indicating that DCAR is an activating receptor for mycobacteria. DCAR-expressing reporter cells are similarly active against several *Mycobacterium* species, suggesting that a component commonly contained in mycobacteria is a ligand for DCAR. Bioassay-guided fractionation experiments have revealed that acylated Phosphatidyl-inositol mannosides (PIMs) are present in the mycobacterial cell wall as DCAR ligands (Fig. 3) (Toyonaga et al. 2016). PIMs are unique phosphoglycolipids in mycobacteria and other related bacteria, and abundant components of the mycobacterial cell wall (Nigou et al. 2003; Guerin et al. 2010; Crellin et al. 2013). PIMs are known mannosylated forms of phosphatidyl-*myo*-inositol (PI) and carry one to six mannose residues (PIMx; x = number of mannoses) (Fig. 4). Most mycobacteria have a remarkably consistent family of four PIMs; these comprise monoacyl-PIM2 (AcPIM2), diacyl-PIM2 (Ac_2PIM2), monoacyl-PIM6 (AcPIM6), and diacyl-PIM6 (Ac_2PIM6) (Gilleron et al. 2001). By using purified AcPIM2, Ac_2PIM2, AcPIM4, AcPIM6, and Ac_2PIM6 separately, we found that the acylated forms of PIM2 is the most potent DCAR ligand, followed by AcPIM4, whereas the least potent is the acylated forms of PIM6. To verify that the interaction between DCAR and PIMs is a direct one, we prepared a soluble DCAR fusion protein (DCAR-Ig). DCAR-Ig, but not other CLR-Ig, bound to plate-coated AcPIM2. We also performed DCAR-Ig binding assays in the presence of EGTA and mannose.

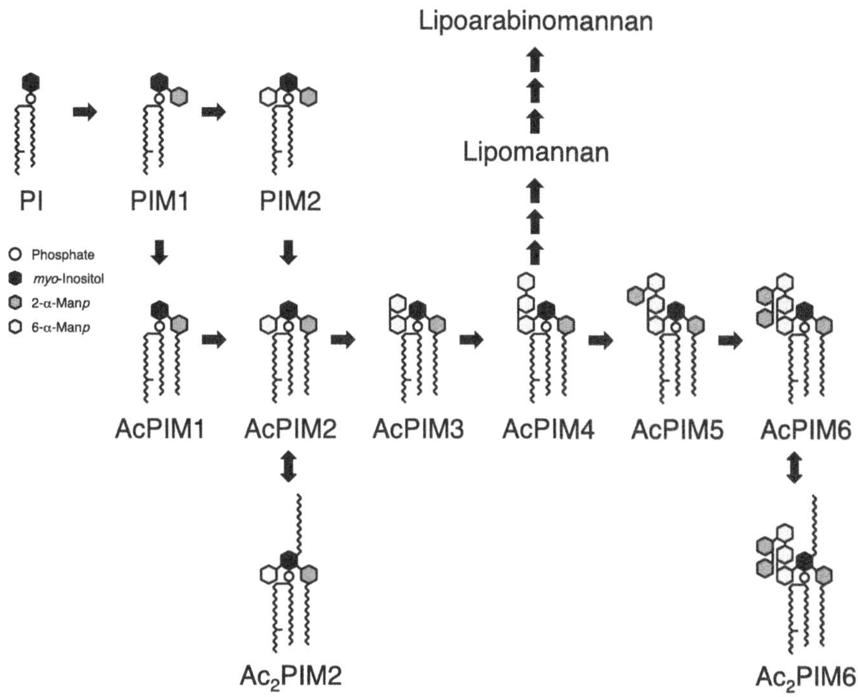

Fig. 4 Proposed pathway for PIM biosynthesis in mycobacteria. PI is converted to PIMs via substitution with one to six mannose units. PIM2 and PIM6 are the two most abundant classes found in mycobacteria, whereas PIM1, PIM3, PIM4, and PIM5 are observed in small amounts. AcPIM4 is proposed to be a branch point of LM/LAM and polar PIM synthesis

EGTA abolished the binding, suggesting that Ca^{2+} is required for the DCAR-PIM interaction. However, high concentration of mannose did not block the binding, suggesting that the PIM mannose moiety alone is not the determinant recognized by DCAR.

In addition to PIMs, we also detected very weak ligand activity in mycobacterial PI and cardiolipin (Fig. 3). As all DCAR ligands identified in mycobacteria contain phosphate moieties, this may be important for ligand recognition by DCAR.

5 PIMs Induce Chemokine Production via DCAR–FcRγ

Acylated PIM2 is further modified by extensive glycosylation (Nigou et al. 2003; Guerin et al. 2010; Crellin et al. 2013). The resultant lipoglycans, termed lipomannan (LM), lipoarabinomannan (LAM), and acylated PIMs are essential structural components of mycobacterial cell walls. PIMs have been suggested to perform additional roles in host immune response modulation such as the production of proinflammatory mediators in myeloid cells (Sprott et al. 2004; Takimoto et al. 2006; Bansal et al. 2009). Several candidates for immune receptors that recognize PIMs have been reported (Jones et al. 2001; Gilleron et al. 2003; Fischer et al. 2004; Torrelles et al. 2006; Driessen et al. 2009). However, the relative contributions of these molecules to PIM-induced immunostimulatory activities remain controversial.

To analyze the innate immune response induced by PIM2, peritoneal exudate cells (PECs) containing SPMs were stimulated with purified acylated PIM2. PECs from wild-type (WT) mice-induced chemokine production (e.g. MCP-1 and MIP-2) upon stimulation with AcPIM2 or Ac_2PIM2, but such production was partially suppressed in PECs from DCAR-deficient mice (Toyonaga et al. 2016). This type of suppression was also observed in FcRγ-deficient PECs, suggesting that acylated PIM2 induces chemokine production via DCAR–FcRγ. Furthermore, intraperitoneal administration of Ac_2PIM2 induced MCP-1 and MIP-2 production in the peritoneal cavities of WT mice. Similar to the in vitro results, chemokine production was partially suppressed in the DCAR-deficient mice. These results suggest that acylated PIM2 may induce innate immune responses such as chemokine production via DCAR–FcRγ.

6 DCAR Contributes to the Th1 Response Against Mycobacterial Infection

We investigated the role of DCAR in mycobacterial infection. Because DCAR is highly expressed on SPMs in the peritoneal cavity, we initially set up intraperitoneal infections with *M. bovis* BCG. When infected with mycobacteria, inflammatory monocytes rapidly accumulate at the infection sites, a known requirement for early immune responses to mycobacteria (Peters et al. 2001; Gonzalez-Juarrero et al. 2003). Two weeks after infection, DCAR-deficient mice had reduced numbers of inflammatory monocytes in their peritoneal cavities, when compared with WT mice. Infiltrating inflammatory monocytes express C-C chemokine receptor type 2 (CCR2), which is an MCP-1 receptor, and are thought to accumulate as a result of local production of MCP-1 upon infection (Monin and Khader 2014; Domingo-Gonzalez et al. 2016). The MCP-1 concentration in the peritoneal cavity was reduced in DCAR-deficient mice. It is also well known that the Th1 response, which mainly consists of IFNγ production from $CD4^+$ T cells, is important for immune protection against mycobacteria (Cooper 2009). IFNγ and IL-12 levels increased in the peritoneal cavities of WT mice after infection, but their production was attenuated in DCAR-deficient mice (Toyonaga et al. 2016). We also examined the effect of DCAR deficiency on the elimination of mycobacteria and found that DCAR-deficient mice had higher bacterial burdens in the peritoneal cavity. We next performed intratracheal infections with *M. bovis* BCG, found that the DCAR-deficient mice showed reduced antigen-specific IFNγ production from mediastinal lymph node cells, and had higher bacterial loads in the lungs two weeks after infection. Mice were also infected with the pathogenic *M. tuberculosis* H37Rv strain. The DCAR-deficient mice infected with this strain produced less MCP-1 and had higher average colony-forming units in their lungs than the WT mice, suggesting that DCAR contributes to host defenses against mycobacteria by promoting the Th1 response.

7 Conclusions

In recent years, the ligand and functions of CLRs have been well studied (Ishikawa et al. 2017; Brown et al. 2018). Among the CLRs present on the mouse chromosome 6F2 locus, Mincle, MCL, and Dectin-2 are known as the receptors for mycobacteria (Ishikawa et al. 2009; Miyake et al. 2013; Yonekawa et al. 2014). DCAR is the fourth CLR within this locus that recognizes mycobacteria. These CLRs recognize the different glycolipid components present in the mycobacterial cell wall. Although it is thought that activation signals are transduced through FcRγ, a common signaling subunit, the response elicited differs for each ligand (Fig. 5). Trehalose dimycolate (TDM), a ligand for Mincle and MCL, strongly induces the production of proinflammatory cytokines such as TNF, whereas the Dectin-2 ligand LAM induces the production of IL-2 and IL-10, as well as proinflammatory cytokines (Ishikawa et al. 2009; Yonekawa et al. 2014). Unlike these glycolipids, DCAR ligand PIMs induces minimal proinflammatory cytokines, but do induce chemokine production, such as MCP-1 (Toyonaga et al. 2016). It is therefore necessary to clarify whether the factors that cause these differences are related to either of all of the following: the expression pattern, cell context, and affinity with the ligand.

DCAR and DCIR1 share 91% amino acid sequence identity in the CRD region, suggesting that they may bind to the same ligand. In contrast to DCAR-deficient mice, it was reported that ITIM-bearing DCIR1-deficient mice can eliminate *M. tuberculosis* H37Rv more efficiently than WT mice (Troegeler et al. 2017). The authors showed that DCIR1 deficiency impairs STAT1-mediated type I IFN signaling in DCs, leading to increased IL-12 production and increased differentiation of T lymphocytes toward a Th1 response during infection. However, it is not yet clear whether these results are related to DCIR1 recognizing an endogenous or mycobacterial ligand. Therefore, further investigation is needed to determine the role of DCIR in infections with mycobacterial pathogens.

Acylated PIM2 is also a known adjuvant that can enhance the acquired immune response (Sprott et al. 2004), and the activities of various analogues of it have been synthesized and studied (Ainge et al. 2006; Parlane et al. 2012; Patil et al. 2015; Zheng et al. 2015). We have found that acylated PIM2 induced the expression of particular inflammatory genes in human DCs derived from peripheral blood monocytes (Toyonaga et al. 2016). So far, the DCAR human homolog has not been clearly identified. In terms of its primary amino acid sequence and chromosomal

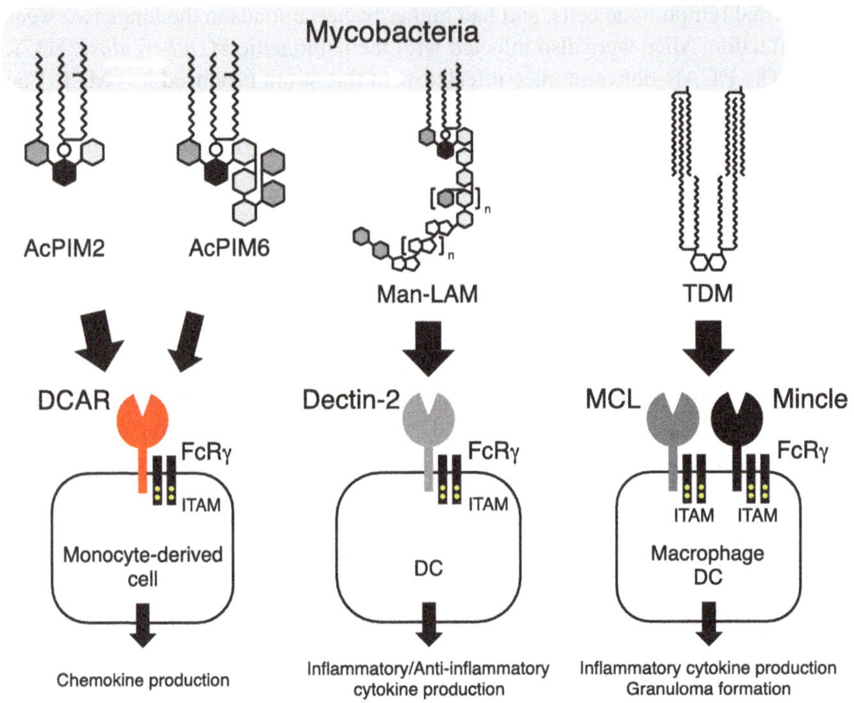

Fig. 5 Recognition of mycobacterial lipid components by FcRγ-coupled CLRs. DCAR recognizes PIMs, Dectin-2 recognizes mannose-capped LAM (Man-LAM), and Mincle and MCL recognize TDM. Interactions between CLRs and mycobacterial ligands trigger cytoplasmic signaling cascades and a number of cellular responses through FcRγ

localization, blood DC antigen 2 (BDCA-2), a human FcRγ-coupled CLR expressed on plasmacytoid DCs, has been thought to be a candidate (Fig. 1) (Kanazawa 2007). However, Zheng et al. reported that no ligands were identified for BDCA-2 when a novel array of synthetic mycobacterial glycans was screened (Zheng et al. 2017). Therefore, further studies are needed to clarify whether there are functional molecules that can substitute for murine DCAR functions in humans. Identification of a human receptor for acylated PIM2 is expected to provide additional insights into the inflammatory pathways in host defenses, with valuable implications for the development of new therapeutic and vaccine strategies targeting immune receptors.

Acknowledgements We thank Sandra Cheesman, Ph.D., from Edanz Group (www.edanzediting.com/ac) for editing a draft of this manuscript.

References

Ainge GD, Parlane NA, Denis M et al (2006) Phosphatidylinositol mannosides: synthesis and adjuvant properties of phosphatidylinositol di- and tetramannosides. Bioorg Med Chem 14 (22):7615–7624. https://doi.org/10.1016/j.bmc.2006.07.003

Bain CC, Hawley CA, Garner H et al (2016) Long-lived self-renewing bone marrow-derived macrophages displace embryo-derived cells to inhabit adult serous cavities. Nat Commun 7: ncomms11852. https://doi.org/10.1038/ncomms11852

Bansal K, Kapoor N, Narayana Y et al (2009) PIM2 Induced COX-2 and MMP-9 expression in macrophages requires PI3K and Notch1 signaling. PLoS ONE 4(3):e4911. https://doi.org/10.1371/journal.pone.0004911

Bates EE, Fournier N, Garcia E et al (1999) APCs express DCIR, a novel C-type lectin surface receptor containing an immunoreceptor tyrosine-based inhibitory motif. J Immunol 163 (4):1973–1983

Brown GD, Willment JA, Whitehead L (2018) C-type lectins in immunity and homeostasis. Nat Rev Immunol 18(6):374–389. https://doi.org/10.1038/s41577-018-0004-8

Cain DW, O'Koren EG, Kan MJ et al (2013) Identification of a tissue-specific, C/EBPbeta-dependent pathway of differentiation for murine peritoneal macrophages. J Immunol 191(9):4665–4675. https://doi.org/10.4049/jimmunol.1300581

Cooper AM (2009) Cell-mediated immune responses in tuberculosis. Annu Rev Immunol 27:393–422. https://doi.org/10.1146/annurev.immunol.021908.132703

Crellin PK, Luo C-Y, Morita YS (2013) Metabolism of plasma membrane lipids in mycobacteria and corynebacteria. Lipid Metabolism 6:119–148

Dambuza IM, Brown GD (2015) C-type lectins in immunity: recent developments. Curr Opin Immunol 32:21–27. https://doi.org/10.1016/j.coi.2014.12.002

Domingo-Gonzalez R, Prince O, Cooper A et al (2016) Cytokines and chemokines in *Mycobacterium tuberculosis* infection. Microbiol Spectr 4(5). https://doi.org/10.1128/microbiolspec.tbtb2-0018-2016

Drickamer K, Taylor ME (2015) Recent insights into structures and functions of C-type lectins in the immune system. Curr Opin Struct Biol 34:26–34. https://doi.org/10.1016/j.sbi.2015.06.003

Driessen NN, Ummels R, Maaskant JJ et al (2009) Role of phosphatidylinositol mannosides in the interaction between mycobacteria and DC-SIGN. Infect Immun 77(10):4538–4547. https://doi.org/10.1128/IAI.01256-08

Fischer K, Scotet E, Niemeyer M et al (2004) Mycobacterial phosphatidylinositol mannoside is a natural antigen for CD1d-restricted T cells. Proc Natl Acad Sci USA 101(29):10685–10690. https://doi.org/10.1073/pnas.0403787101

Gautier EL, Ivanov S, Lesnik P et al (2013) Local apoptosis mediates clearance of macrophages from resolving inflammation in mice. Blood 122(15):2714–2722. https://doi.org/10.1182/blood-2013-01-478206

Ghosn EE, Cassado AA, Govoni GR et al (2010) Two physically, functionally, and developmentally distinct peritoneal macrophage subsets. Proc Natl Acad Sci USA 107 (6):2568–2573. https://doi.org/10.1073/pnas.0915000107

Gilleron M, Ronet C, Mempel M et al (2001) Acylation state of the phosphatidylinositol mannosides from *Mycobacterium bovis* bacillus Calmette Guerin and ability to induce granuloma and recruit natural killer T cells. J Biol Chem 276(37):34896–34904. https://doi.org/10.1074/jbc.M103908200

Gilleron M, Quesniaux VF, Puzo G (2003) Acylation state of the phosphatidylinositol hexamannosides from *Mycobacterium bovis* bacillus Calmette Guerin and mycobacterium tuberculosis H37Rv and its implication in toll-like receptor response. J Biol Chem 278 (32):29880–29889. https://doi.org/10.1074/jbc.M303446200

Gonzalez-Juarrero M, Shim TS, Kipnis A et al (2003) Dynamics of macrophage cell populations during murine pulmonary tuberculosis. J Immunol 171(6):3128–3135

Guerin ME, Kordulakova J, Alzari PM et al (2010) Molecular basis of phosphatidyl-myo-inositol mannoside biosynthesis and regulation in mycobacteria. J Biol Chem 285(44):33577–33583. https://doi.org/10.1074/jbc.R110.168328

Hardison SE, Brown GD (2012) C-type lectin receptors orchestrate antifungal immunity. Nat Immunol 13(9):817–822. https://doi.org/10.1038/ni.2369

Ishikawa E, Ishikawa T, Morita YS et al (2009) Direct recognition of the mycobacterial glycolipid, trehalose dimycolate, by C-type lectin Mincle. J Exp Med 206(13):2879–2888. https://doi.org/10.1084/jem.20091750

Ishikawa E, Mori D, Yamasaki S (2017) Recognition of mycobacterial lipids by immune receptors. Trends Immunol 38(1):66–76. https://doi.org/10.1016/j.it.2016.10.009

Jackson M (2014) The mycobacterial cell envelope-lipids. Cold Spring Harb Perspect Med 4(10). https://doi.org/10.1101/cshperspect.a021105

Jones BW, Heldwein KA, Means TK et al (2001) Differential roles of toll-like receptors in the elicitation of proinflammatory responses by macrophages. Ann Rheum Dis 60 Suppl 3:iii6–12

Kanazawa N, Okazaki T, Nishimura H et al (2002) DCIR acts as an inhibitory receptor depending on its immunoreceptor tyrosine-based inhibitory motif. J Invest Dermatol 118(2):261–266. https://doi.org/10.1046/j.0022-202x.2001.01633.x

Kanazawa N, Tashiro K, Inaba K et al (2003) Dendritic cell immunoactivating receptor, a novel C-type lectin immunoreceptor, acts as an activating receptor through association with Fc receptor gamma chain. J Biol Chem 278(35):32645–32652. https://doi.org/10.1074/jbc.M304226200

Kanazawa N, Tashiro K, Miyachi Y (2004) Signaling and immune regulatory role of the dendritic cell immunoreceptor (DCIR) family lectins: DCIR, DCAR, dectin-2 and BDCA-2. Immunobiology 209(1–2):179–190. https://doi.org/10.1016/j.imbio.2004.03.004

Kanazawa N (2007) Dendritic cell immunoreceptors: C-type lectin receptors for pattern-recognition and signaling on antigen-presenting cells. J Dermatol Sci 45(2):77–86. https://doi.org/10.1016/j.jdermsci.2006.09.001

Kawai T, Akira S (2010) The role of pattern-recognition receptors in innate immunity: update on Toll-like receptors. Nat Immunol 11(5):373–384. https://doi.org/10.1038/ni.1863

Kerscher B, Willment JA, Brown GD (2013) The Dectin-2 family of C-type lectin-like receptors: an update. Int Immunol 25(5):271–277. https://doi.org/10.1093/intimm/dxt006

Kim KW, Williams JW, Wang YT et al (2016) MHC II^+ resident peritoneal and pleural macrophages rely on IRF4 for development from circulating monocytes. J Exp Med 213 (10):1951–1959. https://doi.org/10.1084/jem.20160486

Kishimoto A, Watanabe M, Terauchi K et al (2015) Ubiquitous versus restricted expression of the two mouse dendritic cell C-type lectin receptors, DCIR1 and DCAR2, among myeloid cells. Biochem Biophys Res Commun. https://doi.org/10.1016/j.bbrc.2015.09.146

Liao CT, Rosas M, Davies LC et al (2016) IL-10 differentially controls the infiltration of inflammatory macrophages and antigen-presenting cells during inflammation. Eur J Immunol 46(9):2222–2232. https://doi.org/10.1002/eji.201646528

Marakalala MJ, Ndlovu H (2017) Signaling C-type lectin receptors in antimycobacterial immunity. PLoS Pathog 13(6):e1006333. https://doi.org/10.1371/journal.ppat.1006333

Minnikin DE, Lee OY, Wu HH et al (2015) Pathophysiological implications of cell envelope structure in *Mycobacterium tuberculosis* and related taxa. In: Tuberculosis-expanding knowledge, pp 145–175

Miyake Y, Toyonaga K, Mori D et al (2013) C-type lectin MCL is an FcRgamma-coupled receptor that mediates the adjuvanticity of mycobacterial cord factor. Immunity 38(5):1050–1062. https://doi.org/10.1016/j.immuni.2013.03.010

Monin L, Khader SA (2014) Chemokines in tuberculosis: the good, the bad and the ugly. Semin Immunol 26(6):552–558. https://doi.org/10.1016/j.smim.2014.09.004

Nigou J, Gilleron M, Puzo G (2003) Lipoarabinomannans: from structure to biosynthesis. Biochimie 85(1–2):153–166

Parlane NA, Compton BJ, Hayman CM et al (2012) Phosphatidylinositol di-mannoside and derivates modulate the immune response to and efficacy of a tuberculosis protein vaccine against *Mycobacterium bovis* infection. Vaccine 30(3):580–588. https://doi.org/10.1016/j.vaccine.2011.11.055

Patil PS, Cheng TJ, Zulueta MM et al (2015) Total synthesis of tetraacylated phosphatidylinositol hexamannoside and evaluation of its immunomodulatory activity. Nat Commun 6:7239. https://doi.org/10.1038/ncomms8239

Peters W, Scott HM, Chambers HF et al (2001) Chemokine receptor 2 serves an early and essential role in resistance to *Mycobacterium tuberculosis*. Proc Natl Acad Sci USA 98(14):7958–7963. https://doi.org/10.1073/pnas.131207398

Sancho D, Reis e Sousa C (2012) Signaling by myeloid C-type lectin receptors in immunity and homeostasis. Annu Rev Immunol 30:491–529. https://doi.org/10.1146/annurev-immunol-031210-101352

Shiokawa M, Yamasaki S, Saijo S (2017) C-type lectin receptors in anti-fungal immunity. Curr Opin Microbiol 40:123–130. https://doi.org/10.1016/j.mib.2017.11.004

Sprott GD, Dicaire CJ, Gurnani K et al (2004) Activation of dendritic cells by liposomes prepared from phosphatidylinositol mannosides from *Mycobacterium bovis* bacillus Calmette-Guerin and adjuvant activity in vivo. Infect Immun 72(9):5235–5246. https://doi.org/10.1128/IAI.72.9.5235-5246.2004

T'Jonck W, Guilliams M, Bonnardel J (2018) Niche signals and transcription factors involved in tissue-resident macrophage development. Cell Immunol 330:43–53. https://doi.org/10.1016/j.cellimm.2018.02.005

Takeuchi O, Akira S (2010) Pattern recognition receptors and inflammation. Cell 140(6):805–820. https://doi.org/10.1016/j.cell.2010.01.022

Takimoto H, Maruyama H, Shimada KI et al (2006) Interferon-gamma independent formation of pulmonary granuloma in mice by injections with trehalose dimycolate (cord factor), lipoarabinomannan and phosphatidylinositol mannosides isolated from *Mycobacterium tuberculosis*. Clin Exp Immunol 144(1):134–141. https://doi.org/10.1111/j.1365-2249.2006.03043.x

Torrelles JB, Azad AK, Schlesinger LS (2006) Fine discrimination in the recognition of individual species of phosphatidyl-myo-inositol mannosides from *Mycobacterium tuberculosis* by C-type lectin pattern recognition receptors. J Immunol 177(3):1805–1816

Toyonaga K, Torigoe S, Motomura Y et al (2016) C-Type lectin receptor DCAR recognizes mycobacterial phosphatidyl-inositol Mannosides to promote a Th1 response during infection. Immunity 45(6):1245–1257. https://doi.org/10.1016/j.immuni.2016.10.012

Troegeler A, Mercier I, Cougoule C et al (2017) C-type lectin receptor DCIR modulates immunity to tuberculosis by sustaining type I interferon signaling in dendritic cells. Proc Natl Acad Sci USA 114(4):E540–E549. https://doi.org/10.1073/pnas.1613254114

Yonekawa A, Saijo S, Hoshino Y et al (2014) Dectin-2 is a direct receptor for mannose-capped lipoarabinomannan of mycobacteria. Immunity 41(3):402–413. https://doi.org/10.1016/j.immuni.2014.08.005

Zelensky AN, Gready JE (2005) The C-type lectin-like domain superfamily. FEBS J 272 (24):6179–6217. https://doi.org/10.1111/j.1742-4658.2005.05031.x

Zheng RB, Jegouzo SAF, Joe M et al (2017) Insights into interactions of mycobacteria with the host innate immune system from a novel array of synthetic mycobacterial glycans. ACS Chem Biol 12(12):2990–3002. https://doi.org/10.1021/acschembio.7b00797

Zheng T, Luo D, Compton B et al (2015) Adjuvanticity of a synthetic phosphatidylinositol dimannoside to a subvirion influenza vaccine in an influenza mouse model. J Vaccines Vaccin 6(2):277

Sensing Tissue Damage by Myeloid C-Type Lectin Receptors

Carlos Del Fresno, Francisco J. Cueto and David Sancho

Contents

1 Introduction 118
2 Molecular Signatures Associated to Microorganisms and Self-damage 119
3 Types of Cell Death 120
3.1 Necrosis 120
3.2 Apoptosis 121
3.3 Pyroptosis 122
3.4 Necroptosis 123
3.5 NETosis 123
3.6 Relevance of Oxidative Stress in Cell Death 124
4 C-Type Lectin Receptors in Myeloid Cells 125
4.1 DNGR-1 (CLEC9A) 126
4.2 Mincle (*CLEC4E*) 127
4.3 Cll-1 (*CLEC12A*) 129
4.4 LOX-1 (*OLR1*) 131
4.5 CD301 (*CLEC10A*) 133
4.6 Dec-205 (*LY75*) 134
5 Conclusions 135
References 136

Abstract After both sterile and infectious insults, damage is inflicted on tissues leading to accidental or programmed cell death. In addition, events of programmed cell death also take place under homeostatic conditions, such as in embryo development or in the turnover of hematopoietic cells. Mammalian tissues are seeded

C. Del Fresno (✉) · F. J. Cueto · D. Sancho (✉)
Immunobiology Laboratory, Centro Nacional de Investigaciones Cardiovasculares (CNIC), Madrid 28029, Spain
e-mail: carlos.delfresno@cnic.es

D. Sancho
e-mail: dsancho@cnic.es

Current Topics in Microbiology and Immunology (2020) 429: 117–145
https://doi.org/10.1007/82_2019_194

Published Online: 24 January 2020

with myeloid immune cells, which harbor a plethora of receptors that allow the detection of cell death, modulating immune responses. The myeloid C-type lectin receptors (CLRs) are one of the most prominent families of receptors involved in tailoring immunity after sensing dead cells. In this chapter, we will cover a diversity of signals arising from different forms of cell death and how they are recognized by myeloid CLRs. We will also explore how myeloid cells develop their sentinel function, exploring how some of these CLRs identify cell death and the type of responses triggered thereof. In particular, we will focus on DNGR-1 (*CLEC9A*), Mincle (*CLEC4E*), CLL-1 (*CLEC12A*), LOX-1 (*OLR1*), CD301 (*CLEC10A*) and DEC-205 (*LY75*) as paradigmatic death-sensing CLRs expressed by myeloid cells. The molecular processes triggered after cell death recognition by myeloid CLRs contribute to the regulation of immune responses in pathologies associated with tissue damage, such as infection, autoimmunity and cancer. A better understanding of these processes may help to improve the current approaches for therapeutic intervention.

Abbreviations

ATP	Adenosine Triphosphate
CAR-T	Chimeric Antigen Receptor T Cell
CLEC	C-Type LECtin
CLR	C-Type Lectin Receptor
DC	Dendritic Cell
IFN	Interferon
IRF	Interferon Response Factor
ITAM	Immunoreceptor Tyrosine-based Activating Motif
ITIM	Immunoreceptor Tyrosine-based Inhibitory Motif
LDL	Low-Density Lipoprotein
LPS	Lipopolysaccharide
PRR	Pattern Recognition Receptor
ROS	Reactive Oxygen Species
SHP-1/-2	Src-Homology 2 domain-containing Phosphatase-1/-2
SYK	Spleen Tyrosine Kinase
TLR	Toll-Like Receptor

1 Introduction

Quoting the famous movie "Forrest Gump" on its twenty-fifth anniversary, "*Momma always said dyin' was a part of life.*" Independently of the meaning of this quote in our daily life, this is an important assertion when we take a closer look at our bodies. Based on a cycle of cell death and regeneration, our cells are in a constant turnover under physiological conditions (Pellettieri and Alvarado 2007). This process is even more patent in pathological situations where massive tissue damage can be caused by both sterile and infectious insults. Immune cells populating all tissues constitute a sentinel network that surveys and reacts to cell death (Kroemer 2017).

Pattern recognition receptors (PRRs) allow the sensing of conserved structures released or exposed during cell death (Shekarian et al. 2017). Importantly, cell death can be triggered through diverse programs, which deliver distinct signals. Sensing these signals by myeloid cells originates different immune responses depending on the PRR involved in the recognition. Among the PRRs, C-type lectin receptors (CLRs) exert a prominent function in death recognition (Sancho and Reis e Sousa 2013). Herein, we will review the type of signals generated during cell death and how these signals are produced depending on particular cell death programs. All this information will be integrated from an immune perspective, analyzing how myeloid cells recognize these signals through a panel of CLRs and their subsequent inflammatory outcomes.

2 Molecular Signatures Associated to Microorganisms and Self-damage

Immune cells express different combinations of pattern recognition receptors (PRRs) that allow them to sense pathogen-associated molecular patterns (PAMPs), conserved molecular entities present in microorganisms (Janeway 1989). However, this does not fully explain the initiation of immunity in the apparent absence of infections. The danger theory defends that the immune system acts not only discriminating foreign molecules from the host's own, but also by identifying evolutionarily conserved endogenous molecular patterns associated with threats to homeostasis that are exposed, released or produced in response to stress (Matzinger 2002). These damage-associated molecular patterns (DAMPs) are self-molecules that can be detected by immune cells, inducing either immunity or promoting wound healing. The interplay between DAMPs and PAMPs and their PRRs may instruct immune responses.

Cell death comprises a variety of modalities whereby cells cease to carry out their functions in an irreversible manner. Cell death generates DAMPs, and it is an excellent example of a situation where immune cells have to discriminate homeostatic death from menacing situations. The nature of released DAMPs upon cell death embraces a wide range of molecules that can be preformed or newly synthesized (Yatim et al. 2017). For instance, constitutive intracellular components released upon loss of membrane integrity, such as nucleic acids or cytoskeleton proteins. On the other hand, tissue injury can induce the specific modification of endogenous molecules, such as the glycation of ligands recognized by the receptor for advanced glycation end products (RAGE) or the production of cytokines, such as IL1α or IFN-α, which can act as immunomodulators (Roh and Sohn 2018).

There is evidence that the nature of DAMPs is well conserved along evolution. For instance, the hydrophobic portions (*hyppos*) of different molecules are embedded within biological membranes or hydrophobic cores in macromolecules, and their exposure can be linked to cell stress or death (Seong and Matzinger 2004). Another example of ancient DAMPs may be nucleic acids, which are usually restricted to the nuclear or cytoplasmic space, and their exposure can be indicative of cell death or viral infection. Mammalian cells are armed with an ample repertoire of receptors that detect nucleic acids, both at the intracellular and extracellular levels (Roers et al. 2016). The detection of nucleic acids is also a common feature of invertebrate animals, yeasts and bacteria. Additionally, other properties of the microenvironment can condition the binding of defined DAMPs to those receptors (Rajamäki et al. 2013; Santoni et al. 2015). For instance, bacterial infections and tumors can acidify the cellular microenvironment, impacting sensing of DAMPs and signaling.

3 Types of Cell Death

Cell death occurs both under steady-state conditions (e.g., embryo development) and upon infection or sterile tissue damage. Cell death can be broadly categorized either as an accidental phenomenon (accidental cell death—ACD) or as a regulated cellular process (regulated cell death—RCD). This is actually the current classification system updated by the Nomenclature Committee on Cell Death (NCCD) (Galluzzi et al. 2015), which provides guidelines to define all aspects of cell death. According to the classic paradigm, ACD occurs only under pathological conditions, and alternatively, RCD mostly happens during homeostatic situations (Green et al. 2009). However, this paradigm has been challenged by the discovery of different forms of RCD. We next describe the triggers, morphological features, underlying molecular mechanisms and relationship with immune responses of the main forms of cell death.

3.1 Necrosis

Defined as uncontrolled cell death following severe injury (or ACD), necrosis is accompanied by plasma membrane rupture. This results in the release of the intracellular content out to surrounding tissues, which may generate further damage (D'Arcy 2019). Under this view, the trigger and molecular mechanisms underlying necrosis are unspecific and involve any form of chemical or physical insult that overwhelms control mechanisms and trigger this indiscriminate cell death. However, cells can die displaying necrotic features under a number of RCD processes such as necroptosis, pyroptosis or NETosis among others, which are driven by defined molecular mechanisms (Tang et al. 2019).

Necrosis was classically considered an immunogenic process compared with RCD, due to the random exposition of cellular components/antigens that otherwise should be sheltered under the plasma membrane (Kroemer et al. 2009). However, this notion has been challenged by several observations where certain types of RCD can be immunogenic, while ACD is still considered a pro-inflammatory event associated with uncontrolled tissue damage.

3.2 Apoptosis

Apoptosis represents the first described RCD program (Kerr et al. 1972). It is characterized by cell rounding, nuclear condensation, DNA fragmentation, membrane blebbing and the eventual formation of membrane-surrounded apoptotic bodies (Nagata 2018). Apoptosis can be triggered extrinsically by the engagement of membrane receptors such as death receptors (e.g., FAS, also known as CD95) or intrinsically by mitochondrial outer membrane permeabilization due to exposition to different toxic stimuli (irradiation, toxins and hypoxia) (Reed and Pellecchia 2005). Additionally, granzymes can trigger apoptosis in cells targeted by cytotoxic T and NK cells (Elmore 2007).

Apoptosis is a finely regulated process based on the sequential activation of the caspase family of cysteine proteases (McIlwain et al. 2013). These enzymes exist in steady-state as inactive zymogens. Stimuli that trigger apoptosis activate initiator caspases. Caspase-8 and -9 are prototypical initiator caspases of the extrinsic and intrinsic apoptosis, respectively. The proteolytic action of these caspases activates downstream executor caspases such as Caspase-3. Executor caspases cleave a wide range of proteins, giving rise to destruction of subcellular structures, thus compromising the cellular integrity (Galluzzi et al. 2016).

The immune consequences of apoptosis have been classically considered as silent or anti-inflammatory, such as the induction of T cell tolerance to antigens associated with apoptotic cells (Steinman et al. 2000). In fact, the removal of apoptotic cells by phagocytosis, namely efferocytosis, is considered an anti-inflammatory process (Morioka et al. 2019). However, certain types of caspase-mediated cell death induced upon anthracycline-based antitumor chemotherapies are immunogenic (Casares et al. 2005; Obeid et al. 2007). In addition, apoptotic cells that are not efficiently cleared undergo a process termed secondary necrosis defined as dissolution of the cell following apoptosis (Green et al. 2009). It is difficult to distinguish between secondary and primary necrosis, as both imply loss of plasma membrane integrity and are typically immunogenic due to the release of intracellular pro-inflammatory cell components (Bell et al. 2006). However, secondary necrotic cells have gone through various modifications during the process of apoptosis, and consequently, molecules released during secondary necrosis may drive distinct immunologic responses (Sachet et al. 2017).

Therefore, assigning a silent or tolerogenic role to apoptosis may not be accurate. A systematic characterization of the molecular mechanisms, the clearance of apoptotic bodies and their location may help to characterize the immunogenicity of specific apoptotic processes. Importantly, apoptosis and its dysregulation underlie both physiological and pathological processes, including cell homeostasis, tissue remodeling, organ transplantation and cancer (Linkermann et al. 2014; Singh et al. 2019). In addition, a number of pathologies have been linked to defective clearance of apoptotic cells (Morioka et al. 2019).

3.3 *Pyroptosis*

Pyroptosis represents an alternative caspase-dependent RCD driven by the activation of the inflammasome, a cytosolic multiprotein complex responsible for the release of IL-1α/β and IL-18 (Fink et al. 2008; Tang et al. 2019). Pyroptosis is characterized by the formation of cell membrane pores, leading to membrane rupture and release of cytosolic contents to the extracellular environment (Wang et al. 2019). Pyroptosis can be triggered by many different PAMPs (e.g., bacterial peptidoglycans, viral dsRNA) and DAMPs (ATP, elevated intracellular ROS) through the canonical Caspase-1 pathway (Liu and Lieberman 2017) or after direct recognition of LPS by Caspase-4/5 in humans or Caspase-11 in mice, in a non-canonical pathway (Shi et al. 2014). In addition, pyroptosis can be activated by alternative caspases such as Caspase-8 during *Yersinia* infection (Orning et al. 2018) or Caspase-3 in response to chemotherapy in certain cancer cells (Wang et al. 2017).

The final step for the development of pyroptosis involves the cleavage of gasdermins, a family of proteins that comprises functionally diverse enzymes with pore-forming potential. In particular, Gasdermin-D (GSDMD) is cleaved in most models of pyroptosis (Shi et al. 2015). GSDMD is a proteolytic substrate of Caspases-1/4/5/8/11 and its cleavage releases the N-terminal effector domain from the inhibitory domain. The N-terminal domain of GSDMD oligomerizes in the cell membrane and forms cytotoxic pores, through which caspase substrates of small sizes, such as IL-1β/-18 and the alarmin IL-1α are spread to the extracellular milieu. The accumulation of membrane pores eventually leads to membrane rupture and releasing the entire cellular content (Feng et al. 2018). Additionally, chemotherapy drugs induce pyroptosis by a similar mechanism through Caspase-3 cleavage of Gasdermin-E (Wang et al. 2017).

Due to the involvement of pro-inflammatory cytokines IL-1α/-1β/-18 and the eventual release of the intracellular content, pyroptosis is a highly inflammatory form of cell death (Tang et al. 2019). The main function of pyroptosis is the defense and rapid clearance of bacterial and viral infections (Doitsh et al. 2014; Robinson et al. 2019). However, persistent inflammasome activation can produce excessive and chronic inflammation, eventually contributing to metabolic disorders, autoinflammatory diseases and cancer (Davis et al. 2011).

3.4 Necroptosis

Necroptosis is a form of necrotic RCD sharing morphological features to necrosis, with the rupture of the plasma membrane as the main characteristic (Pasparakis and Vandenabeele 2015). It can be triggered by multiple inflammatory stimuli, both sterile and infectious (Tang et al. 2019). Of note, molecules that induce apoptosis, such as TRAIL or Fas ligand (FasL), can also ignite necroptosis when apoptosis is chemically inhibited (Vercammen et al. 1998).

Mechanistically, necroptosis relies on the phosphorylation of mixed lineage kinase domain-like pseudokinase (MLKL) by the receptor-interacting serine/threonine kinase 3 (RIPK3) conforming the critical RIPK3-MLKL axis for necroptosis induction (Sun et al. 2012). Phosphorylated MLKL can bind phosphatidylinositol phosphates present in the inner face of the plasma membrane (Dondelinger et al. 2014). Once translocated to the membrane, MLKL inserts into it and multimerizes with the final consequence of membrane permeabilization (Galluzzi et al. 2017).

Current findings support that necroptosis is an immunogenic cell death important for the activation of innate and adaptive immunity. For instance, RIPK3-deficient mice are highly sensitive to viral infections (Cho et al. 2009; Huang et al. 2015) and the presence of RIPK3 in tumor cells is required for eliciting cytotoxic antitumor responses (Yatim et al. 2015). In fact, necroptosis in tumor cells promotes antitumor immunogenicity. Thus, *Ripk3* expression in colorectal carcinoma patients is reduced compared with adjacent healthy tissue and high *Ripk3* expression constitutes an independent good prognostic factor (Feng et al. 2015). In line with these findings, the oncogenes *Braf* and *Axl* drive the loss of *Ripk3* expression (Najafov et al. 2018). Consequently, tumor cells dampen necroptosis to become poorly immunogenic and hence escape immunosurveillance; therefore, promoting necroptosis in cancer cells has a great potential as an adjuvant for antitumor therapy (Cho 2018). Necroptosis has not only been involved in infections and malignancies, but it is also implicated in a variety of pathologies such as myocardial infarction and stroke, atherosclerosis or inflammatory bowel disease (Linkermann and Green 2014).

3.5 NETosis

NETosis is an RCD process driven by neutrophil extracellular traps (NETs) extrusion. Despite its name, coined after its first characterization (Brinkmann et al. 2004), NETs are not only produced by neutrophils but by other granulocytes as well (Araźna et al. 2015). NETs are web-like DNA structures formed after chromatin decondensation and release to the extracellular environment following plasma membrane rupture; in addition, several proteins adhere to these DNA-based nets, including histones and components of cytoplasmic granules (Delgado-Rizo et al.

2017). Interestingly, NETosis has also been described without involving cell death, a process termed vital-NETosis (Timp et al. 2015).

The molecular mechanisms implicated in the induction of NETosis are quite diverse (Tang et al. 2019), with notorious differences between cell death-based NETosis, vital-NETosis or even NET formation through release of mitochondrial DNA (Yousefi et al. 2019). In fact, the pyroptosis driver GSDMD (Sollberger et al. 2018) and pyroptosis-related IL-1β (Mitroulis et al. 2011) have been involved in the triggering of NETosis, indicating the complexity of this process. However, production of reactive oxygen species (ROS) is of capital relevance for triggering most forms of the NETotic processes (Araźna et al. 2015).

NETosis is triggered in response to infection but also in sterile conditions, such as autoimmune disorders, ischemia-reperfusion injury, and cancer (Branzk and Papayannopoulos 2013). In fact, the diversity of the molecular mechanisms involved in NETosis initiation is associated with the plethora of stimuli that ignite this process.

A main function of NETosis is to content the spread of infections by trapping pathogenic microorganisms (Mesa and Vasquez 2013), thus contributing to control the infection-mediated inflammation. However, in the development of this process, DAMPs are also released, which can be sensed by surrounding innate immune cells, triggering further inflammation. Due to this feedback loop, NETosis becomes a highly inflammatory process (Tang et al. 2019), which has clinical implications in autoimmune diseases (Lee et al. 2017b).

3.6 Relevance of Oxidative Stress in Cell Death

In previous sections, we have explored in detail forms of PCD with well-described pathophysiological roles such as apoptosis, pyroptosis, necroptosis and NETosis (Jorgensen et al. 2017). Nevertheless, there are some other forms of cell death whose importance needs to be studied in further detail (Tang et al. 2019). However, a common feature of all the cell death programs is that, to some extent, the oxidative stress contributes to their triggering. This oxidative stress results from the imbalance between ROS production and the antioxidant capacity of the cell (Ghosh et al. 2018). One of the main consequences of an excessive ROS presence in a cell is the peroxidation of lipids, the main components of membranes which are indispensable for maintaining the structure and the functionality of the cell (Gaschler and Stockwell 2017). Lipid peroxidation drives different forms of cell death, such as ferroptosis (Dixon et al. 2012), pyroptosis (Kang et al. 2018), necroptosis (Canli et al. 2016) or NETosis (Palladino et al. 2018), among others.

As a consequence of all these death processes, cell rupture occurs, which would drive the activation of the immune system. However, ROS activity can regulate immune activation by altering the structure of otherwise pro-inflammatory components. This is the case for the high mobility group B1 (HMGB1) protein. HMGB1 is a nuclear component that acts as a prototypic DAMP that can be recognized by TLR4, triggering inflammation (Andersson and Tracey 2011). However, once oxidized by mitochondrial ROS during apoptosis, HMGB1 induces immunological tolerance (Kazama et al. 2008). Therefore, each specific cell death-associated situation needs to be deeply characterized in order to know its immunological impact.

4 C-Type Lectin Receptors in Myeloid Cells

The myeloid lineage comprises a group of cells that derive from common myeloid progenitors in the bone marrow. Common myeloid progenitors give rise to diverse cell types, including mast cells, basophils, neutrophils, eosinophils, monocytes, macrophages and dendritic cells (DCs), which belong to the innate immune system, and erythrocytes and megakaryocytes, which can contribute to the immune response, but have a different primary function (Weiskopf et al. 2016). Immune myeloid cells comprise the main core (or classical core) of the innate immune system, characterized by their readiness to respond to insults and for not being antigen-specific. Among myeloid cells, DCs and macrophages continuously surveil tissues having a dual role in the modulation of both innate and adaptive immunity. They are equipped with a diverse arsenal of PRRs composed by toll-like receptors (TLRs), nucleotide-binding domain, leucine-rich repeat-containing (NOD)-like receptors (NLRs), C-type lectin receptors (CLRs), AIM2-like receptors (ALRs) and RIG-I-like receptors (RLRs) (Brubaker et al. 2015). These receptors allow to specifically detect signals from their microenvironment, acting as central hubs that integrate information about environmental cues and possible threats. Thus, myeloid cells orchestrate immune function, collaborating in tissue homeostasis or healing, or instructing both innate and adaptive immunity.

The family of CLRs is of prominent relevance in sensing cell death by myeloid cells (Sancho and Reis e Sousa 2013). CLRs are characterized for bearing C-type lectin-like domains (CTLD), which in many cases include a carbohydrate recognition domain (CRD). Their domain architecture is the base for grouping CLRs in up to 17 groups (Zelensky and Gready 2005). The CRD domain allows CLRs to recognize a wide range of glycans exposed to self- and non-self-ligands (Iborra and Sancho 2014). In addition, we proposed a classification of myeloid CLRs based on their intracellular signaling motifs, in an attempt to classify this complex receptor family based on functional criteria (Sancho and Reis e Sousa 2012). Thus, myeloid CLRs can be broadly classified as immunoreceptor tyrosine-based activating motif (ITAM)-coupled CLRs, hemi-ITAM-(hemITAM)-bearing CLRs, immunoreceptor tyrosine-based inhibitory motif (ITIM)-containing CLRs and a group of CLRs lacking typical signaling motifs (del Fresno et al. 2018a, b). Here, we will explore some selected myeloid CLRs that sense cell death, modulating inflammatory responses. The selection of these CLRs has not been done based on structural criteria, but on their paradigmatic roles in the detection of tissue damage by myeloid cells.

4.1 DNGR-1 (CLEC9A)

Dendritic cell natural killer lectin group receptor-1 (DNGR-1, CLEC9A) is a CLR encoded in chromosomes 12 in humans and 6 in mice (Sancho et al. 2008; Huysamen et al. 2008). The expression pattern of DNGR-1 is very restricted to the dendritic cell lineage and more specifically to type I conventional DCs (cDC1s) and common DC progenitors (Sancho et al. 2008; Schraml et al. 2013). Additionally, DNGR-1 is also expressed by mouse but not human, plasmacytoid dendritic cells (http://www.immgen.org; Sancho et al. 2008; Zilionis et al. 2019).

Structurally, DNGR-1 is a transmembrane protein that belongs to the group V of CLRs. These CLRs show an NK cell receptor-like architecture, having no CRD motifs and whose binding capacity to their ligands is calcium-independent (Sancho and Reis e Sousa 2012). In its intracellular region, DNGR-1 features a hemITAM signaling module that can signal through the spleen tyrosine kinase (SYK) (Sancho et al. 2009; Zelenay et al. 2012). DNGR-1 acts as a homodimer through a cysteine in its neck region. Under conditions of low pH or ionic strength, the neck region of DNGR-1 undergoes changes in its tertiary structure, driving resistance to disassembly of the homodimer (Hanč et al. 2016).

The only known ligand for DNGR-1 is filamentous actin (F-actin), a major component of the cellular cytoskeleton that can be accessible to immune cells upon loss of plasma membrane integrity, as occurs in different forms of cell death (Ahrens et al. 2012; Zhang et al. 2012). The docking site for DNGR-1 homodimers in F-actin involves two contiguous monomers from one protofilament and the adjoining one from the other (Hanč et al. 2015). Moreover, myosin II helps to stabilize the interaction between F-actin and DNGR-1 (Schulz et al. 2018).

Upon recognition of F-actin, SYK is recruited to the hemITAM of DNGR-1. DNGR-1/SYK signaling promotes cross-presentation of dead cell-associated antigens in MHC-I, contributing to priming of $CD8^+$ T cells. However, contrary to Dectin-1, the recruitment of SYK to DNGR-1 hemITAM does not result in NF-κB activation or increased production of inflammatory cytokines (Zelenay et al. 2012). Also, DNGR-1 does not act as a scavenger receptor promoting the uptake of dead cells (Sancho et al. 2009). Instead, DNGR-1 signaling promotes the diversion of phagocytic cargo into non-acidic, non-lysosomal compartments, which prevents antigen degradation and promotes cross-presentation (Zelenay et al. 2012; Iborra et al. 2012). This phenomenon participates in shaping the repertoire of primary antiviral immune responses, without affecting the global effector response (Zelenay et al. 2012; Iborra et al. 2012). However, cDC1s and cross-priming of antiviral $CD8^+$ T cells through DNGR-1 contribute to the generation of tissue-resident memory $CD8^+$ T cells specific for the viral threat (Iborra et al. 2016b).

Moreover, DNGR-1 can additionally limit inflammation in situations of tissue damage, a phenomenon termed disease tolerance (Soares et al. 2017). Recently, cDC1s have been involved in the recruitment of neutrophils to inflammation foci (Janela et al. 2019). In different models of tissue-specific injuries, such as acute pancreatitis induced by caerulein and kidney aggression during systemic candidiasis, sensing of tissue damage through DNGR-1 in cDC1s restricts immunopathology (del Fresno et al. 2018b). Mechanistically, engagement of F-actin by DNGR-1 activates the phosphatase SHP-1, which restrains NFκB activation by heterologous receptors and subsequent expression of pro-inflammatory mediators, such as TNF and CXCL2/MIP-2α that contribute to immunopathology (del Fresno et al. 2018b).

To date, whether SYK and SHP-1 signaling occur simultaneously or alternatively remain obscure. Intriguingly, the inhibitory ITAM configuration by which SHP-1 binds to the hypophosphorylated ITAM requires the transient binding and activation of Syk to the ITAM (Mkaddem et al. 2014; Iborra et al. 2016a). The conditions that determine the overall outcome of F-actin recognition in DNGR-1 signaling remain unknown. Here, the ionic composition of the microenvironment, which may fine-tune the conformation of DNGR-1 homodimers (Hanč et al. 2016), may be an important cofactor, providing versatility to DNGR-1 signaling. In this context, F-actin constitutes the DAMP recognized by DNGR-1, while the environmental ionic strength would represent a modulating factor.

Considering the lack of activating signaling after DNGR-1 engagement and its regulatory role in heterologous pathways, the net inflammatory outcome after death recognition by DNGR-1 may be regulatory. However, surrounding signals to the DNGR-1-mediated dead cell sensing, such as adjuvants in the context of infection or vaccination, may switch these responses toward a pro-inflammatory result, considering the role of this receptor in antigen cross-priming.

4.2 *Mincle* (CLEC4E)

Macrophage-inducible-C-type lectin (Mincle, CLEC4E) is a CLR encoded in chromosome 12 in humans and in chromosome 6 in mice. Mincle is expressed by neutrophils (Lee et al. 2012), macrophages (Yamasaki et al. 2008), DCs (Martínez-López et al. 2019) and monocytes (http://www.immgen.org; Heng et al. 2008).

From the structural point of view, Mincle is a transmembrane protein that belongs to the group II of CLRs, which comprises CLRs with a single CRD relying on calcium for their ligand binding (Sancho and Reis e Sousa 2012). Mincle forms heterodimers with macrophage C-type lectin (MCL, CLEC4D), with which it shares a very similar structure (Furukawa et al. 2013; Lobato-Pascual et al. 2013). Mincle coordinates two Ca^{2+} ions that are responsible for the recognition of sugar moieties in its ligands (Furukawa et al. 2013). Neighboring its sugar-binding sites, Mincle features a surface hydrophobic groove that binds long fatty acids (Feinberg et al. 2013; Decout et al. 2017). In the intracellular region, Mincle presents a short domain that does not contain any tyrosine-based regulatory motif. Instead, Mincle bears an arginine residue in its transmembrane region that mediates its association with immunoreceptor tyrosine-based activation motif (ITAM)-containing FcRγ chain (Yamasaki et al. 2008; Ishikawa et al. 2009, 2013).

Mincle was identified as the long-sought receptor for chord factor, a pro-inflammatory molecule from *Mycobacterium* spp. (Ishikawa et al. 2009; Schoenen et al. 2010). Since then, multiple microbial components have been shown to engage and activate Mincle, driving activation of myeloid cells (Lu et al. 2018). For instance, it promotes protective immune responses against bacterial sepsis after cecal ligation (Lee et al. 2017a) and also drives the activation of macrophages upon sensing *Malassezia* spp. (Yamasaki et al. 2009), an infectious skin fungus whose surface glycolipids were defined as the specific Mincle ligand (Ishikawa et al. 2013). Furthermore, recognition of gut microbiota through Mincle contributes to reinforce the immune response in the gut barrier (Martínez-López et al. 2019). In all these cases, engagement of Mincle induces the phosphorylation of SYK and downstream signaling through CARD9 and NFκB, leading to increased production of pro-inflammatory mediators, such as CXCL2/MIP-2α and TNF, promoting the recruitment of neutrophils to inflammation foci (Werninghaus et al. 2009).

However, Mincle also binds the endogenous spliceosome-associated protein 130 (SAP130) (Yamasaki et al. 2008). This protein is part of the U2 small nuclear ribonucleoprotein (snRNP) complex, implicated in the removal of introns, and with a nuclear localization under steady-state conditions (Kiss 2004). The recognition of SAP130 by Mincle has been characterized in different settings. Upon SAP130 ligation, Mincle induces SYK-dependent production of the pro-inflammatory mediators TNF and CXCL2/MIP-2α. In irradiated thymi, Mincle promotes the production of CXCL2/MIP-2α in macrophages, driving inflammation (Yamasaki et al. 2008). In more physiologically relevant scenarios, the pro-inflammatory signals triggered by Mincle can be observed in mouse models of alcoholic liver injury. During alcoholic liver disease, microbiota-derived LPS can be detected by liver macrophages (Parlesak et al. 2000). Sensing of LPS induces the expression of Mincle in macrophages in an Interleukin-1 receptor-associated kinase M (IRAK-M)-dependent manner (Zhou et al. 2016). Mincle can sense endogenous SAP130 in alcoholic livers, amplifying SYK- and inflammasome-mediated IL1β production (Zhou et al. 2016; Kim et al. 2018). Thus, Mincle aggravates alcoholic liver disease, whereas Mincle-deficient mice display reduced neutrophilia, steatosis and fibrosis in their livers (Zhou et al. 2016; Kim et al. 2018).

In a tumor context, pancreatic ductal adenocarcinomas display stable expression of the necroptotic components RIP1 and RIP3. The genetic deletion of RIP3 or the chemical inhibition of RIP1 protected against tumor development, which was associated with an immunogenic infiltrate. Mechanistically, pancreatic adenocarcinoma cells exploited necroptosis induction to promote their growth by promoting the recruitment of tumor-associated macrophages and myeloid-derived suppressor cells (MDSCs) in a CXCL1/KC-dependent manner, limiting the efficacy of chemotherapeutic agents. But additionally, necroptosis of pancreatic ductal adenocarcinoma cells also promotes the release of SAP130, with a parallel Mincle upregulation in tumor-infiltrating myeloid cells. Interestingly, Mincle-deficient mice phenocopied RIP3-deficient mice and were also protected against oncogenesis (Seifert et al. 2016). Furthermore, administration of trehalose dimycolate, a synthetic analog of the *Mycobacterium tuberculosis* ligand for Mincle, accelerates the growth of both RIP3-competent and -deficient tumors, suggesting that both SAP130 and the bacterial ligands for Mincle induce similar signaling and immune responses (Seifert et al. 2016).

Recently, two new endogenous ligands have been identified for Mincle, β-glucosylceramide (Nagata et al. 2017) and cholesterol sulfate (Kostarnoy et al. 2017). β-glucosylceramide is a ubiquitous metabolite released upon cell death. β-glucosylceramide induces the activation of DCs, promoting the production of pro-inflammatory mediators, such as TNF and CXCL2/MIP-2α. Thus, in a mouse model, thymic irradiation or mice deficient for the enzyme in charge of β-glucosylceramide degradation in the hematopoietic compartment show an increased recruitment of neutrophils to irradiated thymi, resulting in increased destruction of the organ. Of note, β-glucosylceramide can also induce the expression of costimulatory molecules in DCs in vitro and boost priming of adaptive immune responses (Nagata et al. 2017). Similarly, Mincle can sense cholesterol sulfate, an abundant molecule in barrier epithelia. Sensing of cholesterol sulfate by bone marrow DCs promotes their maturation and the production of pro-inflammatory mediators such as IL1α, IL1β, KC, CCL3/MIP1α and CCL4/MIP1β. This is important in models of contact dermatitis, where Mincle plays a detrimental role, aggravating inflammation and increasing tissue abnormalities (Kostarnoy et al. 2017).

The discovery of β-glucosylceramide and cholesterol sulfate as ligands for Mincle with pro-inflammatory properties similar to SAP130 makes it difficult to discriminate their individual contribution to the inflammatory response induced by Mincle during tissue injury, as SAP130 is fundamental for the cell machinery and the others are widely produced metabolites. Overall, it could be concluded that cell death recognition by Mincle generates pro-inflammatory responses. Whether these responses eventually regulate inflammation by a third-party component, as described in the pancreatic adenocarcinoma context, illustrate the complexity of the outcome after cell death recognition. In this regard, sensing of *Leishmania* through Mincle can induce SHP-1 activation through an inhibitory ITAM configuration, restricting DC activation (Iborra et al. 2016a). Whether any endogenous ligand is capable of inducing Mincle signaling through an inhibitory ITAM configuration remains unknown. In any case, Mincle is a great example of a receptor that identifies *hyppos* as danger cues and modulates immunity accordingly.

*4.3 Cll-1 (*CLEC12A*)*

CLEC12A is a receptor also known as MICL, CD371, KLRL1 or DCAL-2, but based on the usage frequency of these names, we proposed the consensus alias CLL-1 together with its gene name, *CLEC12A* (del Fresno et al. 2019). It is codified in chromosome 12 in humans and chromosome 6 in mice. CLEC12A is expressed by human monocytes, macrophages, polymorphonuclear cells (Marshall et al. 2004) and DCs (Han et al. 2004; Gurka et al. 2015), while in the mouse, it is expressed by both $CD8^+$ and $CD8^-$ spleen DCs, pDCs, B cells, thymic DN, DP and $CD8^+$ T cells, most of the granulocytes and myeloid cells, along with bone marrow NK cells (Kasahara and Clark 2012) (http://www.immgen.org; Heng et al. 2008). CLEC12A surface expression is downregulated in myeloid cells upon activation, both in humans (Marshall et al. 2006) and mice (Pyz et al. 2008), although receptor levels were increased upon experimental autoimmune encephalomyelitis (EAE), a mouse model for multiple sclerosis (MS) (Sagar et al. 2017), suggesting a complex implication of CLEC12A in the control of myeloid activation during inflammation.

Structurally, CLEC12A is a transmembrane protein that belongs to the group V of CLRs (Sancho and Reis e Sousa 2012). It bears an ITIM motif in its intracellular domain that couples to the phosphatases SHP-1 and SHP-2, responsible for the regulatory role of CLEC12A upon inflammation (Marshall et al. 2004).

Early studies indicated that CLEC12A recognizes a non-microbial endogenous ligand (Pyz et al. 2008). Although it is now known that the receptor also senses *Plasmodium*-derived hemozoin (Raulf et al. 2019), CLEC12A is a well-established receptor for dead cells that recognizes monosodium urate (MSU) crystals and delivers regulatory signals. Consequently, *Clec12a*-deficient mice show exacerbated inflammatory responses when challenged with MSU, necrotic cells or sub-lethal thymic irradiation (Neumann et al. 2014). In this line, CLEC12A-lacking mice are more sensitive to collagen antibody-induced arthritis (CAIA), which was reproduced by administering CLEC12A-blocking antibodies (Redelinghuys et al. 2016). Indeed, genetic variants of CLEC12A associate with rheumatoid arthritis in humans (Michou et al. 2012). These data support the regulatory role of CLEC12A in inflammation.

However, there are a number of reports showing a pro-inflammatory role for CLEC12A. Indeed, both antibody-based targeting and genetic deletion of CLEC12A protect mice from EAE development, where CLEC12A facilitates the binding and transmigration of DCs across the blood–brain barrier (Sagar et al. 2017). Considering that CLEC12A recognizes MSU from dying cells (Neumann et al. 2014) and that serum uric acid reverses EAE progression (Scott et al. 2002), the maintenance of high uric acid levels upon CLEC12A-deficient conditions was argued as an explanation for the beneficial effect of its loss (Sagar et al. 2017). In any case, the distinct underlying mechanisms triggering different autoimmunity models (Lyons et al. 1999) may explain these apparently controversial results. Thus, during EAE, CLEC12A could show a “hijacking” effect over uric acid plus its contribution to DC infiltration, while for CAIA, CLEC12A may act in myeloid cells as an intrinsic regulator of the inflammatory response after sensing tissue damage. In addition, MSU recognition by CLEC12A amplifies cytosolic RNA-mediated IFN-I production, and therefore, *Clec12a*-deficient mice are sensitive to lymphocytic choriomeningitis virus (LCMV) infection. In this case, the molecular mechanism relies on Src kinase activation downstream CLEC12A, which promotes TBK1-IRF3 signaling and IFN-I production (Li et al. 2019).

Interestingly, CLEC12A expression is one of the most differentially upregulated surface markers in leukemic stem cells (Daga et al. 2019). Its combined expression with some other markers has been proposed as an efficient tool for detecting minimal residual disease in acute myeloid leukemia (AML), and high CLEC12A levels along with CD123 are a strong prognostic marker for leukemia relapse (Roug et al. 2014). In fact, CLEC12A has been proposed as a therapeutic target in human AML (Williams et al. 2019), with successful antibody-drug conjugate (Jiang et al. 2018), bispecific antibody (van Loo et al. 2019) or CAR-T cell (Laborda et al. 2017) strategies. The relevance of cell death recognition in this context is a matter of debate. However, it is tempting to speculate that CLEC12A expression by tumor cells may help them to boost an anti-inflammatory environment after the recognition of surrounding dead cells, thus facilitating the scape from immunosurveillance. Overall, these data indicate that the sensing of tissue damage by CLEC12A can give rise to both pro-inflammatory and regulatory responses.

4.4 *LOX-1 (*OLR1*)*

The CLR LOX-1 is encoded by the *OLR1* gene and it is also known as LOXIN (del Fresno et al. 2019). It is codified in chromosome 12 in humans and chromosome 6 in mice. LOX-1 was first described as an endothelial receptor (Sawamura et al. 1997), but it is also expressed by human macrophages (Yoshida et al. 1998), monocytes (Draude et al. 1999) and DCs (Nickel et al. 2009), with a similar expression pattern in the mouse, where it is also expressed by bone marrow neutrophils (http://www.immgen.org; Heng et al. 2008). LOX-1 expression is low under steady-state conditions, but diverse inflammatory stimuli induce the surface expression of this CLR (Kattoor et al. 2019).

From the structural point of view, similarly to CLEC12A, LOX-1 is a transmembrane protein that belongs to the group V of CLRs (Sancho and Reis e Sousa 2012). LOX-1 does not bear any distinguishable ITAM or ITIM motif in its intracellular domain, what it is usually associated with CLRs with endocytic capacity that mediate antigen capture for further processing and presentation to T cells (Geijtenbeek and Gringhuis 2009). Notably, the lack of a defined intracellular module facilitates the involvement of this type of CLRs in flexible signaling pathways triggered by themselves or heterologous receptors (del Fresno et al. 2018a).

The first described ligand of LOX-1 is oxidized-low-density lipoproteins (LDLs) (oxLDL) (Sawamura et al. 1997). The involvement of this receptor in atherosclerosis was uncovered in atherosclerosis-prone mice (*LDLR*$^{-/-}$), which showed reduced pathology when crossed with *Olr1*-deficient mice, indicating that LOX-1 is pro-atherogenic (Mehta et al. 2007). OxLDL internalization after recognition by LOX-1 is a key event in the development of coronary artery disease, with LOX-1 genetic variants associated with increased risk (Tian et al. 2019). In fact, a serum-soluble cleaved form of LOX-1 is used as a diagnostic and prognostic tool in this pathology (Tian et al. 2019).

In addition to oxLDL, LOX-1 recognizes other endogenous ligands such as modified lipoproteins, activated platelets, heat shock proteins (HSP) (Huysamen and Brown 2009) and apoptotic cells (Oka et al. 1998). The recognition of apoptotic cells may occur by two alternative mechanisms. On the one side, LOX-1 directly recognizes phosphatidylserine on the membrane of cells undergoing apoptosis (Murphy et al. 2006). On the other side, the recognition and internalization of apoptotic cells may occur through LOX-1 binding to HSPs. HSPs are cytoplasmic compounds identified as stress-responsive proteins, which are also translocated to the cell surface during different processes of cell death (Goh et al. 2011). In myeloid cells, LOX-1 binds different members of this family such as HSP-60 (Liu et al. 2019) or HSP-70 (Parlato et al. 2010). In this sense, HSPs facilitate the uptake of dying cells (Zhu et al. 2016). The functional consequences of HSP-60 recognition by LOX-1 are the activation of macrophages resulting in cytokine production (Liu et al. 2019) and the induction of antigen presentation in DCs after binding HSP-70 (Delneste et al. 2002), being the HSP-70-LOX1 interaction an underlying mechanism for the uptake of apoptotic cells (Parlato et al. 2010).

LOX-1 expression is upregulated in circulating and tumor-infiltrating neutrophils, while its expression is negligible in healthy volunteers. Interestingly, LOX-1^+ neutrophils show transcriptional and suppressive characteristics of myeloid-derived suppressor cells (Condamine et al. 2016). Indeed, the expression of LOX-1 on polymorphonuclear cells defines highly immunosuppressive cells that correlate with the recurrence of glioblastoma and disease progression (Chai et al. 2019). LOX-1-mediated delivery of immunosuppressive signals in a cell death-rich environment such as a tumor could be a strategy to escape from immunosurveillance. In any case, these data suggest that LOX-1 could be a marker for human MDSCs. The regulatory role for LOX-1 observed in these settings and whether it is neutrophil-specific deserves further investigation.

LOX-1-deficient mice show improved survival in a cecal ligation and puncture (CLP) sepsis model. This phenotype is accompanied by a dampened systemic production of pro-inflammatory cytokines, while neutrophil infiltration into the peritoneum was boosted, favoring bacterial clearance (Wu et al. 2011). In this model, LOX-1 plays an intriguing role, promoting inflammatory responses but suppressing neutrophil mobilization, although indirect effects cannot be ruled out. LOX-1 has also been described as a surface receptor for both gram-positive and gram-negative bacteria (Shimaoka et al. 2001), where a bacterial HSP-60 homolog could be the specific LOX-1 ligand (Zhu et al. 2013). Therefore, it is difficult to interpret in this context whether the observed effects are dependent on bacteria or cell death recognition.

In summary, the immune balance after cell death sensing by LOX-1 could be described as pro-inflammatory due to its repercussion in myeloid activation and antigen presentation. However, although the implication of cell death recognition is not fully characterized in LOX-1^+ intratumor MDSCs or after infection, the triggering of LOX-1-mediated immunosuppressive signals cannot be discarded. An interesting hypothesis would be that different LOX-1 ligands trigger diverse inflammatory responses, either regulatory or pro-inflammatory.

4.5 CD301 (CLEC10A)

CLEC10A can be also named CD301, CD301a, MGL-1, HML or CLECSF14 but based on their usage frequency, we proposed the consensus alias CD301 together with its gene name, *CLEC10A* (del Fresno et al. 2019). It is encoded in chromosome 17 in humans. In mice, there are two homologous genes, *Clec10a*/CD301 and *Mgl2*/CD301b (Tsuiji 2003), both encoded in chromosome 11. CLEC10A is expressed by human monocyte-derived DCs (Higashi et al. 2002), $CD1c^+$ DCs (Heger et al. 2018) and macrophages, specifically in alternatively activated macrophages (Raes et al. 2005), with a similar expression pattern in mouse DCs and macrophages (http://www.immgen.org; Heng et al. 2008). CLEC10A surface expression is downregulated upon DC maturation while anti-inflammatory agents such as dexamethasone enhance CLEC10A expression both in macrophages and DCs (van Vliet et al. 2006). Structurally, CLEC10A is a transmembrane protein that belongs to the group II of CLRs (Sancho and Reis e Sousa 2012). Similar to LOX-1, it does not bear any distinguishable ITAM or ITIM motif in its intracellular domain.

CLEC10A recognizes N-acetylgalactosamine (GalNAc) residues present in glycoproteins and glycosphingolipids, which can be found on the surface of helminth parasites (van Vliet et al. 2005), tumors (Mortezai et al. 2013) and apoptotic cells (Malagolini et al. 2009). CLEC10A mediates the internalization of these cells and diverts the cargo to the phagolysosomal route (van Vliet et al. 2007).

Clec10a-deficient mice were generated to explore the role of this CLR in tissue homeostasis. The only noticeable defect of *Clec10a*-deficient mice is a mild increase in blood erythrocytes, which is caused by an impaired scavenging of aged erythrocytes by macrophages (Onami et al. 2002). However, the implication of CLEC10A in the sensing and removal of apoptotic cells is evident upon X-ray irradiation of pregnant mice, where massive apoptosis is induced in the developing embryos. Here, the number of pups was reduced in the absence of *Clec10a*, with a defective clearance of apoptotic cells in tissues suffering a vast apoptosis generation, such as the neural tube (Yuita et al. 2005).

In the same line, the lack of *Clec10a* originated evident phenotypes under pathological conditions running in the presence of massive cell death. This is the case for multiple sclerosis (MS). *Clec10a* expression is upregulated in MS lesions at the brain, particularly in $P2Y12R^+$ M2-polarized microglia (Ilarregui et al. 2019). In accordance with an immunosuppressive function for CD301 in this model, *Clec10a*$^{-/-}$ mice display a worsened outcome during EAE development at the resolution phase of the pathology (Ilarregui et al. 2019). Despite ex vivo experiments showed that CLEC10A induces the apoptosis of T cells, whether a defect in apoptotic cell removal accounts for the observed phenotype cannot be ruled out. Similarly, *Clec10a*-deficient mice are more susceptible to pulmonary infections by *Klebsiella pneumoniae* due to a severe lung immunopathology, despite a comparable bacterial burden (Jondle et al. 2016). In this case, also, the phenotype could rely on a defective clearance of dead cells due to the lack of CLEC10A.

Overall, CLEC10A acts as a regulatory CLR after sensing apoptotic events. This is in line with the anti-inflammatory signals generated by efferocytosis, i.e., the engulfment and clearance of apoptotic cells (Morioka et al. 2019). In fact, CLEC10A also regulates inflammation by recognizing stimuli different from apoptotic cells, such as commensal bacteria during colitis (Saba et al. 2009) or CD45 on the surface of effector T cells (van Vliet et al. 2006).

4.6 Dec-205 (LY75)

DEC-205 receptor is also known as Gp200-MR6 or CD205, but after analyzing the usage frequency of these names, we proposed the consensus alias DEC-205 together with its gene name *LY75* (del Fresno et al. 2019). It is encoded in the second chromosome in both humans and mice. While essentially any human hematopoietic cell express DEC-205 (Kato et al. 2006), in the mouse, it is expressed at low levels by B and T cells and granulocytes (Sancho and Reis e Sousa 2012), with a high expression in DCs (Jiang et al. 1995), in particular $CD8^+$ DCs (Heath et al. 2004) (http://www.immgen.org; Heng et al. 2008). From the structural point of view, DEC-205 belongs to the group VI of transmembrane CLRs, which includes CLRs bearing multiple CRD whose ligand binding capacity is calcium-dependent. The intracellular domain of DEC-205 is characterized by the absence of any identifiable ITAM or ITIM motif (Sancho and Reis e Sousa 2012).

DEC-205 can bind bacterial ligands (Zhang et al. 2008), oxLDL (Nickel et al. 2009) and apoptotic cells (Shrimpton et al. 2009). Using antibodies that bind DEC-205, early studies showed that this receptor internalizes antigens to the endo/lysosome in thymic epithelial cells and DCs (Jiang et al. 1995). Afterward, it was demonstrated that this DEC-205-mediated endocytic capacity participated in the clearance of apoptotic thymocytes by thymic epithelial cells (Small and Kraal 2003). Eventually, DEC-205 was defined as a death-sensing receptor that recognizes cells undergoing apoptosis and secondary necrosis (Shrimpton et al. 2009). This binding is pH-dependent, requiring an acidic environment (Cao et al. 2015). This is evolutionarily consistent with the function of DEC-205 as a death receptor, because extracellular acidification usually associates with inflammation, constituting a danger signal (Rajamäki et al. 2013). Eventually, keratins have been described as the actual ligand for DEC-205 (Cao et al. 2016).

Most of the latest studies on DEC-205 are focused on its use as a target in antigen delivery strategies due to its expression pattern and endocytic capacity leading to antigen presentation (Iberg and Hawiger 2019). In this sense, the immune outcome can be modulated to promote immunity against viral infections (Padilla-Quirarte et al. 2019) or cancer (Johnson et al. 2008), or tolerance for diabetes treatment (Petzold et al. 2012).

There is not a clear picture of the inflammatory responses triggered after natural recognition of dead cells by DEC-205. As indicated for CLEC10A, the clearance of apoptotic cells may be silent for the immune system, but further studies are required to clarify this point. In addition, lessons learnt from targeting studies suggest that, depending on the surrounding context, DEC-205 may be quite plastic in the signaling delivered after its engagement, which constitutes a core property of CLRs (del Fresno et al. 2018a).

5 Conclusions

During virtually any stress conditions, tissue damage and cell death take place, and the recognition of associated molecular cues by myeloid CLRs in these circumstances is key for tailoring immune responses. However, programmed cell death is also a relevant process that occurs under homeostatic conditions for the maintenance of tissue fitness. Here, we have dissected different forms of cell death and how these events condition the inflammatory responses triggered after their sensing.

Myeloid CLRs are main players in tissue damage sensing, displaying great functional plasticity, both in terms of ligand recognition and the inflammatory outcomes of this sensing. As CLRs are prompted to cross talk, their specific ligand, the surrounding environment and other receptors simultaneously activated are important factors to consider. All these variables can affect the functional consequences of the engagement of a particular CLR.

As described in this chapter, the recognition of cell death by myeloid CLRs can generate either regulatory or inflammatory reactions, impacting both innate and adaptive immune responses. Therefore, understanding how myeloid CLRs (and other receptors) recognize certain cell death programs is fundamental to harness these responses for therapeutic purposes, as CLRs are surface molecules whose function can be easily targeted.

Acknowledgements We thank all members of the David Sancho laboratory at CNIC for scientific discussions. Carlos Del Fresno is supported by AECC Foundation (INVES192DELF). Francisco Javier Cueto is the recipient of a Ph.D. "La Caixa" fellowship (LCF/BQ/ES14/10320011). Work in the DS laboratory is funded by the CNIC; by the European Research Council (ERC-2016-Consolidator Grant 725091); by the European Commission (635122-PROCROP H2020); by Ministerio de Ciencia, Innovación e Universidades (MICINN), Agencia Estatal de Investigación and Fondo Europeo de Desarrollo Regional (FEDER) (SAF2016-79040-R); by Comunidad de Madrid (B2017/BMD-3733 Immunothercan-CM); by FIS-Instituto de Salud Carlos III, MICINN and FEDER (RD16/0015/0018-REEM); by Acteria Foundation; by Atresmedia (Constantes y Vitales prize) and by Fundació La Marató de TV3 (201723). The CNIC is supported by the Instituto de Salud Carlos III (ISCIII), the MICINN and the Pro-CNIC Foundation and is a Severo Ochoa Center of Excellence (SEV-2015-0505).

References

Ahrens S, Zelenay S, Sancho D et al (2012) F-actin is an evolutionarily conserved damage-associated molecular pattern recognized by DNGR-1, a receptor for dead cells. Immunity 36:635–645. https://doi.org/10.1016/j.immuni.2012.03.008

Andersson U, Tracey KJ (2011) HMGB1 is a therapeutic target for sterile inflammation and infection. Annu Rev Immunol 29:139–162. https://doi.org/10.1146/annurev-immunol-030409-101323

Araźna M, Pruchniak MP, Demkow U (2015) Reactive oxygen species, granulocytes, and netosis. Adv Exp Med Biol 836:1–7. https://doi.org/10.1007/5584_2014_12

Bell CW, Jiang W, Reich CF, Pisetsky DS (2006) The extracellular release of HMGB1 during apoptotic cell death. Am J Physiol—Cell Physiol 291:C1318–C1325. https://doi.org/10.1152/ajpcell.00616.2005

Branzk N, Papayannopoulos V (2013) Molecular mechanisms regulating NETosis in infection and disease. Semin Immunopathol 35:513–530. https://doi.org/10.1007/s00281-013-0384-6

Brinkmann V, Reichard U, Goosmann C et al (2004) Neutrophil extracellular traps kill bacteria. Science 303(80):1532–1535. https://doi.org/10.1126/science.1092385

Brubaker SW, Bonham KS, Zanoni I, Kagan JC (2015) Innate immune pattern recognition: a cell biological perspective. Annu Rev Immunol 33:257–290. https://doi.org/10.1146/annurev-immunol-032414-112240

Canli Ö, Alankus Y, Grootjans S et al (2016) Glutathione peroxidase 4 prevents necroptosis in mouse erythroid precursors. Blood 127:139–148. https://doi.org/10.1182/blood-2015-06

Cao L, Shi X, Chang H et al (2015) PH-dependent recognition of apoptotic and necrotic cells by the human dendritic cell receptor DEC205. Proc Natl Acad Sci USA 112:7237–7242. https://doi.org/10.1073/pnas.1505924112

Cao L, Chang H, Shi X et al (2016) Keratin mediates the recognition of apoptotic and necrotic cells through dendritic cell receptor DEC205/CD205. Proc Natl Acad Sci USA 113:13438–13443. https://doi.org/10.1073/pnas.1609331113

Casares N, Pequignot MO, Tesniere A et al (2005) Caspase-dependent immunogenicity of doxorubicin-induced tumor cell death. J Exp Med 202:1691–1701. https://doi.org/10.1084/jem.20050915

Chai E, Zhang L, Li C (2019) LOX-1$^+$ PMN-MDSC enhances immune suppression which promotes glioblastoma multiforme progression. Cancer Manag Res 11:7307–7315. https://doi.org/10.2147/cmar.s210545

Cho YS (2018) The role of necroptosis in the treatment of diseases. BMB Rep 51:219–224. https://doi.org/10.5483/BMBRep.2018.51.5.074

Cho YS, Challa S, Moquin D et al (2009) Phosphorylation-driven assembly of the RIP1-RIP3 complex regulates programmed necrosis and virus-induced inflammation. Cell 137:1112–1123. https://doi.org/10.1016/j.cell.2009.05.037

Condamine T, Dominguez GA, Youn JI et al (2016) Lectin-type oxidized LDL receptor-1 distinguishes population of human polymorphonuclear myeloid-derived suppressor cells in cancer patients. Sci Immunol 1. https://doi.org/10.1126/sciimmunol.aaf8943

D'Arcy MS (2019) Cell death: a review of the major forms of apoptosis, necrosis and autophagy. Cell Biol Int 43:582–592. https://doi.org/10.1002/cbin.11137

Daga S, Rosenberger A, Quehenberger F et al (2019) High GPR56 surface expression correlates with a leukemic stem cell gene signature in CD34-positive AML. Cancer Med 8:1771–1778. https://doi.org/10.1002/cam4.2053

Davis BK, Wen H, Ting JP-Y (2011) The inflammasome NLRs in immunity, inflammation, and associated diseases. Annu Rev Immunol 29:707–735. https://doi.org/10.1146/annurev-immunol-031210-101405

Decout A, Silva-Gomes S, Drocourt D et al (2017) Rational design of adjuvants targeting the C-type lectin Mincle. Proc Natl Acad Sci 114:2675–2680. https://doi.org/10.1073/pnas.1612421114

del Fresno C, Iborra S, Saz-Leal P et al (2018a) Flexible signaling of Myeloid C-type lectin receptors in immunity and inflammation. Front Immunol 9. https://doi.org/10.3389/fimmu.2018.00804

del Fresno C, Saz-Leal P, Enamorado M et al (2018b) DNGR-1 in dendritic cells limits tissue damage by dampening neutrophil recruitment. Science 362(80):351–356. https://doi.org/10.1126/science.aan8423

del Fresno C, Cueto FJ, Sancho D (2019) A proposal for nomenclature in myeloid C-type lectin receptors. Front Immunol 10. https://doi.org/10.3389/fimmu.2019.02098

Delgado-Rizo V, Martínez-Guzmán MA, Iñiguez-Gutierrez L et al (2017) Neutrophil extracellular traps and its implications in inflammation: an overview. Front Immunol 8. https://doi.org/10.3389/fimmu.2017.00081

Delneste Y, Magistrelli G, Gauchat J-FO et al (2002) Involvement of LOX-1 in dendritic cell-mediated antigen cross-presentation cross-priming is currently a challenge in antitumor im-munotherapy. Heat shock proteins (HSP) are molecular chaperones that control the folding and prevent the aggregation of prote. Immunity 17:353–362

Dixon SJ, Lemberg KM, Lamprecht MR et al (2012) Ferroptosis: an iron-dependent form of nonapoptotic cell death. Cell 149:1060–1072. https://doi.org/10.1016/j.cell.2012.03.042

Doitsh G, Galloway NLK, Geng X et al (2014) Cell death by pyroptosis drives CD4 T-cell depletion in HIV-1 infection. Nature 505:509–514. https://doi.org/10.1038/nature12940

Dondelinger Y, Declercq W, Montessuit S et al (2014) MLKL compromises plasma membrane integrity by binding to phosphatidylinositol phosphates. Cell Rep 7:971–981. https://doi.org/10.1016/j.celrep.2014.04.026

Draude G, Hrboticky N, Lorenz RL (1999) The expression of the lectin-like oxidized low-density lipoprotein receptor (LOX-1) on human vascular smooth muscle cells and monocytes and its down-regulation by lovastatin. Biochem Pharmacol 57:383–386. https://doi.org/10.1016/S0006-2952(98)00313-X

Elmore S (2007) Apoptosis: a review of programmed cell death. Toxicol Pathol 35:495–516. https://doi.org/10.1080/01926230701320337

Feinberg H, Jégouzo SAF, Rowntree TJW et al (2013) Mechanism for recognition of an unusual mycobacterial glycolipid by the macrophage receptor Mincle. J Biol Chem 288:28457–28465. https://doi.org/10.1074/jbc.M113.497149

Feng X, Song Q, Yu A et al (2015) Receptor-interacting protein kinase 3 is a predictor of survival and plays a tumor suppressive role in colorectal cancer. Neoplasma 62:592–601. https://doi.org/10.4149/neo_2015_071

Feng S, Fox D, Man SM (2018) Mechanisms of gasdermin family members in inflammasome signaling and cell death. J Mol Biol 430:3068–3080. https://doi.org/10.1016/j.jmb.2018.07.002

Fink SL, Bergsbaken T, Cookson BT (2008) Anthrax lethal toxin and Salmonella elicit the common cell death pathway of caspase-1-dependent pyroptosis via distinct mechanisms. Proc Natl Acad Sci USA 105:4312–4317. https://doi.org/10.1073/pnas.0707370105

Furukawa A, Kamishikiryo J, Mroi D et al (2013) Structural analysis for glycolipid recognition by the. Proc Natl Acad Sci USA 110:17438–17443. https://doi.org/10.1073/pnas.1312649110/-/DCSupplemental.www.pnas.org/cgi/doi/10.1073/pnas.1312649110

Galluzzi L, Bravo-San Pedro JM, Vitale I et al (2015) Essential versus accessory aspects of cell death: recommendations of the NCCD 2015. Cell Death Differ 22:58–73. https://doi.org/10.1038/cdd.2014.137

Galluzzi L, López-Soto A, Kumar S, Kroemer G (2016) Caspases connect cell-death signaling to organismal homeostasis. Immunity 44:221–231. https://doi.org/10.1016/j.immuni.2016.01.020

Galluzzi L, Kepp O, Chan FK-M, Kroemer G (2017) Necroptosis: mechanisms and relevance to disease. Annu Rev Pathol Mech Dis 12:103–130. https://doi.org/10.1146/annurev-pathol-052016-100247

Gaschler MM, Stockwell BR (2017) Lipid peroxidation in cell death. Biochem Biophys Res Commun 482:419–425. https://doi.org/10.1016/j.bbrc.2016.10.086

Geijtenbeek TBH, Gringhuis SI (2009) Signalling through C-type lectin receptors: shaping immune responses. Nat Rev Immunol 9:465–479. https://doi.org/10.1038/nri2569

Ghosh N, Das A, Chaffee S et al (2018) Reactive oxygen species, oxidative damage and cell death. Immun Inflamm Heal Dis 45–55. https://doi.org/10.1016/b978-0-12-805417-8.00004-4
Goh YC, Yap CT, Huang BH et al (2011) Heat-shock protein 60 translocates to the surface of apoptotic cells and differentiated megakaryocytes and stimulates phagocytosis. Cell Mol Life Sci 68:1581–1592. https://doi.org/10.1007/s00018-010-0534-0
Green DR, Ferguson T, Zitvogel L, Kroemer G (2009) Immunogenic and tolerogenic cell death. Nat Rev Immunol 9:353–363. https://doi.org/10.1038/nri2545
Gurka S, Dirks S, Photiadis J, Kroczek RA (2015) Expression analysis of surface molecules on human thymic dendritic cells with the 10th HLDA Workshop antibody panel. Clin Transl Immunol 4:e47. https://doi.org/10.1038/cti.2015.21
Han Y, Zhang M, Li N et al (2004) KLRL1, a novel killer cell lectinlike receptor, inhibits natural killer cell cytotoxicity. Blood 104:2858–2866. https://doi.org/10.1182/blood-2004-03-0878
Hanč P, Fujii T, Iborra S et al (2015) Structure of the complex of F-actin and DNGR-1, a C-type lectin receptor involved in dendritic cell cross-presentation of dead cell-associated antigens. Immunity 42:839–849. https://doi.org/10.1016/j.immuni.2015.04.009
Hanč P, Schulz O, Fischbach H et al (2016) A pH- and ionic strength-dependent conformational change in the neck region regulates DNGR-1 function in dendritic cells. EMBO J 35:2484–2497. https://doi.org/10.15252/embj.201694695
Heath WR, Belz GT, Behrens GMN et al (2004) Cross-presentation, dendritic cell subsets, and the generation of immunity to cellular antigens. Immunol Rev 199:9–26. https://doi.org/10.1111/j.0105-2896.2004.00142.x
Heger L, Balk S, Lühr JJ et al (2018) CLEC10A Is a specific marker for human $CD1c^+$ dendritic cells and enhances their toll-like receptor 7/8-Induced cytokine secretion. Front Immunol 9. https://doi.org/10.3389/fimmu.2018.00744
Heng TSP, Painter MW, Elpek K et al (2008) The immunological genome project: networks of gene expression in immune cells. Nat Immunol 9:1091–1094. https://doi.org/10.1038/ni1008-1091
Higashi N, Fujioka K, Denda-Nagai K et al (2002) The macrophage C-type lectin specific for galactose/N-acetylgalactosamine is an endocytic receptor expressed on monocyte-derived immature dendritic cells. J Biol Chem 277:20686–20693. https://doi.org/10.1074/jbc.M202104200
Huang Z, Wu SQ, Liang Y et al (2015) RIP1/RIP3 binding to HSV-1 ICP6 initiates necroptosis to restrict virus propagation in mice. Cell Host Microbe 17:229–242. https://doi.org/10.1016/j.chom.2015.01.002
Huysamen C, Brown GD (2009) The fungal pattern recognition receptor, Dectin-1, and the associated cluster of C-type lectin-like receptors. FEMS Microbiol Lett 290:121–128. https://doi.org/10.1111/j.1574-6968.2008.01418.x
Huysamen C, Willment JA, Dennehy KM, Brown GD (2008) CLEC9A is a novel activation C-type lectin-like receptor expressed on $BDCA3^+$ dendritic cells and a subset of monocytes. J Biol Chem 283:16693–16701. https://doi.org/10.1074/jbc.M709923200
Iberg CA, Hawiger D (2019) Advancing immunomodulation by in vivo antigen delivery to DEC-205 and other cell surface molecules using recombinant chimeric antibodies. Int Immunopharmacol 73:575–580. https://doi.org/10.1016/j.intimp.2019.05.037
Iborra S, Sancho D (2014) Signalling versatility following self and non-self sensing by myeloid C-type lectin receptors. Immunobiology
Iborra S, Izquierdo HM, Martínez-López M et al (2012) The DC receptor DNGR-1 mediates cross-priming of CTLs during vaccinia virus infection in mice. J Clin Invest 122:1628–1643. https://doi.org/10.1172/JCI60660
Iborra S, Martínez-López M, Cueto FJ et al (2016a) Leishmania uses Mincle to target an inhibitory ITAM signaling pathway in dendritic cells that dampens adaptive immunity to infection. Immunity 45:788–801. https://doi.org/10.1016/j.immuni.2016.09.012
Iborra S, Martínez-López M, Khouili SC et al (2016b) Optimal generation of tissue-resident but not circulating memory T cells during viral infection requires crosspriming by $DNGR\text{-}1^+$ dendritic cells. Immunity 45:847–860. https://doi.org/10.1016/j.immuni.2016.08.019

Ilarregui JM, Kooij G, Rodríguez E et al (2019) Macrophage galactose-type lectin (MGL) is induced on M2 microglia and participates in the resolution phase of autoimmune neuroinflammation. J Neuroinflammation 16. https://doi.org/10.1186/s12974-019-1522-4
Ishikawa E, Ishikawa T, Morita YS et al (2009) Direct recognition of the mycobacterial glycolipid, trehalose dimycolate, by C-type lectin Mincle. J Exp Med 206:2879–2888. https://doi.org/10.1084/jem.20091750
Ishikawa T, Itoh F, Yoshida S et al (2013) Identification of distinct ligands for the C-type lectin receptors Mincle and Dectin-2 in the pathogenic fungus Malassezia. Cell Host Microbe 13:477–488. https://doi.org/10.1016/j.chom.2013.03.008
Janela B, Patel AA, Lau MC et al (2019) A subset of type I conventional dendritic cells controls cutaneous bacterial infections through VEGFα-mediated recruitment of neutrophils. Immunity 50:1069–1083.e8. https://doi.org/10.1016/j.immuni.2019.03.001
Janeway CA (1989) Approaching the asymptote? Evolution and revolution in immunology. Cold Spring Harb Symp Quant Biol 54(Pt 1):4475–4487
Jiang W, Swiggard WJ, Heufler C et al (1995) The receptor DEC-205 expressed by dendritic cells and thymic epithelial cells is involved in antigen processing. Nature 375:151–155. https://doi.org/10.1038/375151a0
Jiang Y-P, Liu BY, Zheng Q et al (2018) CLT030, a leukemic stem cell–targeting CLL1 antibody-drug conjugate for treatment of acute myeloid leukemia. Blood Adv 2:1738–1749. https://doi.org/10.1182/bloodadvances.2018020107
Johnson TS, Mahnke K, Storn V et al (2008) Inhibition of melanoma growth by targeting of antigen to dendritic cells via an anti-DEC-205 single-chain fragment variable molecule. Clin Cancer Res 14:8169–8177. https://doi.org/10.1158/1078-0432.CCR-08-1474
Jondle CN, Sharma A, Simonson TJ et al (2016) Macrophage galactose-type lectin-1 deficiency is associated with increased neutrophilia and hyperinflammation in gram-negative pneumonia. J Immunol 196:3088–3096. https://doi.org/10.4049/jimmunol.1501790
Jorgensen I, Rayamajhi M, Miao EA (2017) Programmed cell death as a defence against infection. Nat Rev Immunol 17:151–164. https://doi.org/10.1038/nri.2016.147
Kang R, Zeng L, Zhu S et al (2018) Lipid peroxidation drives gasdermin D-mediated pyroptosis in lethal polymicrobial sepsis. Cell Host Microbe 24:97–108.e4. https://doi.org/10.1016/j.chom.2018.05.009
Kasahara S, Clark EA (2012) Dendritic cell-associated lectin 2 (DCAL2) defines a distinct CD8α—dendritic cell subset. J Leukoc Biol 91:437–448. https://doi.org/10.1189/jlb.0711384
Kato M, McDonald KJ, Khan S et al (2006) Expression of human DEC-205 (CD205) multilectin receptor on leukocytes. Int Immunol 18:857–869. https://doi.org/10.1093/intimm/dxl022
Kattoor AJ, Goel A, Mehta JL (2019) LOX-1: regulation, signaling and its role in atherosclerosis. Antioxidants 8:218. https://doi.org/10.3390/antiox8070218
Kazama H, Ricci JE, Herndon JM et al (2008) Induction of immunological tolerance by apoptotic cells requires caspase-dependent oxidation of high-mobility group box-1 protein. Immunity 29:21–32. https://doi.org/10.1016/j.immuni.2008.05.013
Kerr JFR, Wyllie AH, Currie AR (1972) Apoptosis: a basic biological phenomenon with wide-ranging implications in tissue kinetics. Br J Cancer 26:239–257. https://doi.org/10.1038/bjc.1972.33
Kim JW, Roh YS, Jeong H et al (2018) Spliceosome-associated protein 130 exacerbates alcohol-induced liver injury by inducing NLRP3 inflammasome-mediated IL-1β in mice. Am J Pathol 188:967–980. https://doi.org/10.1016/j.ajpath.2017.12.010
Kiss T (2004) Biogenesis of small nuclear RNPs. J Cell Sci 117:5949–5951. https://doi.org/10.1242/jcs.01487
Kostarnoy AV, Gancheva PG, Lepenies B et al (2017) Receptor Mincle promotes skin allergies and is capable of recognizing cholesterol sulfate. Proc Natl Acad Sci USA 114:E2758–E2765. https://doi.org/10.1073/pnas.1611665114
Kroemer G (2017) Death, danger & immunity: fundamental mechanisms linking pathogenic or iatrogenic cell death events to immune responses. Immunol Rev 280:5–7. https://doi.org/10.1111/imr.12604

Kroemer G, Galluzzi L, Vandenabeele P et al (2009) Classification of cell death: recommendations of the nomenclature committee on cell death 2009. Cell Death Differ 16:3–11. https://doi.org/10.1038/cdd.2008.150

Laborda E, Mazagova M, Shao S et al (2017) Development of a chimeric antigen receptor targeting c-type lectin-like molecule-1 for human acute myeloid leukemia. Int J Mol Sci 18:1–8. https://doi.org/10.3390/ijms18112259

Lee WB, Kang JS, Yan JJ et al (2012) Neutrophils promote mycobacterial trehalose dimycolate-induced lung inflammation via the Mincle pathway. PLoS Pathog. https://doi.org/10.1371/journal.ppat.1002614

Lee WB, Yan JJ, Kang JS et al (2017a) Mincle activation enhances neutrophil migration and resistance to polymicrobial septic peritonitis. Sci Rep 7:41106. https://doi.org/10.1038/srep41106

Lee KH, Kronbichler A, Park DDY et al (2017b) Neutrophil extracellular traps (NETs) in autoimmune diseases: a comprehensive review. Autoimmun Rev 16:1160–1173. https://doi.org/10.1016/j.autrev.2017.09.012

Li K, Neumann K, Duhan V et al (2019) The uric acid crystal receptor Clec12A potentiates type I interferon responses. Proc Natl Acad Sci 116:18544–18549. https://doi.org/10.1073/pnas.1821351116

Linkermann A, Green DR (2014) Necroptosis. N Engl J Med 370:455–465. https://doi.org/10.1056/NEJMra1310050

Linkermann A, Stockwell BR, Krautwald S, Anders H-J (2014) Regulated cell death and inflammation: an auto-amplification loop causes organ failure. Nat Rev Immunol 14:759–767. https://doi.org/10.1038/nri3743

Liu X, Lieberman J (2017) A mechanistic understanding of pyroptosis: the fiery death triggered by invasive infection. Adv Immunol 135:81–117. https://doi.org/10.1016/bs.ai.2017.02.002

Liu B, Li S, Xiu B et al (2019) C-terminus of heat shock protein 60 can activate macrophages by lectin-like oxidized low-density lipoprotein receptor 1. Biochem Biophys Res Commun 508:1113–1119. https://doi.org/10.1016/j.bbrc.2018.12.008

Lobato-Pascual A, Saether PC, Fossum S et al (2013) Mincle, the receptor for mycobacterial cord factor, forms a functional receptor complex with MCL and FcεRI-γ. Eur J Immunol 43:3167–3174. https://doi.org/10.1002/eji.201343752

Lu X, Nagata M, Yamasaki S (2018) Mincle: 20 years of a versatile sensor of insults. Int Immunol 30:233–239. https://doi.org/10.1093/intimm/dxy028

Lyons JA, San M, Happ MP, Cross AH (1999) B cells are critical to induction of experimental allergic encephalomyelitis by protein but not by a short encephalitogenic peptide. Eur J Immunol 29:3432–3439. https://doi.org/10.1002/(SICI)1521-4141(199911)29:11%3c3432:AID-IMMU3432%3e3.0.CO;2-2

Malagolini N, Chiricolo M, Marini M, Dall'Olio F (2009) Exposure of α2,6-sialylated lactosaminic chains marks apoptotic and necrotic death in different cell types. Glycobiology 19:172–181. https://doi.org/10.1093/glycob/cwn122

Marshall ASJ, Willmen JA, Lin HH et al (2004) Identification and characterization of a novel human myeloid inhibitory C-type lectin-like receptor (MICL) that is predominantly expressed on granulocytes and monocytes. J Biol Chem 279:14792–14802. https://doi.org/10.1074/jbc.M313127200

Marshall ASJ, Willment JA, Pyz E et al (2006) Human MICL (CLEC12A) is differentially glycosylated and is down-regulated following cellular activation. Eur J Immunol 36:2159–2169. https://doi.org/10.1002/eji.200535628

Martínez-López M, Iborra S, Conde-Garrosa R et al (2019) Microbiota sensing by Mincle-syk axis in dendritic cells regulates interleukin-17 and -22 production and promotes intestinal barrier integrity. Immunity. https://doi.org/10.1016/j.immuni.2018.12.020

Matzinger P (2002) The danger model: a renewed sense of self. Science 296(80):301–305. https://doi.org/10.1126/science.1071059

McIlwain DR, Berger T, Mak TW (2013) Caspase functions in cell death and disease. Cold Spring Harb Perspect Biol 5:1–28. https://doi.org/10.1101/cshperspect.a008656

Mehta JL, Sanada N, Hu CP et al (2007) Deletion of LOX-1 reduces atherogenesis in LDLR knockout mice fed high cholesterol diet. Circ Res 100:1634–1642. https://doi.org/10.1161/CIRCRESAHA.107.149724

Mesa MA, Vasquez G (2013) NETosis. Autoimmune Dis 2013:1–7. https://doi.org/10.1155/2013/651497

Michou L, Cornélis F, Levesque JM et al (2012) A genetic association study of the CLEC12A gene in rheumatoid arthritis. Jt Bone Spine 79:451–456. https://doi.org/10.1016/j.jbspin.2011.12.012

Mitroulis I, Kambas K, Chrysanthopoulou A et al (2011) Neutrophil extracellular trap formation is associated with IL-1β and autophagy-related signaling in gout. PLoS One 6. https://doi.org/10.1371/journal.pone.0029318

Mkaddem SB, Hayem G, Jönsson F et al (2014) Shifting Fc RIIA-ITAM from activation to inhibitory configuration ameliorates arthritis. J Clin Invest 1–15. https://doi.org/10.1172/JCI74572

Morioka S, Maueröder C, Ravichandran KS (2019) Living on the edge: efferocytosis at the interface of homeostasis and pathology. Immunity 50:1149–1162. https://doi.org/10.1016/j.immuni.2019.04.018

Mortezai N, Behnken HN, Kurze AK et al (2013) Tumor-associated Neu5Ac-Tn and Neu5Gc-Tn antigens bind to C-type lectin CLEC10A (CD301, MGL). Glycobiology 23:844–852. https://doi.org/10.1093/glycob/cwt021

Murphy JE, Tacon D, Tedbury PR et al (2006) LOX-1 scavenger receptor mediates calcium-dependent recognition of phosphatidylserine and apoptotic cells. Biochem J 393:107–115. https://doi.org/10.1042/BJ20051166

Nagata S (2018) Apoptosis and clearance of apoptotic cells. Annu Rev Immunol 36:489–517. https://doi.org/10.1146/annurev-immunol-042617-053010

Nagata M, Izumi Y, Ishikawa E et al (2017) Intracellular metabolite β-glucosylceramide is an endogenous Mincle ligand possessing immunostimulatory activity. Proc Natl Acad Sci USA 114:E3285–E3294. https://doi.org/10.1073/pnas.1618133114

Najafov A, Zervantonakis IK, Mookhtiar AK et al (2018) BRAF and AXL oncogenes drive RIPK3 expression loss in cancer. PLoS Biol 16:1–22. https://doi.org/10.1371/journal.pbio.2005756

Neumann K, Castiñeiras-Vilariño M, Höckendorf U et al (2014) Clec12a is an inhibitory receptor for uric acid crystals that regulates inflammation in response to cell death. Immunity 40:389–399. https://doi.org/10.1016/j.immuni.2013.12.015

Nickel T, Schmauss D, Hanssen H et al (2009) oxLDL uptake by dendritic cells induces upregulation of scavenger-receptors, maturation and differentiation. Atherosclerosis 205:442–450. https://doi.org/10.1016/j.atherosclerosis.2009.01.002

Obeid M, Tesniere A, Ghiringhelli F et al (2007) Calreticulin exposure dictates the immunogenicity of cancer cell death. Nat Med 13:54–61. https://doi.org/10.1038/nm1523

Oka K, Sawamura T, Kikuta KI et al (1998) Lectin-like oxidized low-density lipoprotein receptor 1 mediates phagocytosis of aged/apoptotic cells in endothelial cells. Proc Natl Acad Sci USA 95:9535–9540. https://doi.org/10.1073/pnas.95.16.9535

Onami TM, Lin M-Y, Page DM et al (2002) Generation of mice deficient for macrophage galactose- and N-acetylgalactosamine-specific lectin: limited role in lymphoid and erythroid homeostasis and evidence for multiple lectins. Mol Cell Biol 22:6661. https://doi.org/10.1128/mcb.22.18.6661.2002

Orning P, Weng D, Starheim K et al (2018) Pathogen blockade of TAK1 triggers caspase-8–dependent cleavage of gasdermin D and cell death. Science 362(80):1064–1069. https://doi.org/10.1126/science.aau2818

Padilla-Quirarte HO, Badillo-Godinez O, Gutierrez-Xicotencatl L et al (2019) Targeting M2e to DEC-205 induces an enhanced serum antibody-dependent heterosubtypic protection against influenza A virus infection. Vaccine 37:2624–2633. https://doi.org/10.1016/j.vaccine.2019.02.050

Palladino END, Katunga LA, Kolar GR, Ford DA (2018) 2-Chlorofatty acids: Lipid mediators of neutrophil extracellular trap formation. J Lipid Res 59:1424–1432. https://doi.org/10.1194/jlr.M084731

Parlato S, Romagnoli G, Spadaro F et al (2010) LOX-1 as a natural IFN-α-mediated signal for apoptotic cell uptake and antigen presentation in dendritic cells. Blood 115:1554–1563. https://doi.org/10.1182/blood-2009-07-234468

Parlesak A, Schäfer C, Schütz T et al (2000) Increased intestinal permeability to macromolecules and endotoxemia in patients with chronic alcohol abuse in different stages of alcohol-induced liver disease. J Hepatol 32:742–747. https://doi.org/10.1016/S0168-8278(00)80242-1

Pasparakis M, Vandenabeele P (2015) Necroptosis and its role in inflammation. Nature 517:311–320. https://doi.org/10.1038/nature14191

Pellettieri J, Alvarado AS (2007) Cell turnover and adult tissue homeostasis: from humans to planarians. Annu Rev Genet 41:83–105. https://doi.org/10.1146/annurev.genet.41.110306.130244

Petzold C, Schallenberg S, Stern JNH, Kretschmer K (2012) Targeted antigen delivery to DEC-205$^+$ dendritic cells for tolerogenic vaccination. Rev Diabet Stud 9:305–318. https://doi.org/10.1900/RDS.2012.9.305

Pyz E, Huysamen C, Marshall ASJ et al (2008) Characterisation of murine MICL (CLEC12A) and evidence for an endogenous ligand. Eur J Immunol 38:1157–1163. https://doi.org/10.1002/eji.200738057

Raes G, Brys L, Dahal BK et al (2005) Macrophage galactose-type C-type lectins as novel markers for alternatively activated macrophages elicited by parasitic infections and allergic airway inflammation. J Leukoc Biol 77:321–327. https://doi.org/10.1189/jlb.0304212

Rajamäki K, Nordström T, Nurmi K et al (2013) Extracellular acidosis is a novel danger signal alerting innate immunity via the NLRP3 inflammasome. J Biol Chem 288:13410–13419. https://doi.org/10.1074/jbc.M112.426254

Raulf MK, Johannssen T, Matthiesen S et al (2019) The C-type lectin receptor CLEC12A recognizes plasmodial hemozoin and contributes to cerebral malaria development. Cell Rep 28:30–38.e5. https://doi.org/10.1016/j.celrep.2019.06.015

Redelinghuys P, Whitehead L, Augello A et al (2016) MICL controls inflammation in rheumatoid arthritis. Ann Rheum Dis 75:1386–1391. https://doi.org/10.1136/annrheumdis-2014-206644

Reed JC, Pellecchia M (2005) Apoptosis-based therapies for hematologic malignancies. Blood 106:408–418. https://doi.org/10.1182/blood-2004-07-2761

Robinson N, Ganesan R, Hegedűs C et al (2019) Programmed necrotic cell death of macrophages: focus on pyroptosis, necroptosis, and parthanatos. Redox Biol 26. https://doi.org/10.1016/j.redox.2019.101239

Roers A, Hiller B, Hornung V (2016) Recognition of endogenous nucleic acids by the innate immune system. Immunity 44:739–754. https://doi.org/10.1016/j.immuni.2016.04.002

Roh JS, Sohn DH (2018) Damage-associated molecular patterns in inflammatory diseases. Immune Netw 18:e27. https://doi.org/10.4110/in.2018.18.e27

Roug AS, Larsen HO, Nederby L et al (2014) HMICL and CD123 in combination with a CD45/CD34/CD117 backbone—a universal marker combination for the detection of minimal residual disease in acute myeloid leukaemia. Br J Haematol 164:212–222. https://doi.org/10.1111/bjh.12614

Saba K, Denda-Nagai K, Irimura T (2009) A C-type lectin MGL1/CD301a plays an anti-inflammatory role in murine experimental colitis. Am J Pathol 174:144–152. https://doi.org/10.2353/ajpath.2009.080235

Sachet M, Liang YY, Oehler R (2017) The immune response to secondary necrotic cells. Apoptosis 22:1189–1204. https://doi.org/10.1007/s10495-017-1413-z

Sagar Di, Singh NP, Ginwala R et al (2017) Antibody blockade of CLEC12A delays EAE onset and attenuates disease severity by impairing myeloid cell CNS infiltration and restoring positive immunity. Sci Rep 7. https://doi.org/10.1038/s41598-017-03027-x

Sancho D, Reis e Sousa C (2012) Signaling by myeloid C-type lectin receptors in immunity and homeostasis. Annu Rev Immunol 30:491–529. https://doi.org/10.1146/annurev-immunol-031210-101352

Sancho D, Reis e Sousa C (2013) Sensing of cell death by myeloid C-type lectin receptors. Curr Opin Immunol 25:46–52

Sancho D, Mourão-Sá D, Joffre OP et al (2008) Tumor therapy in mice via antigen targeting to a novel, DC-restricted C-type lectin. J Clin Invest 118:2098–2110. https://doi.org/10.1172/JCI34584DS1

Sancho D, Joffre OP, Keller AM et al (2009) Identification of a dendritic cell receptor that couples sensing of necrosis to immunity. Nature 458:899–903. https://doi.org/10.1038/nature07750

Santoni G, Cardinali C, Morelli BB et al (2015) Danger- and pathogen-associated molecular patterns recognition by pattern-recognition receptors and ion channels of the transient receptor potential family triggers the inflammasome activation in immune cells and sensory neurons. J Neuroinflammation 12. https://doi.org/10.1186/s12974-015-0239-2

Sawamura T, Kume N, Aoyama T et al (1997) An endothelial receptor for oxidized low-density lipoprotein. Nature 386:73–77. https://doi.org/10.1038/386073a0

Schoenen H, Bodendorfer B, Hitchens K et al (2010) Cutting edge: Mincle is essential for recognition and adjuvanticity of the mycobacterial cord factor and its synthetic analog trehalose-dibehenate. J Immunol 184:2756–2760. https://doi.org/10.4049/jimmunol.0904013

Schraml BU, Van Blijswijk J, Zelenay S et al (2013) Genetic tracing via DNGR-1 expression history defines dendritic cells as a hematopoietic lineage. Cell 154:843–858. https://doi.org/10.1016/j.cell.2013.07.014

Schulz O, Hanč P, Böttcher JP et al (2018) Myosin II synergizes with F-actin to promote DNGR-1-dependent cross-presentation of dead cell-associated antigens. Cell Rep 24:419–428. https://doi.org/10.1016/j.celrep.2018.06.038

Scott GS, Spitsin SV, Kean RB et al (2002) Therapeutic intervention in experimental allergic encephalomyelitis by administration of uric acid precursors. Proc Natl Acad Sci USA 99:16303–16308. https://doi.org/10.1073/pnas.212645999

Seifert L, Werba G, Tiwari S et al (2016) The necrosome promotes pancreatic oncogenesis via CXCL1 and Mincle-induced immune suppression. Nature 532:245–249. https://doi.org/10.1038/nature17403

Seong S-YY, Matzinger P (2004) Opinion: hydrophobicity: an ancient damage-associated molecular pattern that initiates innate immune responses. Nat Rev Immunol 4:469–478. https://doi.org/10.1038/nri1372

Shekarian T, Valsesia-Wittmann S, Brody J et al (2017) Pattern recognition receptors: immune targets to enhance cancer immunotherapy. Ann Oncol 28:1756–1766. https://doi.org/10.1093/annonc/mdx179

Shi J, Zhao Y, Wang Y et al (2014) Inflammatory caspases are innate immune receptors for intracellular LPS. Nature 514:187–192

Shi J, Zhao Y, Wang K et al (2015) Cleavage of GSDMD by inflammatory caspases determines pyroptotic cell death. Nature 526:660–665. https://doi.org/10.1038/nature15514

Shimaoka T, Kume N, Minami M et al (2001) LOX-1 supports adhesion of gram-positive and gram-negative bacteria. J Immunol 166:5108–5114. https://doi.org/10.4049/jimmunol.166.8.5108

Shrimpton RE, Butler M, Morel AS et al (2009) CD205 (DEC-205): a recognition receptor for apoptotic and necrotic self. Mol Immunol 46:1229–1239. https://doi.org/10.1016/j.molimm.2008.11.016

Singh R, Letai A, Sarosiek K (2019) Regulation of apoptosis in health and disease: the balancing act of BCL-2 family proteins. Nat Rev Mol Cell Biol 20:175–193. https://doi.org/10.1038/s41580-018-0089-8

Small M, Kraal G (2003) In vitro evidence for participation of DEC-205 expressed by thymic cortical epithelial cells in clearance of apoptotic thymocytes. Int Immunol 15:197–203. https://doi.org/10.1093/intimm/dxg024

Soares MP, Teixeira L, Moita LF (2017) Disease tolerance and immunity in host protection against infection. Nat Rev Immunol 17:83–96. https://doi.org/10.1038/nri.2016.136

Sollberger G, Choidas A, Burn GL et al (2018) Gasdermin D plays a vital role in the generation of neutrophil extracellular traps. Sci Immunol 3:6689. https://doi.org/10.1126/sciimmunol.aar6689

Steinman RM, Turley S, Mellman I, Inaba K (2000) The induction of tolerance by dendritic cells that have captured apoptotic cells. J Exp Med 191:411–416. https://doi.org/10.1084/jem.191.3.411

Sun L, Wang H, Wang Z et al (2012) Mixed lineage kinase domain-like protein mediates necrosis signaling downstream of RIP3 kinase. Cell 148:213–227. https://doi.org/10.1016/j.cell.2011.11.031

Tang D, Kang R, Vanden Berghe T et al (2019) The molecular machinery of regulated cell death. Cell Res 29:347–364. https://doi.org/10.1038/s41422-019-0164-5

Tian K, Ogura S, Little PJ et al (2019) Targeting LOX-1 in atherosclerosis and vasculopathy: current knowledge and future perspectives. Ann NY Acad Sci 1443:34–53. https://doi.org/10.1111/nyas.13984

Timp JF, Braekkan SK, Versteeg HH, Cannegieter SC (2015) NETosis: how vital is it? Bryan. Blood 122:1–10. https://doi.org/10.1182/blood-2013-04

Tsuiji M (2003) Molecular cloning and characterization of a novel mouse macrophage C-type lectin, mMGL2, which has a distinct carbohydrate specificity from mMGL1. Trends Glycosci Glycotechnol 15:153–157. https://doi.org/10.4052/tigg.15.153

van Loo PF, Hangalapura BN, Thordardottir S et al (2019) MCLA-117, a CLEC12AxCD3 bispecific antibody targeting a leukaemic stem cell antigen, induces T cell-mediated AML blast lysis. Expert Opin Biol Ther 19:721–733. https://doi.org/10.1080/14712598.2019.1623200

van Vliet SJ, van Liempt E, Saeland E et al (2005) Carbohydrate profiling reveals a distinctive role for the C-type lectin MGL in the recognition of helminth parasites and tumor antigens by dendritic cells. Int Immunol 17:661–669. https://doi.org/10.1093/intimm/dxh246

van Vliet SJ, Gringhuis SI, Geijtenbeek TBH, van Kooyk Y (2006) Regulation of effector T cells by antigen-presenting cells via interaction of the C-type lectin MGL with CD45. Nat Immunol 7:1200–1208. https://doi.org/10.1038/ni1390

van Vliet SJ, Aarnoudse CA, Broks-van den Nerg VCM et al (2007) MGL-mediated internalization and antigen presentation by dendritic cells: a role for tyrosine-5. Eur J Immunol 37:2075–2081. https://doi.org/10.1002/eji.200636838

Vercammen D, Beyaert R, Denecker G et al (1998) Inhibition of caspases increases the sensitivity of L929 cells to necrosis mediated by tumor necrosis factor. J Exp Med 187:1477–1485. https://doi.org/10.1084/jem.187.9.1477

Wang Y, Gao W, Shi X et al (2017) Chemotherapy drugs induce pyroptosis through caspase-3 cleavage of a gasdermin. Nature 547:99–103

Wang Y-Y, Liu X-L, Zhao R (2019) Induction of pyroptosis and its implications in Cancer Management. Front Oncol 9:1–10. https://doi.org/10.3389/fonc.2019.00971

Weiskopf K, Schnorr PJ, Pang WW et al (2016) Myeloid cell origins, differentiation, and clinical implications. Microbiol Spectr 4. https://doi.org/10.1128/microbiolspec.mchd-0031-2016

Werninghaus K, Babiak A, Gross O et al (2009) Adjuvanticity of a synthetic cord factor analogue for subunit Mycobacterium tuberculosis vaccination requires FcRgamma-Syk-Card9-dependent innate immune activation. J Exp Med 206:89–97. https://doi.org/10.1084/jem.20081445

Williams BA, Law A, Hunyadkurti J et al (2019) Antibody therapies for acute myeloid leukemia: unconjugated, toxin-conjugated, radio-conjugated and multivalent formats. J Clin Med 8:1261. https://doi.org/10.3390/jcm8081261

Wu Z, Sawamura T, Kurdowska AK et al (2011) Lox-1 deletion improves neutrophil responses, enhances bacterial clearance, and reduces lung injury in a murine polymicrobial sepsis model. Infect Immun 79:2865–2870. https://doi.org/10.1128/IAI.01317-10

Yamasaki S, Ishikawa E, Sakuma M et al (2008) Mincle is an ITAM-coupled activating receptor that senses damaged cells. Nat Immunol 9:1179–1188. https://doi.org/10.1038/ni.1651

Yamasaki S, Matsumoto M, Takeuchi O et al (2009) C-type lectin Mincle is an activating receptor for pathogenic fungus, Malassezia. Proc Natl Acad Sci USA 106:1897–1902. https://doi.org/10.1073/pnas.0805177106

Yatim N, Jusforgues-Saklani H, Orozco S et al (2015) RIPK1 and NF-κB signaling in dying cells determines cross-priming of $CD8^+$ T cells. Science (80-) 350:328–334

Yatim N, Cullen S, Albert ML (2017) Dying cells actively regulate adaptive immune responses. Nat Rev Immunol 17:262–275

Yoshida H, Kondratenko N, Green S et al (1998) Identification of the lectin-like receptor for oxidized low-density lipoprotein in human macrophages and its potential role as a scavenger receptor. Biochem J 334:9–13. https://doi.org/10.1042/bj3340009

Yousefi S, Stojkov D, Germic N et al (2019) Untangling "NETosis" from NETs. Eur J Immunol 49:221–227. https://doi.org/10.1002/eji.201747053

Yuita H, Tsuiji M, Tajika Y et al (2005) Retardation of removal of radiation-induced apoptotic cells in developing neural tubes in macrophage galactose-type C-type lectin-1-deficient mouse embryos. Glycobiology 15:1368–1375. https://doi.org/10.1093/glycob/cwj028

Zelenay S, Keller AM, Whitney PG et al (2012) The dendritic cell receptor DNGR-1 controls endocytic handling of necrotic cell antigens to favor cross-priming of CTLs in virus-infected mice. J Clin Invest 122:1615–1627. https://doi.org/10.1172/JCI60644.viral

Zelensky AN, Gready JE (2005) The C-type lectin-like domain superfamily. FEBS J 272:6179–6217. https://doi.org/10.1111/j.1742-4658.2005.05031.x

Zhang SS, Chae GP, Zhang P et al (2008) Plasminogen activator Pla of Yersinia pestis utilizes murine DEC-205 (CD205) as a receptor to promote dissemination. J Biol Chem 283:31511–31521. https://doi.org/10.1074/jbc.M804646200

Zhang J-GG, Czabotar PE, Policheni AN et al (2012) The dendritic cell receptor Clec9A binds damaged cells via exposed actin filaments. Immunity 36:646–657. https://doi.org/10.1016/j.immuni.2012.03.009

Zhou H, Yu M, Zhao J et al (2016) IRAKM-Mincle axis links cell death to inflammation: Pathophysiological implications for chronic alcoholic liver disease. Hepatology 64:1978–1993. https://doi.org/10.1002/hep.28811

Zhu H, Lee C, Zhang D et al (2013) Surface-associated GroEL facilitates the adhesion of Escherichia coli to macrophages through lectin-like oxidized low-density lipoprotein receptor-1. Microbes Infect 15:172–180. https://doi.org/10.1016/j.micinf.2012.10.001

Zhu H, Fang X, Zhang D et al (2016) Membrane-bound heat shock proteins facilitate the uptake of dying cells and cross-presentation of cellular antigen. Apoptosis 21:96–109. https://doi.org/10.1007/s10495-015-1187-0

Zilionis R, Engblom C, Pfirschke C et al (2019) Single-cell transcriptomics of human and mouse lung cancers reveals conserved myeloid populations across individuals and species. Immunity 50:1317–1334.e10. https://doi.org/10.1016/j.immuni.2019.03.009

Structural Aspects of Carbohydrate Recognition Mechanisms of C-Type Lectins

Masamichi Nagae and Yoshiki Yamaguchi

Contents

1 Introduction 148
2 C-Type Lectin Fold 150
3 Sugar Binding Motifs: EPN and QPD 150
3.1 Other Monosaccharide Binding Modes 152
4 Oligosaccharide Recognition 153
4.1 Oligosaccharide Recognition via EPN Motif 156
4.2 Oligosaccharide Recognition of QPD Motif-Containing C-Type Lectins 167
4.3 Oligosaccharide Recognition of C-Type Lectin-like Domains 169
4.4 Genetic Variants of C-Type Lectins 170
5 Functional Oligomerization of C-Type Lectin Domains 171
6 Conclusion and Future Perspective 172
References 172

Abstract Carbohydrate recognition is an essential function occurring in all living organisms. Lectins are carbohydrate-binding proteins and are classified into several families. In mammals, Ca^{2+}-dependent C-type lectins, such as β-galactoside-binding galectin and sialic acid-binding siglec, play crucial roles in the immune response and homeostasis. C-type lectins are abundant and diverse in animals. Their immunological activities include lymphocyte homing, pathogen recognition, and clearance of apoptotic bodies. C-type lectin domains are composed of 110–130 amino acid residues with highly conserved structural folds. Remarkably, individual lectins can accept a wide variety of sugar ligands and can distinguish subtle

M. Nagae (✉)
Department of Pharmaceutical Sciences, The University of Tokyo, Hongo 7-3-1, Bunkyo-Ku, Tokyo 113-0033, Japan
e-mail: mnagae@mol.f.u-tokyo.ac.jp

Y. Yamaguchi (✉)
Faculty of Pharmaceutical Sciences, Tohoku Medical and Pharmaceutical University, Miyagi 981-8558, Japan
e-mail: yyoshiki@tohoku-mpu.ac.jp

Current Topics in Microbiology and Immunology (2020) 429: 147–176
https://doi.org/10.1007/82_2019_181

Published Online: 29 November 2019

structural differences in closely related ligands. In addition, several C-type lectin-like proteins specifically bind to carbohydrate ligands in Ca^{2+}-independent ways. The accumulated 3D structural evidence clarifies the unexpected structural versatility of C-type lectins underlying the variety of ligand binding modes. In this issue, we focus on the structural aspects of carbohydrate recognition mechanisms of C-type lectins and C-type lectin-like proteins.

1 Introduction

Carbohydrates, most often linked to proteins or lipids, cover the surface of all living cells and are fundamental determinants of health and disease. Carbohydrate-binding proteins or lectins interact with glycans in specific ways, eliciting many important cellular responses including those involved in immune cell homeostasis. The Ca^{2+}-dependent or C-type lectins are the largest and most diverse lectin family found in animals (Zelensky and Gready 2005; Drickamer and Taylor 2015). Well-studied members include galectin and siglec in mammals. C-type lectins share highly homologous structural modules in their carbohydrate recognition domains. These domains are ubiquitously found both in soluble and membrane proteins. More than 100 human proteins contain at least one C-type lectin domain. The C-type lectin family is subdivided into 16 groups (Groups 1–16) based on phylogenetic relationships and domain architecture (Cummings and McEver 2015) (Fig. 1a). Most of these groups have a single C-type lectin domain. Exceptions with 8–10 C-type lectin domains in their polypeptides (e.g. macrophage mannose receptor and DEC-205) are found in Group 6. Also, many proteins have C-type lectin domains that lack the conserved Ca^{2+} binding site, designated as "C-type lectin-like domains," and thus do not always bind carbohydrate ligands (e.g., Group 5 in Fig. 1a). This has allowed for distinction between sugar-binding C-type lectins and the broader family of C-type lectin-like domains (Zelensky and Gready 2005).

C-type lectins accept a wide variety of carbohydrate and non-carbohydrate ligands such as lipids and proteins via their lectin or lectin-like domains. Some C-type lectins play specific roles in glycoconjugate recognition, with the aglycon moiety of the ligand often contributing to the interaction. From an immunological point of view, various C-type lectins work as pattern recognition receptors (PRRs) which recognize highly conserved specific molecular signatures called pathogen-associated molecular patterns (PAMPs) and damage-associated molecular patterns (DAMPs) that are crucial for discrimination of self from non-self (Varki 2017; Sancho and Reis e Sousa 2013). In order to understand the physiological roles of C-type lectins in detail, 3D structural information is essential. We describe our current knowledge of the carbohydrate recognition mechanism of C-type lectins at monosaccharide, oligosaccharide, and polysaccharide levels.

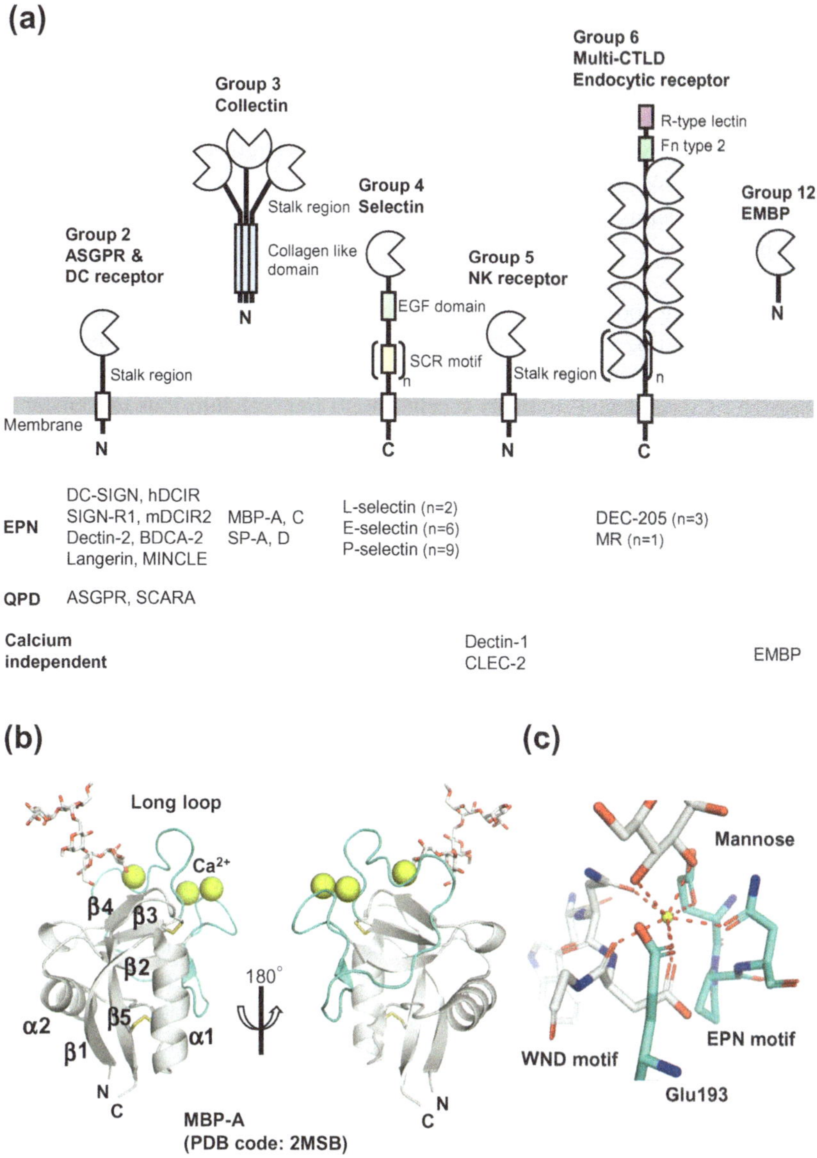

Fig. 1 **a** Domain architecture of representative C-type lectin-containing proteins described in this issue. This figure is prepared from the review (Zelensky and Gready 2005) with modifications. C-type lectins are classified into three sugar-binding motifs (EPN, QPD, and calcium independent). Fn type 2: Fibronectin type 2, SCR: Short consensus repeat, EGF: Epidermal Growth Factor, ASGPR: Asialoglycoprotein receptor, SRACLA: scavenger receptor C-type lectin. **b** Overall structure of C-type lectin domain (Rat mannose binding protein A (MBP-A), *PDB code* 2MSB). Polypeptide, carbohydrate and calcium ions are shown in ribbon, stick, and sphere models, respectively. The long loop region is colored in cyan. Disulfide bonds are shown in yellow stick model. **c** Close-up view of the primary binding site of MBP-A. The coordination and hydrogen bonds are depicted with red dotted lines

2 C-Type Lectin Fold

Since the pioneering work on the mannose-binding protein was reported (Weis et al. 1992), hundreds of atomic structures have been elucidated for C-type lectins. As of July 2019, over 250 atomic structures of C-type lectins were deposited in the Protein Data Bank. Overall fold and disulfide bond pattern are highly conserved among C-type lectins and C-type lectin-like domains. The C-type lectin domain is typically composed of 110–130 amino acid residues and the overall fold is formed by two α-helices and six or seven β-strands forming two antiparallel β-sheets (Fig. 1b). A long loop is found around the "top face" and inserted between two β-strands (β2 and β3 in typical cases). This loop is characteristic of C-type lectins, playing crucial roles in calcium coordination and sugar recognition (colored in cyan in Fig. 1b). C-type lectin domains accept one to four calcium ions (Zelensky and Gready 2005). The sugar-binding calcium ion is located at the top face of the C-type lectin domain. In the sugar-binding site, the calcium ion makes coordination bonds with both the lectin domain and bound monosaccharide (Fig. 1c). The other calcium ions mainly stabilize the 3D structure and occasionally form a secondary sugar-binding site.

3 Sugar Binding Motifs: EPN and QPD

A calcium ion forms multiple coordination bonds with amino acid residues that are well conserved among C-type lectins. The residues that define the sugar binding are called motifs. The asparagine and aspartate from the WND (Trp-Asn-Asp) motif and one carbonyl side chain (Glu side chain in Fig. 1c) form coordination bonds with Ca^{2+} in all C-type lectins. In addition, the EPN (Glu-Pro-Asn) or QPD (Gln-Pro-Asp) motif contributes to calcium coordination and forms the sugar-binding site. These motifs have two carbonyl groups separated by a proline residue. Two adjacent hydroxyl groups from a monosaccharide make coordination bonds with the calcium ion and hydrogen bonds with the EPN or QPD motif (Zelensky and Gready 2005). C-type lectin-like domains lack these conserved motifs and do not bind a calcium ion.

EPN and QPD motifs define the monosaccharide specificity of C-type lectins. Hence, C-type lectins can be classified into three groups: (i) EPN motif-containing C-type lectins, (ii) QPD motif-containing C-type lectins, and (iii) C-type lectin-like domains without these two motifs. EPN motif-containing C-type lectins usually accept D-mannose, D-glucose, L-fucose, and *N*-acetyl-D-glucosamine (GlcNAc) through equatorial 3-OH and 4-OH groups (left panel in Fig. 2a). In contrast, QPD motif-containing C-type lectins bind *N*-acetyl-D-galactosamine (GalNAc) and D-galactose through equatorial 3-OH and axial 4-OH groups (right panel in Fig. 2a). In both cases, bound monosaccharide is stabilized through hydrogen bonds with the carboxyl ($-COO^-$) and amide (–CONH) groups from these motifs. Importantly,

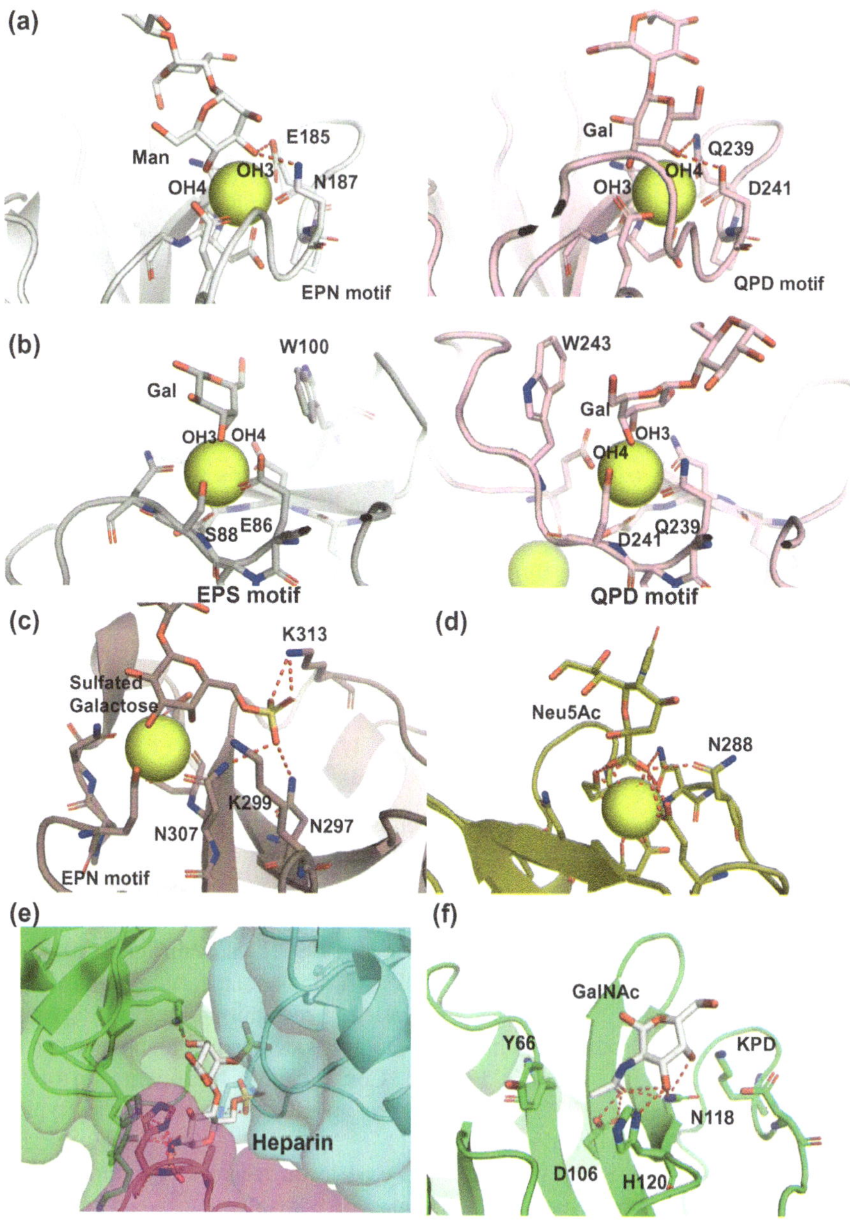

hydrogen donors and acceptors are switched in the two cases. In this way, the hydrogen bond network defines the position and orientation of the bound monosaccharide. In fact, replacing the EPN motif with QPD in mannose binding protein A (MBP-A) changes the binding ability to favor galactose. This result proves the role of these motifs in monosaccharide specificity (Drickamer 1992). In a

◂**Fig. 2** **a** Monosaccharide recognition by EPN (MBP-A, *PDB code* 2MSB, left panel) and QPD motifs (ASGPR, *PDB code* 5JPV, right panel). The amino acid residues which contribute to calcium ion coordination and sugar binding are shown in stick models. The positions of EPN and QPD motifs are indicated. **b** Galactose recognition by TC14 (*PDB code* 1TLG, left panel) and ASGPR (*PDB code* 5JPV, right panel). The tryptophan residues (W100 in TC14 and W243 in ASGPR) are shown in stick model and make stacking interactions with galactose rings. **c** Sulfated galactose recognition by langerin (*PDB code* 3P5I). The amino acid residues interacting with the sulfate group are highlighted. **d** Sialic acid recognition by murine SIGN-R1 (*PDB code* 4CAJ). Coordination and hydrogen bonds are depicted with red dotted lines. An additional hydrogen bond with asparagine (N288) is also shown. **e** Heparin disaccharide recognition by EMBP (*PDB code* 2BRS). One disaccharide unit (colored in white) is surrounded by three EMBP molecules (shown in semitransparent surface models, and colored in green, cyan, and magenta). Direct intermolecular hydrogen bonds are shown in red dotted lines. **f** Calcium-independent GalNAc recognition by SPL-2 (*PDB code* 6A7S). Direct intermolecular hydrogen bonds are shown in red dotted lines. Y66 contacts the GalNAc residue

ligand-free form, two water molecules occupy the positions which will be taken over by 3-OH and 4-OH groups of the binding sugar residue, forming an eight-coordinated calcium ion (Ng et al. 1996; Feinberg et al. 2000). The orientation of the sugar ring at the primary binding site is affected by the surrounding amino acid residues. Occasionally, the bound sugar occurs as a mixture of the two orientations in a single crystal structure (Ng et al. 2002).

3.1 Other Monosaccharide Binding Modes

Monosaccharide specificities of C-type lectins are generally defined by these two motifs, but there are several known exceptions. A tunicate lectin TC14 from *Polyandrocarpa misakiensis* has an EPS motif and a sea cucumber lectin CEL-IV from *Cuvumaria echinate* has the EPN motif. Contrary to the motif rule, they bind galactose at the primary binding site (Poget et al. 1999; Hatakeyama et al. 2011). In each structure, a tryptophan side chain stacks with the apolar face of the galactose ring (left panel in Fig. 2b). In the case of galactose recognition by the QPD motif, the sugar is typically stabilized with the tryptophan side chain, which is located on the opposite side (right panel in Fig. 2b). Consequently, the orientation of the galactose ring observed in TC14 and CEL-IV is inverted compared with the typical galactose-QPD motif interaction.

A second example is the mammalian C-type lectin receptor langerin which is expressed on Langerhans cells and mediates carbohydrate-dependent uptake of pathogens. Langerin has an EPN motif but, atypically, it accepts glycans with terminal 6-sulfated galactose. A crystal structure of langerin complexed with $6SO_4$-Galβ1-4GlcNAc shows that the galactose residue coordinates a calcium ion and the sulfate group forms salt bridges with two lysine residues located close to the primary binding site. This electrostatic interaction appears to compensate for the nonoptimal binding of galactose with the EPN motif (Fig. 2c) (Feinberg et al. 2011).

Another exception is the interaction between SIGN-R1 and sialic acid (Neu5Ac) (Silva-Martin et al. 2014). SIGN-R1, also known as CD209a, is a murine C-type lectin receptor with an EPN motif and is expressed in myeloid cells. A crystal structure of SIGN-R1 in complex with Neu5Ac shows that the carboxylic group, not the hydroxyl groups, of the sialic acid makes coordination bonds with the calcium ion of SIGN-R1 and hydrogen bonds with adjacent amino acid residues (Fig. 2d). One asparagine residue (N288) contributes to sialic acid binding, independent of calcium coordination.

C-type lectin-like domains without an EPN or QPD motif are predicted to lack typical sugar-binding ability. Nevertheless, several do bind carbohydrate ligands. For instance, the eosinophil major basic protein (EMBP) is a constituent of the eosinophil secondary granule and has a C-type lectin-like domain. Surface plasmon resonance assay demonstrated that EMBP directly binds to heparin and heparan sulfate, but not to hyaluronic acid (Swaminathan et al. 2005). A crystal structure of human EMBP in complex with heparin disaccharide has been reported (Swaminathan et al. 2005). The authors introduced the disaccharide ligand by soaking into ligand-free crystals. Three EMBP molecules in the crystal lattice contact one heparin disaccharide unit. The major contact site is located close to the primary binding site of typical C-type lectins and the bound disaccharide unit is mainly stabilized by electrostatic interactions and hydrogen bonds (Fig. 2e).

One more example is the bivalve lectins SPL-1 and SPL-2, which show high affinities for GlcNAc or GalNAc containing glycans. Intriguingly, RPD and KPD motifs are found in SPL-1 and SPL-2, instead of EPN or QPD. Crystal structures of SPL-2 in complex with GalNAc demonstrated that the sugar binds near the putative primary binding site. However, the interaction mode is different from typical sugar binding via an EPN or QPD motif (Unno et al. 2019). 3-OH and 4-OH groups of GalNAc make hydrogen bonds with the putative primary binding site of the C-type lectin, and the acetamido group is sandwiched by tyrosine and histidine side chains via a stacking interaction (Fig. 2f).

A shrimp C-type lectin MjGCTL has a QAP (Gln-Ala-Pro) motif which was predicted to not to have calcium binding ability. However, a recent study shows that the lectin has sugar-dependent hemocyte encapsulation activity (Alenton et al. 2017). The carbohydrate recognition mechanism of this C-type lectin-like domain could be different. Structural analysis of carbohydrate recognition mechanisms with novel motifs will expand our understanding of this lectin family.

4 Oligosaccharide Recognition

C-type lectins have individual specificities despite the high sequence and structural similarities. For instance, langerin binds a diverse range of carbohydrates including high-mannose-type glycan, fucosylated blood group antigens, and glycans with terminal 6-sulfated galactose. Meanwhile, a C-type lectin receptor DCAR specifically recognizes phosphatidylinositol mannoside (PIM), a mycobacteria glycolipid, promoting a Th1 response during infection (Toyonaga et al. 2016). A fundamental question is how highly conserved C-type lectins recognize various types of glycan ligands.

Monosaccharide binding preference is defined by an EPN or QPD motif, while the specificity toward oligo- and polysaccharides is determined by the secondary binding site which is located near the primary binding site. The amino acid residues coordinating calcium ions are highly conserved, while the amino acid residues located within ~15 Å from calcium ion are less so. In particular, the secondary binding site is often formed by amino acid residues on the top face, where three β-strands (β2–β4), the long loop region and a part of two α-helices are located. The specific residues in the secondary binding site usually contribute additional interactions, or conversely, work to discourage binding of certain ligands.

From a structural aspect, it is important to understand how each C-type lectin attains specificity for the target glycan. To get a broader picture, 3D structures of C-type lectin domains in complexes with oligosaccharide ligands were extracted from the PDB and these are summarized in Table 1. In this chapter, we describe the structural basis for the specific oligosaccharide recognition mechanism of C-type lectins.

Table 1 List of 3D structures of C-type lectins and C-type lectin-like domains complexed with glycan ligands. The glycan complex structures were extracted to include more than just the disaccharide units. Crystal structures of C-type lectins in ligand free forms and complexed with monosaccharide or non-carbohydrate ligands were omitted

Receptor	Source	Ligand	PDB ID	References
(i) EPN motif				
High-mannose-type N-glycan				
MBP-A	Rat	Man_5	2MSB	Weis et al. (1992)
MBP-A	Rat	Manα1-3Man	1KWY	Ng et al. (2002)
MBP-A (H189V)	Rat	Manα1-3Man	1KWZ	Ng et al. (2002)
MBP-A	Rat	$Man_6GlcNAc_2$	1KX1	Ng et al. (2002)
SP-D	Human	Manα1-2Man	3G83	Crouch et al. (2009)
SP-D (R343V)	Human	Manα1-2Man	3G84	Crouch et al. (2009)
DC-SIGN	Human	Man_4	1SL4	Guo et al. (2004)
DC-SIGN	Human	Man_6	2IT5	Feinberg et al. (2007)
DC-SIGN	Human	Man_2	2IT6	Feinberg et al. (2007)
DC-SIGN	Human	Pseudo-dimannose mimic	2XR5	Sutkeviciute et al. (2014)

(continued)

Table 1 (continued)

Receptor	Source	Ligand	PDB ID	References
DC-SIGN	Human	Pseudo-trimannose mimic	2XR6	Sutkeviciute et al. (2014)
Dectin-2	Human	Man_9	5VYB	Feinberg et al. (2017)
L-selectin	Human	$Man_5GlcNAc_2$	5VC1	Wedepohl et al. (2017)
L-selectin	Human	$Man_5GlcNAc_2$	3CFW	Mehta-D'souza et al. (2017)
Langerin	Human	Manα1-3Man (Man_5)	3P5D	Feinberg et al. (2011)
Langerin	Human	Manα1-2Man	3P5F	Feinberg et al. (2011)
Complex-type *N*-glycan				
DC-SIGN	Human	$GlcNAc_2Man_3$	1K9I	Feinberg et al. (2001)
DC-SIGNR	Human	$GlcNAc_2Man_3$	1K9 J	Feinberg et al. (2001)
hDCIR	Human	GlcNAcβ1-2Man	5B1X	Nagae et al. (2016)
mDCIR2	Mouse	Bisected glycan	3VYK	Nagae et al. (2013)
BDCA-2	Human	Galβ1-4GlcNAcβ1-2Man	4ZET	Jegouzo et al. (2015)
Codakine	Bivalve	Biantennary glycan	2VUZ	Gourdine et al. (2008)
SP-D	Porcine	Biantennary glycan	6BBE	van Eijk et al. (2018)
Fucose-containing glycan				
P-selectin	Human	PSGL-1	1G1S	Somers et al. (2000)
P-selectin	Human	Sialyl-Lewis^X (soaked)	1G1R	Somers et al. (2000)
E-selectin	Human	Sialyl-Lewis^X (soaked)	1G1T	Somers et al. (2000)
DC-SIGN	Human	Lacto-*N*-fucopentaose III	1SL5	Guo et al. (2004)
DC-SIGNR	Human	Lewis^X	1SL6	Guo et al. (2004)
MBP-A (K3)	Rat	3'-NeuAc-Le^X	2KMB	Ng and Weis (1997)
MBP-A (K3)	Rat	3'-sulfo-Le^X	3KMB	Ng and Weis (1997)
MBP-A (K3)	Rat	4'-sulfo-Le^X	4KMB	Ng and Weis (1997)
E-selectin	Human	Sialyl-Lewis^X	4CSY	Preston et al. (2016)
Langerin	Human	Galα1-3[Fucα1-2]Gal	3P5G	Feinberg et al. (2011)
Glucose-containing glycan				
SP-D	Human	Maltose	1PWB	Shrive et al. (2003)
Mincle	Cow	Trehalose (Glcα1-α1Glc)	4KZV	Feinberg et al. (2013)
Mincle	Cow	Trehalose monobutylate	4ZRV	Feinberg et al. (2016)

(continued)

Table 1 (continued)

Receptor	Source	Ligand	PDB ID	References
Mincle	Cow	Trehalose	4ZRW	Feinberg et al. (2016)
Mincle	Cow	KMJ1	5KTI	Feinberg et al. (2016)
Mincle	Cow	Brartemicin	5KTH	Feinberg et al. (2016)
Langerin	Human	Laminaritriose	3P5H	Feinberg et al. (2011)
Langerin	Human	Maltose (Glcα1-4Glc)	3P7H	Chatwell et al. (2008)
SIGN-R1	Mouse	Dextran sulfate	4C9F	Silva-Martin et al. (2014)
Others				
CEL-IV	Sea cucumber	Raffinose (Galα1-6Glcβ1-2Fru)	3ALU	(Hatakeyama et al. 2011)
CEL-IV	Sea cucumber	Melibiose (Galα1-6Glc)	3ALT	Hatakeyama et al. (2011)
Langerin	Human	$6SO_4$-Galβ1-4GlcNAc	3P5I	Feinberg et al. (2011)
Langerin	Human	GlcNAcβ1-3Galβ1-4Glc	4N33	Feinberg et al. (2013)
Langerin	Human (K313I)	GlcNAcβ1-3Galβ1-4Glc	4N35	Feinberg et al. (2013)
Langerin	Human (N288D/K313I)	GlcNAcβ1-3Galβ1-4GlcNAc	4N38	Feinberg et al. (2013)
(ii) QPD motif				
RSL	Rattlesnake	Lactose (Galβ1-4Glc)	1JZN	Walker et al. (2004)
SCARA4	Mouse	Lewis^X	2OX9	Feinberg et al. (2007)
ASGPR	Human	Lactose (Galβ1-4Glc)	5JPV	Sanhueza et al. (2017)
(iii) C-type lectin-like domains				
EMBP	Human	Heparin	2BRS	Swaminathan et al. (2005)
Dectin-1	Mouse	Glcβ1-3Glcβ1-3Glc	2CL8	Brown et al. (2007)
CLEC-2	Human	*O*-glycosylated podoplanin	3WSR	Nagae et al. (2014)

4.1 Oligosaccharide Recognition via EPN Motif

The structural basis for the specific oligosaccharide recognition of EPN motif-containing C-type lectins has been well studied. Here, we introduce the recognition mechanism of EPN motif-containing C-type lectins toward representative oligosaccharides such as high-mannose type, complex-type *N*-glycans, sialyl-Lewis^X, and glucose-containing glycans. The primary binding sites of corresponding C-type lectins bind mannose, fucose, and glucose residues of these oligosaccharides, and the secondary binding sites define their specificities.

4.1.1 Oligosaccharide Recognition of EPN Motif-Containing C-Type Lectins Through Mannose

(I) Recognition of high-mannose-type *N*-glycan through mannose

High-mannose-type *N*-glycans are often found in viral and fungal glycoproteins as well as in nascent mammalian glycoproteins. A representative structure of mammalian high-mannose-type *N*-glycan is shown as $Man_9GlcNAc_2$ which contains mannose residues of α1-2, α1-3, and α1-6 linkages (Fig. 3a). The Manα1-2 Man unit is a common terminal structure on mannans of yeast and other fungi. Therefore, this disaccharide unit can be a target for several C-type lectins working as pattern recognition receptors in the immune system. Crystal structures of 16 EPN motif-containing C-type lectins have been reported in complexes with high-mannose-type glycans (Table 1). Of these, mannose-binding protein A (MBP-A) and pulmonary surfactant protein D (SP-D) are soluble proteins categorized as the collectin family (Group 3). DC-specific intercellular adhesion molecule-3-grabbing nonintegrin (DC-SIGN), langerin, and DC-associated C-type lectin-2 (Dectin-2) are type II membrane proteins which belong to the NK receptors family (group 5). L-Selectin is an adhesion receptor of the selectin family (Group 4) (Fig. 1a). The interaction modes of these lectins with high-mannose-type glycans are visualized in Fig. 3b. C-type lectins accept α1-2, 1-3 and 1-6 linkages of the mannose disaccharide unit at their primary binding sites with several variations. In this section, the atomic recognition modes of high-mannose-type glycan are discussed for each glycosidic linkage.

(I-1) Manα1-2Man unit recognition

There are two known binding modes for the α1-2 linkage: (a) mannose at the non-reducing end (non-reducing mannose) bound to the primary site (SP-D wild type and MBP-A in Fig. 3b), and (b) mannose at the reducing end (reducing mannose) bound to the primary site (Dectin-2, SP-D R343 V mutant, DC-SIGN and langerin in Fig. 3b). The surrounding amino acid residues determine the choice of the binding modes. In SP-D, Arg313 is located close to the primary site and the replacement of this arginine with valine dramatically changes the disaccharide binding mode (Crouch et al. 2009). The R343 V mutant accepts reducing mannose at the primary binding site and forms additional hydrogen bonds with the non-reducing mannose via the secondary binding site (Fig. 3b). Compared with mode (a), the orientation of mannose in mode (b) is flipped in the primary binding sites of Dectin-2, SP-D R343 V and DC-SIGN. Dectin-2 mainly recognizes inner mannose residues of Man_7 glycan (Man-A and C) and the disaccharide recognition mode seems similar to that of DC-SIGN. However, in DC-SIGN-Man_2 complex,

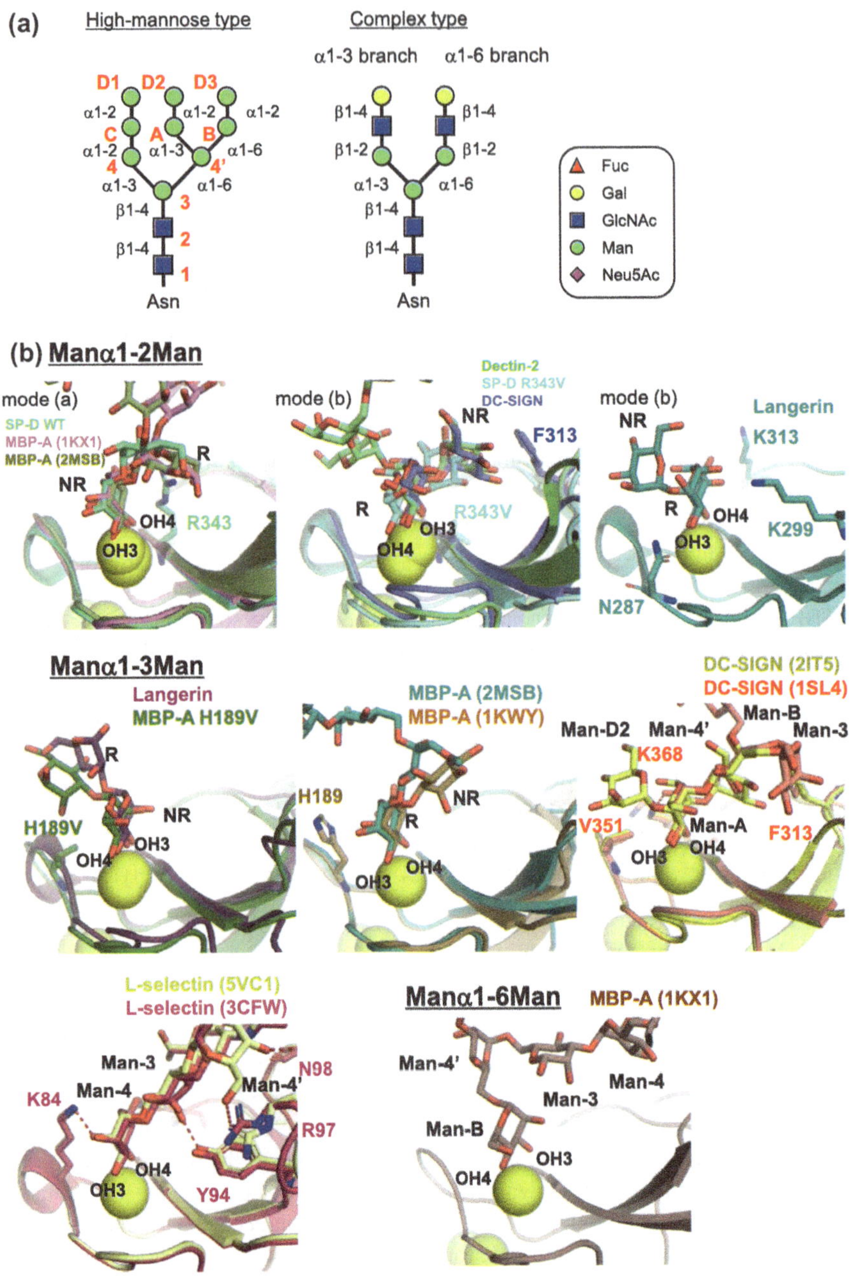
(a)
High-mannose type
Complex type
α1-3 branch
α1-6 branch
D1 D2 D3
C A B
4 4'
3
2
1
Asn
Fuc
Gal
GlcNAc
Man
Neu5Ac
(b) Manα1-2Man
mode (a)
SP-D WT
MBP-A (1KX1)
MBP-A (2MSB)
mode (b)
Dectin-2
SP-D R343V
DC-SIGN
Langerin
Manα1-3Man
Langerin
MBP-A H189V
MBP-A (2MSB)
MBP-A (1KWY)
DC-SIGN (2IT5)
DC-SIGN (1SL4)
L-selectin (5VC1)
L-selectin (3CFW)
Manα1-6Man
MBP-A (1KX1)

◂**Fig. 3** **a** A typical high-mannose-type glycan ($Man_9GlcNAc_2$, left panel) and biantennary complex-type glycan (right panel) linked to asparagine. Monosaccharide symbols follow the SNFG (symbol nomenclature for glycans) system (Varki et al. 2015). The glycosidic linkages and residue numbers are labeled at each position. **b** Summary of the interaction between high-mannose-type glycan and C-type lectins. Structures are classified based on the glycosidic linkages (Manα1-2Man, Manα1-3Man and Manα1-6Man) of the mannose residues. The positions of reducing and non-reducing mannose residues are labeled as "R" and "NR," respectively. The residue names are also labeled. **c** Three cross-linked structures of C-type lectin-high-mannose glycan complexes. Two MBP-A glycan complexes (*PDB code* 2MSB, left and *PDB code* 1KX1, middle panels) and one Dectin-2 complex (*PDB code* 5VYB, right panel) are shown. The mannose residues at the primary binding sites are labeled. **d** Complex-type *N*-glycan recognition by BDCA-2 (*PDB code* 4ZET, left panel), hDCIR (*PDB code* 5B1X, middle panel), and mDCIR2 (*PDB code* 3VYK, right panel). **e** The branch-specific interaction with complex-type *N*-glycan. All four C-type lectins (mDCIR2 (*PDB code* 3VYK), DC-SIGN (*PDB code* 1K9I), Codakine (*PDB code* 2VUZ), and SP-D (*PDB code* 6BBE)) accept mannose residues (Man-4) of the α1-3 branch at the primary binding sites. It should be noted that the relative position of the α1-6 branch of the mDCIR2 complex is different from the other C-type lectins. C-type lectin domains are shown in surface models

2-OH and 3-OH groups of the non-reducing mannose residue are adjacent to phenylalanine (F313 in Fig. 3b), which prevents further extension toward the non-reducing side. A structure of langerin with a Manα1-2Man oligosaccharide has also been described (Fig. 3b right panel), showing that the bound reducing mannose residue is flipped 180° compared with the other complexes in mode (b). Since the reducing mannose residue is clamped between asparagine (N287) and lysine (K299), the 1-OH group cannot be used for extension to a larger glycan. Another conformation has also been reported in which the non-reducing mannose resides in the primary binding site, though the electron density of reducing mannose is missing. Langerin preferentially binds the reducing mannose of the disaccharide, but rather binds the non-reducing end of whole high-mannose glycan.

(I-2) Manα1-3Man unit recognition

C-type lectins can bind the terminal and inner Manα1-3Man unit of *N*-glycan (Fig. 3b). Langerin and MBP-A H189 V mutant recognize the non-reducing mannose residue of Manα1-3Man unit. Langerin Manα1-3Man disaccharide unit complex was obtained by using core Man_5 oligosaccharide (Manα1-3[Manα1-3[Manα1-6]Manα1-6]Man). In this complex, non-reducing mannose tightly interacts with the primary binding site, while the reducing ends are exposed to solvent (Fig. 3b). Wild-type MBP-A recognizes the reducing end of the Manα1-3Man disaccharide unit, and however, introduction of an H189 V mutation causes a change in the bound residue (non-reducing end) as well as a 180° flip of the bound ligand (Ng et al. 2002) (Fig. 3b). The position of this histidine in MBP-A is different from that of R343 in SP-D, indicating that the binding mode is determined by a marginal balance of the surrounding residues.

DC-SIGN recognizes inner α1-3 linked mannose residues. A DC-SIGN-Man_4 (Manα1-3[Manα1-6]Manα1-6Man) complex shows that all four mannose residues (corresponding to Man-A, Man-B, Man-4' and Man-3 in Fig. 3a) are uniquely

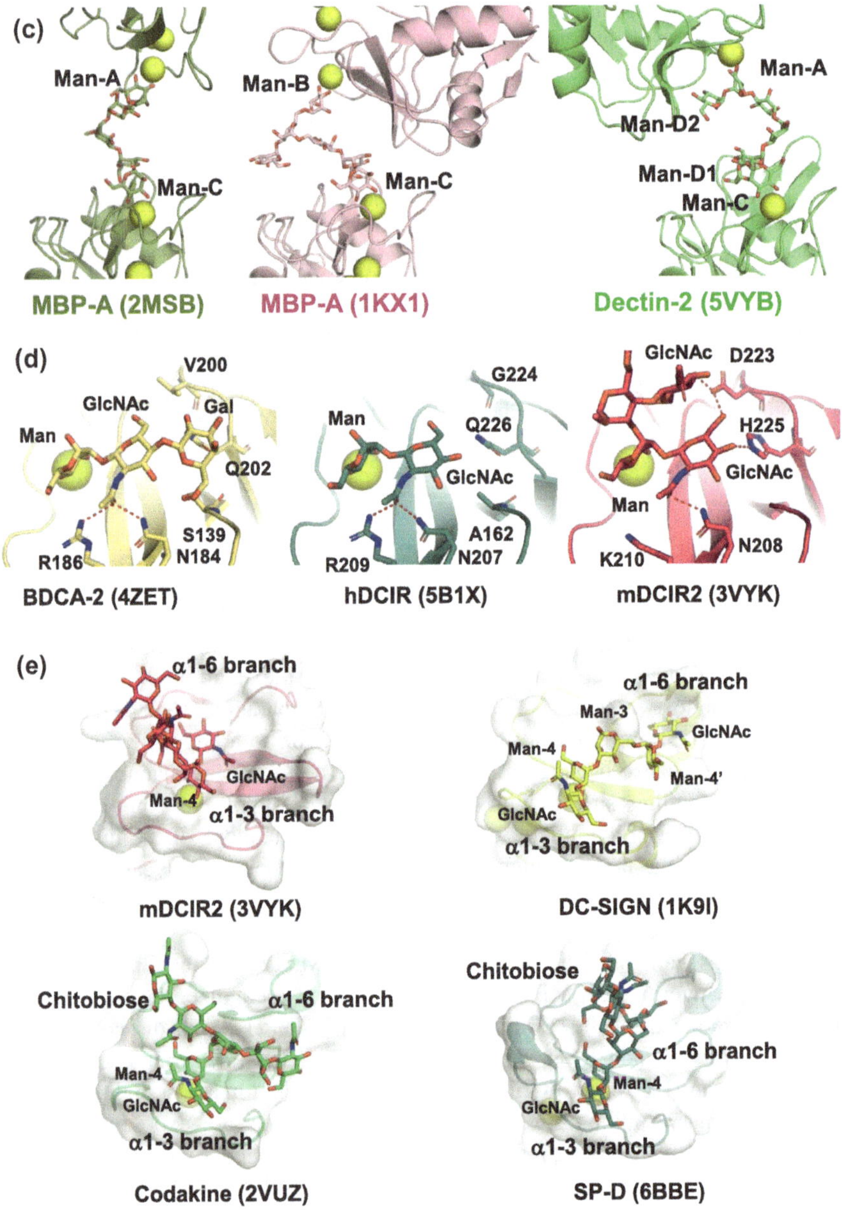

Fig. 3 (continued)

defined. In DC-SIGN-Man_6 (Manα1-2Manα1-3Man[Manα1-2Manα1-6]Man) complex, Manα1-2Manα1-3Man, which is a part of the D2 arm, interacts with DC-SIGN in a major conformation. In addition, only the terminal Manα1-2Man unit (Man-A and Man-D2 residues) binds in a minor conformation. The interaction mode of the major conformation is similar to that of the Man_4 complex and these two glycans are well superimposable (Fig. 3b). This observation indicates that DC-SIGN tightly associates with the inner tetrasaccharide unit (Manα1-2Manα1-3Manα1-6Man) of high-mannose-type glycan. The mannose residue (Man-A in Fig. 3b) coordinates with the calcium ion and both ends of the mannose residues are located at the secondary binding sites. The minor conformation of Man_6 complex is similar to the major conformation of Man_2 complex. Hence, DC-SIGN can recognize the inner α1-3 linked mannose residue as well as terminal α1-2 linked mannose of high mannose glycan. These multiple binding modes of DC-SIGN may enhance its apparent affinity towards glycoproteins carrying high-mannose glycan under physiological conditions.

L-selectin is a member of the selectin family and mediates cell adhesion and signaling in inflammation. A major physiological ligand of L-selectin is thought to be sialyl-$Lewis^X$ (sLe^X) as described in a later section. In addition to this ligand, L-selectin binds inner α1-3 linked mannose residues of high-mannose-type glycan. Interestingly, the binding mode is different from that of DC-SIGN. Crystal structures of *N*-glycosylated L-selectin were reported by two groups (Wedepohl et al. 2017; Mehta-D'souza et al. 2017). In these structures, L-selectin tightly binds to the *N*-glycan ($Man_5GlcNAc_2$) from the symmetry related L-selectin molecule via the sugar-binding site. The mannose at α1-3 branch (Man-4) resides in the primary binding site and the adjacent sugar residues, GlcNAc, β-mannose (Man-3) and α1-6 branched mannose (Man-4'), interact with L-selectin via the secondary binding site (Fig. 3b).

(I-3) Manα1-6Man recognition

To date there is only one example showing the binding mode of α1-6 linked mannose. The 3D structure is obtained from an MBP-A-Man_5 (Manα1-2Manα1-3[Manα1-3Manα1-6]Man) complex (PDB code: 2MSB). Remarkably, one high-mannose -type glycan bridges two MBP-A molecules (discussed in the next section). The Man-B residue at the non-reducing end of the Manα1-6Man unit binds to MBP-A. However, there is no apparent interaction between MBP-A and the mannose at the reducing side (Fig. 3b).

(I-4) Cross-linking by high-mannose-type glycan

High-mannose-type glycan can function as a multivalent ligand for C-type lectins. Three cross-linked structures have been reported so far (Fig. 3c). Two out of three are MBP-A-high-mannose-type glycan complexes, while the other structure is Dectin-2-high-mannose-type glycan complex. Crystal structure of an MBP-A-Man_5 (Manα1-2Manα1-3[Manα1-3Manα1-6]Man) complex (PDB code: 2MSB) shows that one high-mannose-type glycan is simultaneously recognized by two MBP-A molecules (Weis et al. 1992). One MBP-A molecule binds to the Manα1-2Man unit of the α1-3 branch (especially the Man-C residue), and the other MBP-A molecule binds to the Manα1-3Man unit of the α1-6 branch (Man-A, left panel in Fig. 3c). In both molecules, non-reducing mannose residues occupy the primary binding sites. The other complex, MBP-A-$Man_6GlcNAc_2$ complex, shows that one high-mannose glycan bridges two MBP-A trimers (Ng et al. 2002). In this MBP-A complex, one MBP-A molecule interacts with the Manα1-2Man unit, and the other MBP-A molecule interacts with the Manα1-6Man units of the α1-6 branches (especially the Man-B residue, middle panel in Fig. 3c). Comparing the two MBP-A cross-linked structures, all four MBP-A molecules recognize Man-C residues at the non-reducing end.

The Dectin-2-high-mannose-type glycan complex differs from the MBP-A complexes in terms of glycan binding mode. A crystal structure of Dectin-2 in complex with Man_9 glycan shows that two Dectin-2 molecules sandwich one high-mannose-type glycan (Feinberg et al. 2017). One Dectin-2 recognizes the Manα1-2Man unit (Man-C and D1 residues) of the D1 arm, and the other Dectin-2 molecule interacts with the same disaccharide unit (Man-A and D2 residues) of the D2 arm (right panel in Fig. 3c). The interaction modes of the two Dectin-2 molecules are the same. The reducing end of the disaccharide resides in the primary binding site, while the non-reducing end is located at the secondary binding site. The recognition mode of inner mannose residues seems suitable for binding fungal mannans which have variable structures at their non-reducing ends.

(II) Complex-type *N*-glycan recognition via mannose

Complex-type *N*-glycan is synthesized from high-mannose-type glycan by a series of enzymatic processes (Fig. 3a). Several C-type lectin receptors encoded in the Dectin-2 cluster on the natural killer gene complex preferentially interact with complex-type *N*-glycans, such as blood DC antigen 2 (BDCA-2), human DC immunoreceptor (hDCIR), and murine DCIR2 (mDCIR2) (Fig. 1a). These C-type lectins share high amino acid sequence identities; however, their ligand preferences are slightly different.

BDCA-2 binds to galactose-terminated biantennary glycans, defining an epitope found on a limited number of bi- and triantennary glycans (Riboldi et al. 2011). Unusually, BDCA-2 with an EPN motif binds galactosylated glycan. A crystal structure shows why. BDCA-2 primarily recognizes a mannose residue of the trisaccharide unit (Galβ1-4GlcNAcβ1-2Man) with a serine residue (S139) and additional interactions define the specificity between BDCA-2 and Galβ1-4GlcNAc (Fig. 3d (Jegouzo et al. 2015)).

In contrast, human DCIR (hDCIR) binds to GlcNAc-terminated biantennary *N*-glycan. The binding mode of hDCIR toward the disaccharide unit (GlcNAcβ1-2Man) is similar to that of BDCA-2 [Fig. 3d, (Nagae et al. 2016)]. However, the serine residue (S139) interacting with galactose in BDCA-2 is not conserved in hDCIR (A162). This indicates that the specificity is defined by the additional interaction with the terminal glycan residue.

BDCA-2 and hDCIR can interact with both α1-3 and α1-6 arms of biantennary complex-type *N*-glycan. mDCIR2 has unique specificity toward GlcNAc-terminated biantennary glycan with bisecting GlcNAc (Nagae et al. 2013). mDCIR2 shows arm preference, and the galactosylation of the α1-3 branch strongly inhibits binding. A crystal structure of a mDCIR2-bisected glycan complex demonstrated that mDCIR2 strictly recognizes the disaccharide unit (GlcNAcβ1-2Man) of the α1-3 arm as well as bisecting GlcNAc (Fig. 3d). The binding mode of GlcNAcβ1-2Man is similar to those of BDCA-2 and hDCIR. However, aspartate (D223) tightly interacts with the bisecting GlcNAc (Fig. 3d), which is not conserved in BDCA-2 or hDCIR. Due to the simultaneous interaction with both GlcNAcβ1-2Man and bisecting GlcNAc, mDCIR2 selects the α1-3 branch. Such simultaneous interaction is impossible using the α1-6 branch because it is located slightly too far from the bisecting GlcNAc.

As described in the previous paragraph, DC-SIGN (and DC-SIGNR) preferentially interact with high-mannose-type glycans and 3D structures of DC-SIGN have been reported bound to high-mannose glycan. Additionally, DC-SIGN and DC-SIGNR bound to complex-type glycans have been reported (Feinberg et al. 2001). The interaction modes of these receptors are different from those of BDCA-2, hDCIR, and mDCIR2. The mannose residue of the α1-3 branch (Man-4) coordinates a calcium ion, but the orientation is flipped compared with the mDCIR2-bisected glycan complex (Fig. 3e). The GlcNAc residue of the α1-3 branch is located away from the secondary binding site due to this flipping and the α1-6 branch is located on the surface of DC-SIGN. The mannose (Man-4') and GlcNAc residues of the α1-6 branch form hydrogen bonds with DC-SIGN.

The bivalve lectin, codakine, from *Codakia orbicularis* binds a biantennary complex-type glycan (Gourdine et al. 2008). The binding mode of codakine is similar to that of DC-SIGN rather than mDCIR2 (Fig. 3e). The α1-3 branched mannose (Man-4) makes coordination bonds with a calcium ion and the α1-6 branch interacts with codakine via its secondary binding site.

A crystal structure of glycosylated porcine SP-D demonstrates that the sugar-binding site of SP-D accepts complex-type *N*-glycan attached on symmetry-related SP-D (van Eijk et al. 2018). The interaction mode is similar to those of DC-SIGN and codakine. The α1-6 branch is strongly kinked, possibly due to crystal packing (Fig. 3e).

4.1.2 Oligosaccharide Recognition of EPN Motif-Containing C-Type Lectins Through Fucose

C-type lectins primarily recognize mannose residues of high-mannose and complex-type *N*-glycans. In contrast, several C-type lectins such as selectins recognize the OH3 and OH4 groups of the fucose residue in sialyl-Lewis^X (sLe^X, Neu5Acα2-3Galβ1-4[Fucα1-3]GlcNAcβ1-R). sLe^X is a terminal component of *N*- and *O*-glycans on hematopoietic and endothelial cells. L-selectin prefers sLe^X modified with sulfate on the GlcNAc residue. Selectins are expressed on vascular endothelium, platelets, or leukocytes and bind to cell surface glycoproteins harboring sLe^X glycans such as PSGL-1. Upon ligand binding, selectins show the

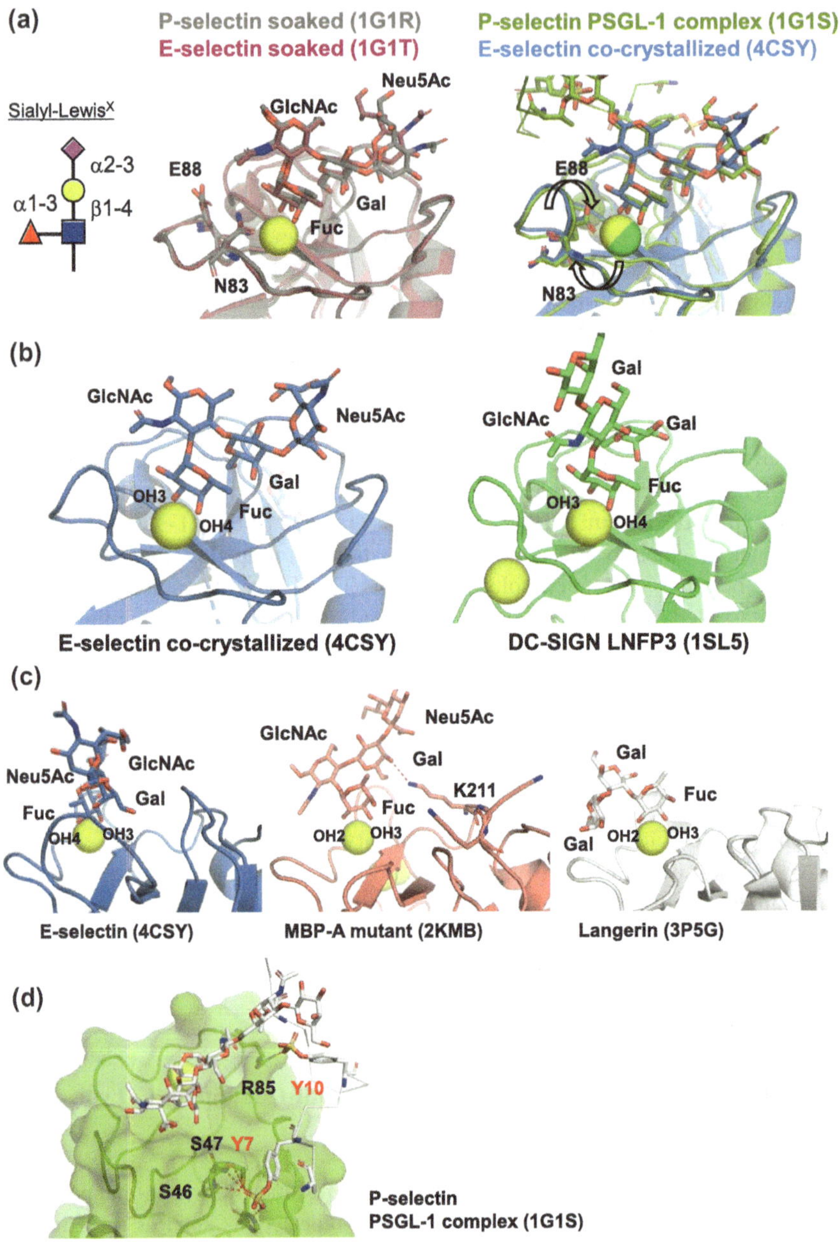

catch-bond behavior which is essential for initial tethering and rolling along the vascular endothelium and subsequent firm adhesion (Kansas 1996).

Crystal structures of E- and P-selectins complexed with sLe^X have been reported (Somers et al. 2000; Preston et al. 2016). These lectins recognize a fucose residue at

◂**Fig. 4** **a** Sialyl-Lewis^X recognition by P- and E-selectins. Bent conformations (low affinity state) of P-selectin (*PDB code* 1G1R) and E-selectin (*PDB code* 1G1T) are shown in left panel. Extend conformations (high affinity state) of P-selectin (*PDB code* 1G1S) and E-selectin (*PDB code* 4CSY) are shown in right panel. Two amino acid residues (N83 and E88) which take different positions in two conformations are shown in stick models. **b** Structural superposition of E-selectin-sialyl-Lewis^X complex (*PDB code* 4CSY, left panel) and DC-SIGN-LNFP3 complex (*PDB code* 1SL5, right panel). **c** Structural comparisons between E-selectin (*PDB code* 4CSY), MBP-A K3 mutant (*PDB code* 2KMB) and Langerin (*PDB code* 3P5G) in complexes with fucose-containing glycans. **d** Aglycon recognition observed in P-selectin-PSGL-1 complex (*PDB code* 1G1S). The interactions between P-selectin and sulfated tyrosine residues of PSGL-1 are highlighted and labeled. The P-selectin molecule is shown in surface model

the primary binding site and form additional interactions with adjacent residues (GlcNAc, Gal, and Neu5Ac). A structural comparison of four structures shows that the positions of fucose, GlcNAc, and Gal are well superimposable, while the position of the terminal Neu5Ac is variable (Fig. 4a). Interestingly, a 3D structural difference is observed between sLe^X co-crystallized and soaked complexes. In the soaked complexes, asparagine (N83) makes coordination bonds with the calcium ion (left panel in Fig. 4a). In contrast, glutamate (E88) makes coordination bonds with the calcium ion in co-crystallized complexes (right panel in Fig. 4a). This difference causes a positional shift of the flexible loop, leading to a global conformational change from bend (low affinity) to extend (high affinity) forms. The glutamate is therefore a key residue in stabilizing the high affinity conformation (Mehta-D'souza et al. 2017).

The structure of L-selectin in complex with high-mannose-type glycan is similar to the co-crystallized complex, rather than the sLe^X soaked complex, even though it assumes a bend conformation (Fig. 3b). The glutamate coordinates with the calcium ion and the flexible loop takes a similar position as in the co-crystallized complex.

Other C-type lectins, such as DC-SIGN, MBP-A mutant, and langerin, also bind to fucose-containing glycans. A crystal structure of DC-SIGN in complex with lacto-*N*-fucopentaose III (Galβ1-4[Fucα1-3]GlcNAcβ1-3Galβ1-4Glc) shows that the position of the fucose coincides well with those of selectins. However, the positions of GlcNAc and galactose are slightly different. This subtle difference may be derived from differences in the secondary binding site of DC-SIGN (Fig. 4b).

The introduction of a triple mutation (K211-K212-K213) in MBP-A (K3 mutant) enables it to accept a series of Lewis^X glycans as P- and E-selectins do (Ng and Weis 1997). However, the binding mode is quite different from those of selectins and DC-SIGN (middle panel in Fig. 4c). In this case, 2-OH (equatorial) and 3-OH (equatorial) groups of the fucose make coordination bonds with the calcium ion. One of three lysine residues (K211) makes a hydrogen bond with the galactose residue.

The fucose recognition mode of langerin is also different from those of selectins and DC-SIGN (Feinberg et al. 2011). In the structure of langerin-blood group B trisaccharide (Galα1-3[Fucα1-2]Gal) complex, 2-OH and 3-OH groups of the fucose residue also coordinate with the calcium ion (right panel in Fig. 4c).

The aglycon moiety of PSGL-1 contributes to the specific interaction with selectins. Of particular note, the physiological interaction between human P-selectin and PSGL-1 requires both sLe^X capped core 2 *O*-glycan and one or more sulfated tyrosine residues in the N-terminal region of PSGL-1. In the crystal structure (Somers et al. 2000), human P-selectin recognizes both the sLe^X attached to threonine and the sulfated tyrosine residues of human PSGL-1 (Fig. 4d). P-selectin also binds to the N-terminus of murine PSGL-1, although the sequence is different from human PSGL-1. Cell-based biochemical assays suggest that sulfation of tyrosine (Y13) and the *O*-glycan on T17 are necessary for murine PSGL-1 to bind optimally to P-selectin (Xia et al. 2003). The spacing of these residues in the sequence is considerably closer than the corresponding residues in the human PSGL-1 sequence (Y7, Y10 and T16). It is likely that murine PSGL-1 binds to P-selectin using a different conformation of the polypeptide.

4.1.3 Oligosaccharide Recognition of EPN Motif-Containing C-Type Lectins Through Glucose

A representative glucose specific C-type lectin is Macrophage inducible calcium-dependent lectin (Mincle), also known as CLEC4E. Mincle is expressed on macrophages and interacts with trehalose-6-6′-dimycolate (TDM), a glycolipid found on the surface of *Mycobacterium tuberculosis* (Matsunaga and Moody 2009; Ishikawa et al. 2009). TDM comprises a trehalose (Glcα1-α1Glc) headgroup and two complex branched and hydroxylated acyl chains. The acyl chains are attached to the 6-OH groups of each of the sugar residues. Crystal structures of bovine Mincle C-type lectin domains in complexes with a series of ligands allow visualization of the interaction modes with glycolipids. A Mincle–trehalose complex structure shows that both glucose residues interact tightly with Mincle. The 3-OH and 4-OH of one glucose coordinate with the calcium ion and the second glucose contacts a second binding site via hydrogen bonds (Feinberg et al. 2013) (Fig. 5a). The additional interaction of the aglycon moiety, such as the lipid part, should greatly improve the specificity. The Mincle–trehalose monobutyrate complex shows that the alkyl chain is located near the hydrophobic groove of Mincle (Feinberg et al. 2016). These hydrophobic residues likely form the extended binding site for the lipid moiety.

In addition to Mincle, langerin, SP-D, and SIGN-R1 can bind glucose residues at their primary binding sites. The structure of a langerin-laminaritoriose (Glcβ1-3Glcβ1-3Glc) complex contrasts markedly with the Mincle–trehalose complex structure. In the langerin complex, the 1-OH and 2-OH of the glucose at the reducing end resides in the primary binding site and the other glucose at the non-reducing end points away towards solvent (Feinberg et al. 2011) (Fig. 5b). The langerin–maltose (Glcα1-4Glc) complex reveals that the 3-OH and 4-OH of glucose at the reducing end also coordinate with the calcium ion, but the orientation is totally flipped compared with that in the Mincle–treharose complex (Chatwell et al. 2008). The glucose at the non-reducing end is also exposed to solvent (Fig. 5c).

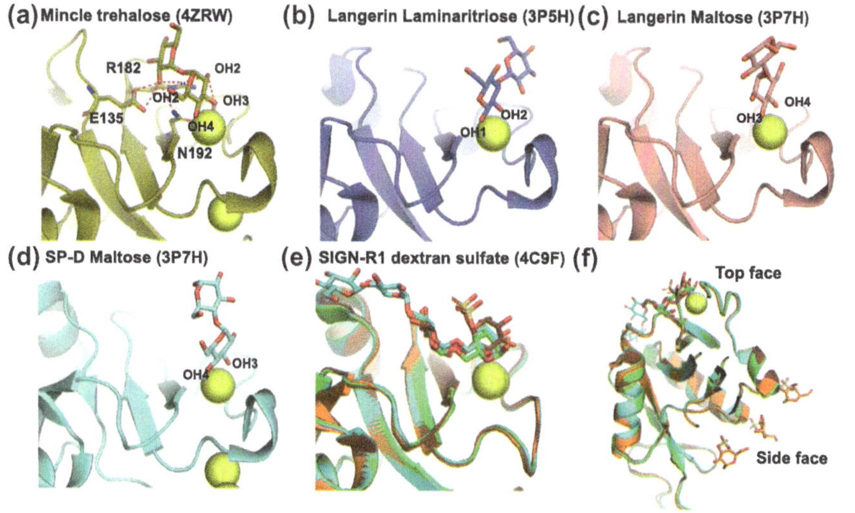

Fig. 5 Glucose recognition mechanism of EPN motif-containing C-type lectins: **a** Mincle-trehalose complex (*PDB code* 4ZRW). **b** Langerin-laminaritriose complex (*PDB code* 3P5H). **c** Langerin-maltose complex (*PDB code* 3P7H). **d** SP-D-maltose complex (*PDB code* 3P7H). **e** Dextran sulfate recognition of SIGN-R1 (*PDB code* 4C9F). Four complexes in the asymmetric unit are superimposed. **f** Two sugar-binding sites (top face and side face) of SIGN-R1 (*PDB code* 4C9F)

In contrast, the orientation of glucose in a SP-D-maltose complex is the same as in the Mincle complex (Fig. 5d). A crystal structure of SIGN-R1 in complex with oligo-dextran sulfate (α1-3 and α1-6 linked glucose polymer with sulfation) demonstrates a somewhat unusual binding mode (Silva-Martin et al. 2014). The glucose at the primary binding site is positioned differently compared with typical C-type lectins. Although the top face of the lectin accepts at least four glucose residues, only 4-OH of Glcα1-6Glc is located within the coordination bond in the primary binding site (Fig. 5e). In the case of sulfated glucose, ring oxygen (O5) seems to make a coordination bond with the calcium ion. It is noteworthy that "side" face of SIGN-R1 can accept the repetitive molecular patterns of the polysaccharide chain (Fig. 5f). This binding mode seems favorable for an interaction of a small globular domain with long polysaccharide chains.

4.2 Oligosaccharide Recognition of QPD Motif-Containing C-Type Lectins

QPD motif-containing C-type lectins have been less studied than those with EPN. Only three structures, Rattlesnake venom lectin (RSL) in complex with lactose (Walker et al. 2004), asialoglycoprotein receptor (ASGPR) lactose complex

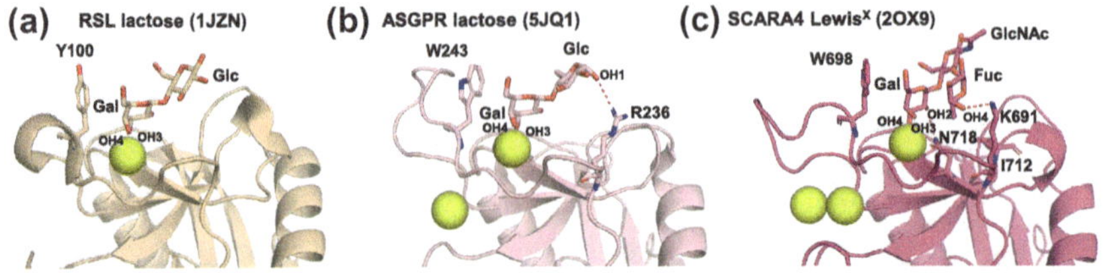

Fig. 6 Carbohydrate recognition of QPD motif-containing C-type lectins: **a** RSL-lactose complex (*PDB code* 1JZN). **b** ASGPR-lactose complex (*PDB code* 5JQ1) and **c** SCARA4-Lewis^{X} complex (*PDB code* 2OX9)

(Sanhueza et al. 2017) and scavenger receptor C-type lectin (SRACLA4) in complex with Lewis^{X} trisaccharide (Fucα1-3[Galβ1-4]GlcNAc) (Feinberg et al. 2007), are deposited in the PDB. In these structures, the 3-OH and 4-OH groups of galactose form coordination bonds with the calcium ion and the apolar face of the galactose is stabilized by a stacking interaction with hydrophobic amino acid residues (Fig. 6a–c). RSL recognizes only the galactose residue of lactose. In contrast, the βOH1 group of the glucose residue makes a hydrogen bond with arginine (R236) in an ASGPR-lactose complex (Fig. 6a and b). In the SRACLA4-Lewis^{X} complex, the fucose residue makes additional hydrophobic contact with isoleucine (I712) and hydrogen bonds with lysine (K691) (Fig. 6c). A structural comparison among the three complexes clarifies that the aromatic residues (Y100 in RSL, W243 in ASGPR, and W698 in SRACLA4) are located in equivalent positions and positively charged residues (R236 in ASGPR and K691 in SRACLA4) occupy similar positions in two of the complexes, evidently engaged in ligand recognition.

It is tempting to compare the recognition modes of Lewis^{X} by the EPN and QPD motifs. In EPN motif-containing C-type lectins, fucose coordinates a calcium ion in the primary binding site and galactose resides in the secondary binding site (Fig. 4a, b). In the case of the QPD motif, by contrast, galactose resides in the primary binding site and fucose is located in the secondary binding site. Interestingly, the conformation of the Lewis^{X} trisaccharide is similar in both complexes. This observation suggests that both lectins recognize a stable conformation of the glycan. Glycan array analysis revealed that SRCL preferentially binds to Lewis^{a} and Lewis^{X}, while DC-SIGN widely accepts various types of glycan such as Lewis^{a}, Lewis^{b}, Lewis^{X}, and Lewis^{Y} (Feinberg et al. 2007; Guo et al. 2004). This may originate from differences in the binding modes of galactose/fucose.

4.3 Oligosaccharide Recognition of C-Type Lectin-like Domains

C-type lectin-like domains do not bind calcium due to the lack of conserved calcium binding motifs. However, some C-type lectin-like domains directly bind to carbohydrates in calcium independent ways. The sugar-binding modes of C-type lectin-like domains are thus expected to be completely different from those of typical C-type lectin domains and can be expected to be diverse.

Dectin-1 is a C-type lectin-like receptor having a single extracellular C-type lectin-like domain, a short stalk region, a single transmembrane helix and a cytoplasmic ITAM. Dectin-1 is a β-glucan receptor and shows preference for the β1-3 linked glucose polymer (Palma et al. 2006). Since the C-type lectin-like domain of Dectin-1 lacks a QPD or EPN motif, it loses its calcium binding ability at the primary site. A crystal structure of Dectin-1 complexed with laminaritriose shows that two C-type lectin domains sandwich one laminaritriose via their lateral faces (Brown et al. 2007) (Fig. 7a). Laminaritiose assumes a planar conformation and is stabilized by several hydrogen bonds. It should be noted that mutational experiments suggested that the top face of the Dectin-1 C-type lectin-like domain contributes to β-glucan binding, which is located far from the binding site observed in the crystal structure (Dulal et al. 2018). Direct evidence, as for example solution NMR analysis, is strongly needed to solve this discrepancy.

Another example is C-type lectin like receptor 2 (CLEC-2), also known as CLEC1B. CLEC-2 is a type II transmembrane receptor with a short N-terminal cytoplasmic tail containing a single tyrosine-based activation motif (hemITAM), a transmembrane segment, an extracellular stalk region and a C-type lectin-like domain which has no calcium binding activity. Podoplanin is a transmembrane *O*-glycoprotein that binds to CLEC-2 in a glycosylation-dependent manner.

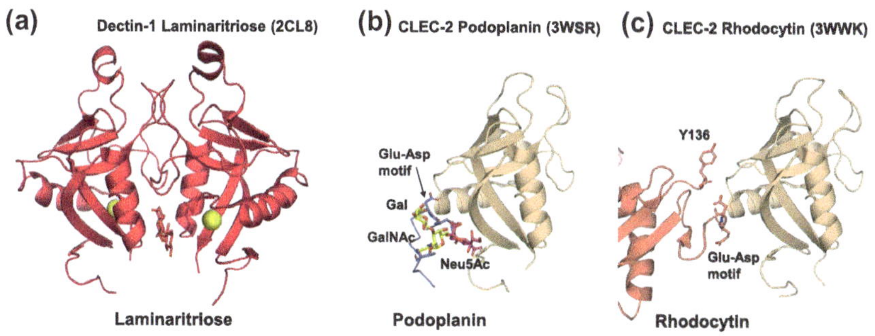

Fig. 7 Ca^{2+}-independent carbohydrate recognition of C-type lectin-like domain: **a** Dectin-1-laminaritoriose complex (*PDB code* 2CL8). **b** CLEC-2-*O*-glycosylated (Neu5Acα2-6[Galβ1-3] GalNAc-*O*-Thr52) podoplanin complex (*PDB code* 3WSR) and **c** CLEC-2-rhodocytin complex (*PDB code* 3WWK). Carbohydrate, Glu-Asp motif, and C-terminal Y136 are shown in stick models

Crystallographic analysis revealed that two consecutive acidic residues (Glu-Asp motif) as well as an α2-6 linked sialic acid residue attached on the *O*-glycan interact with the lateral face of CLEC-2 (Nagae et al. 2014). Four arginine residues of CLEC-2 interact with the sialic acid and the two acidic residues of podoplanin (Fig. 7b). Interestingly, snake venom rhodocytin also binds to CLEC-2, however the interaction mode is somewhat different. Although the Glu-Asp motif is conserved in rhodocytin, rhodocytin is not *O*-glycosylated. Instead, the carboxylate of the rhodocytin C-terminus contributes to an electrostatic interaction with the arginine residue of CLEC-2 (Fig. 7c).

4.4 Genetic Variants of C-Type Lectins

Specific sugar recognition is affected by changes in amino acid residues. Single-nucleotide polymorphisms (SNPs) and genetic variations are potentially involved in the susceptibility for developing disease and disease outcomes. Various SNPs and disease causative mutations of C-type lectins have been reported (Goyal et al. 2016). The effects of SNPs and species-dependent sugar binding are not well studied and are limited to a few examples. In human langerin, the W264R mutation was found in an individual who lacks Birbeck granules in the Langerhans cells (Verdijk et al. 2005). This mutation results in the loss of mannose-binding ability (Ward et al. 2006). W264 is located inside the protein, and thus this mutation destabilizes the local folding of langerin. At present several SNPs, A278 V, N288D, A300P, and K313I, in the human langerin C-type lectin domain are deposited in the SNP database (Ward et al. 2006; Feinberg et al. 2013). Of note, K313 is one of the two lysine residues which are critical for recognition of sulfated galactose (Fig. 2c). The affinity for sulfated galactose is dramatically reduced, instead, the affinity against terminal GlcNAc is increased (Feinberg et al. 2013).

Surfactant protein A (SP-A) consists of two isoforms, SP-A1 and SP-A2, encoded by separate genes. SP-A defends against invading pathogens in the lung. The missense mutations of SP-A2 C-type lectin domain result in idiopathic pulmonary fibrosis (IPF), a serious lung disease affecting older adults (Wang et al. 2009). These mutations, like F198S and G231 V, dramatically reduce the expression level of the protein probably due to the disruption of normal protein folding. In addition, the K223Q mutation in the C-type lectin domain is one of the SNPs which are significantly associated with tuberculosis (Malik et al. 2006). This replacement could affect the glycan recognition of SP-A2, though as yet the mechanism has not been revealed.

For SP-D, three SNPs, M11T, A160T, and S270T, have been reported (Leth-Larsen et al. 2005). These residues are located in the signal peptides of collagen-like and C-type lectin domains. The S270T mutant is located on the opposite side of the sugar binding site and the effect of the mutation on sugar binding is obscure. Another example is Dectin-1. The Y238X mutation of Dectin-1 is found in Africans and western Eurasians (Ferwerda et al. 2009). The Y238X mutant is poorly expressed and lacks β-glucan activity. Y236 in mouse Dectin-1, corresponding to human Y238, is buried inside the C-type lectin domain. Thus, the mutation likely inhibits proper folding. Although the binding of β-glucan is significantly lower in Y238X patients, fungal phagocytosis and fungal killing are normal, suggesting the presence of alternative receptors for phagocytosis.

The species-dependent sugar recognition mechanism has also been investigated. Human and murine langerin share 66% amino acid similarity, but show different ligand preferences for bacterial polysaccharides (Hanske et al. 2017). A crystal structure of murine langerin shows that the different residues map to the secondary binding site, which possibly accounts for the different ligand specificities.

5 Functional Oligomerization of C-Type Lectin Domains

C-type lectin and lectin-like domains often form stable homo- or heterodimers that are critical for their physiological functions. For example, C-type lectin-like domains of NK receptors (Group 5) such as Ly49s and NKGs accept various ligands via homo or heterodimers (Li and Mariuzza 2014). The C-type lectin domain itself forms a monomer in solution. However, many sugar-binding C-type lectins such as collectin and DC receptor families form multimers via their stalk or coiled-coil regions. Biochemical studies of murine Dectin-1 suggests that monomeric C-type lectin-like domain cooperatively forms an oligomer upon β-glucan binding (Dulal et al. 2018).

Domain swapping is a mechanism for two or more protein molecules to form a dimer or higher oligomer by exchanging an identical structural element (Liu and Eisenberg 2002). Oligomerization by domain swapping is often found in snake venom C-type lectin-like proteins (Eble 2019). In these proteins, the long loop between β2 and β3 strands extends away from the core of the protein to form a domain-swapped dimer. Various snake venom C-type lectin-like proteins form ordered heterooligomers via domain-swapping. Interestingly, domain-swapped heterodimers, bitiscetin, botrocetin, and rhodocetin, functionally grab their target platelet receptors such as von Willebrand factor (vWF) A1 domain, GPIbα, GPVI, and α2β1 integrin via an extended surface formed by an extended loop region (Maita et al. 2003; Fukuda et al. 2005; Eble et al. 2017).

In mammalian C-type lectin receptors, domain-swapped dimers can form under crystallization conditions. Crystal structures of several C-type lectin receptors are reported as domain-swapped dimers in a ligand-free state, such as CRD-4 of macrophage mannose receptor (Feinberg et al. 2000), BDCA-2 (Nagae et al. 2014), and NKRp1a (Kolenko et al. 2011). These domain-swapped dimers are similar to those of snake venom C-type lectin-like proteins. An interesting point is that the functional states of BDCA-2 and NKRp1a are monomeric and the domain-swapped dimers are thought to be an inactive state.

Of note, the BDCA-2 C-type lectin domain forms a domain-swapped dimer under three different crystallization conditions (Nagae et al. 2014). Although 1 mM calcium chloride was present during protein purification, no calcium ion was found in the calcium binding site. The side chain of E178, which is critical for calcium coordination, points away from the putative calcium binding site. This indicates that the domain-swapped BDCA-2 cannot accept a carbohydrate ligand. Subsequently, a crystal structure of BDCA-2 in complex with trisaccharide was reported. BDCA-2 is a monomer in the crystal and binds the sugar at the calcium binding site (Jegouzo et al. 2015).

In the case of NKRp1a, the extended loop points away from the central core and mediates formation of a domain-swapped dimer in the crystal (Kolenko et al. 2011). Although the refolding buffer contains a high concentration of calcium chloride, there is no calcium ion in the structure. In contrast, a solution structure determined by NMR is monomeric with the loop tightly anchored to the central region (Rozbesky et al. 2016). Calcium titration analysis suggests monomeric NKRp1a binds a calcium ion weakly with a dissociation constant in the mM range.

Although the physiological relevance of domain-swapped dimer formation is still obscure, the domain-swapped dimer is likely a metastable conformation. Formation of the domain-swapped dimer may be used as a temporary inactive state under certain conditions.

6 Conclusion and Future Perspective

C-type lectins accept various glycans via highly conserved binding sites. Although the primary binding site is the most conserved, there are differences in the secondary binding site that enable various types of glycans to bind. Furthermore, accumulating of 3D structural information demonstrates that C-type lectin-like domains also recognize sugar ligands in a calcium-independent manner. The relationship between SNP and sugar-binding activity has a direct bearing on the susceptibility for pathogens, and however, the mechanism is largely unknown. Structural analysis of C-type lectins from individuals is a promising avenue to our understanding of individual variations in the immune system.

References

Alenton RR, Koiwai K, Miyaguchi K, Kondo H, Hirono I (2017) Pathogen recognition of a novel C-type lectin from *Marsupenaeus japonicus* reveals the divergent sugar-binding specificity of QAP motif. Sci Rep 7:45818

Brown J, O'Callaghan CA, Marshall AS, Gilbert RJ, Siebold C, Gordon S, Brown GD, Jones EY (2007) Structure of the fungal β-glucan-binding immune receptor dectin-1: implications for function. Protein Sci 16:1042–1052

Chatwell L, Holla A, Kaufer BB, Skerra A (2008) The carbohydrate recognition domain of Langerin reveals high structural similarity with the one of DC-SIGN but an additional, calcium-independent sugar-binding site. Mol Immunol 45:1981–1994

Crouch E, Hartshorn K, Horlacher T, McDonald B, Smith K, Cafarella T, Seaton B, Seeberger PH, Head J (2009) Recognition of mannosylated ligands and influenza A virus by human surfactant protein D: contributions of an extended site and residue 343. Biochemistry 48:3335–3345

Cummings RD, McEver RP (2015) C-type lectins. In: Varki A, Cummings RD, Esko JD, Stanley P, Hart GW, Aebi M, Darvill AG, Kinoshita T, Packer NH, Prestegard JH, Schnaar RL, Seeberger PH (eds) Essentials of glycobiology. Cold Spring Harbor, NY, pp 435–452

Drickamer K (1992) Engineering galactose-binding activity into a C-type mannose-binding protein. Nature 360:183–186

Drickamer K, Taylor ME (2015) Recent insights into structures and functions of C-type lectins in the immune system. Curr Opin Struct Biol 34:26–34

Dulal HP, Adachi Y, Ohno N, Yamaguchi Y (2018) β-glucan-induced cooperative oligomerization of Dectin-1 C-type lectin-like domain. Glycobiology 28:612–623

Eble JA (2019) Structurally robust and functionally highly versatile-C-type lectin (-related) proteins in snake venoms. Toxins (Basel) 11:136

Eble JA, McDougall M, Orriss GL, Niland S, Johanningmeier B, Pohlentz G, Meier M, Karrasch S, Estevao-Costa MI, Martins Lima A, Stetefeld J (2017) Dramatic and concerted conformational changes enable rhodocetin to block α2β1 integrin selectively. PLoS Biol 15: e2001492

Feinberg H, Torgersen D, Drickamer K, Weis WI (2000a) Mechanism of pH-dependent *N*-acetylgalactosamine binding by a functional mimic of the hepatocyte asialoglycoprotein receptor. J Biol Chem 275:35176–35184

Feinberg H, Park-Snyder S, Kolatkar AR, Heise CT, Taylor ME, Weis WI (2000b) Structure of a C-type carbohydrate recognition domain from the macrophage mannose receptor. J Biol Chem 275:21539–21548

Feinberg H, Mitchell DA, Drickamer K, Weis WI (2001) Structural basis for selective recognition of oligosaccharides by DC-SIGN and DC-SIGNR. Science 294:2163–2166

Feinberg H, Taylor ME, Weis WI (2007a) Scavenger receptor C-type lectin binds to the leukocyte cell surface glycan Lewis(x) by a novel mechanism. J Biol Chem 282:17250–17258

Feinberg H, Castelli R, Drickamer K, Seeberger PH, Weis WI (2007b) Multiple modes of binding enhance the affinity of DC-SIGN for high mannose *N*-linked glycans found on viral glycoproteins. J Biol Chem 282:4202–4209

Feinberg H, Taylor ME, Razi N, McBride R, Knirel YA, Graham SA, Drickamer K, Weis WI (2011) Structural basis for langerin recognition of diverse pathogen and mammalian glycans through a single binding site. J Mol Biol 405:1027–1039

Feinberg H, Jegouzo SA, Rowntree TJ, Guan Y, Brash MA, Taylor ME, Weis WI, Drickamer K (2013a) Mechanism for recognition of an unusual mycobacterial glycolipid by the macrophage receptor mincle. J Biol Chem 288:28457–28465

Feinberg H, Rowntree TJ, Tan SL, Drickamer K, Weis WI, Taylor ME (2013b) Common polymorphisms in human langerin change specificity for glycan ligands. J Biol Chem 288:36762–36771

Feinberg H, Rambaruth ND, Jegouzo SA, Jacobsen KM, Djurhuus R, Poulsen TB, Weis WI, Taylor ME, Drickamer K (2016) Binding sites for acylated trehalose analogs of glycolipid ligands on an extended carbohydrate recognition domain of the macrophage receptor mincle. J Biol Chem 291:21222–21233

Feinberg H, Jegouzo SAF, Rex MJ, Drickamer K, Weis WI, Taylor ME (2017) Mechanism of pathogen recognition by human dectin-2. J Biol Chem 292:13402–13414

Ferwerda B, Ferwerda G, Plantinga TS, Willment JA, van Spriel AB, Venselaar H, Elbers CC, Johnson MD, Cambi A, Huysamen C, Jacobs L, Jansen T, Verheijen K, Masthoff L, Morre SA, Vriend G, Williams DL, Perfect JR, Joosten LA, Wijmenga C, van der Meer JW, Adema GJ, Kullberg BJ, Brown GD, Netea MG (2009) Human dectin-1 deficiency and mucocutaneous fungal infections. N Engl J Med 361:1760–1767

Fukuda K, Doggett T, Laurenzi IJ, Liddington RC, Diacovo TG (2005) The snake venom protein botrocetin acts as a biological brace to promote dysfunctional platelet aggregation. Nat Struct Mol Biol 12:152–159

Gourdine JP, Cioci G, Miguet L, Unverzagt C, Silva DV, Varrot A, Gautier C, Smith-Ravin EJ, Imberty A (2008) High affinity interaction between a bivalve C-type lectin and a biantennary complex-type N-glycan revealed by crystallography and microcalorimetry. J Biol Chem 283:30112–30120

Goyal S, Klassert TE, Slevogt H (2016) C-type lectin receptors in tuberculosis: what we know. Med Microbiol Immunol 205:513–535

Guo Y, Feinberg H, Conroy E, Mitchell DA, Alvarez R, Blixt O, Taylor ME, Weis WI, Drickamer K (2004) Structural basis for distinct ligand-binding and targeting properties of the receptors DC-SIGN and DC-SIGNR. Nat Struct Mol Biol 11:591–598

Hanske J, Schulze J, Aretz J, McBride R, Loll B, Schmidt H, Knirel Y, Rabsch W, Wahl MC, Paulson JC, Rademacher C (2017) Bacterial polysaccharide specificity of the pattern recognition receptor langerin is highly species-dependent. J Biol Chem 292:862–871

Hatakeyama T, Kamiya T, Kusunoki M, Nakamura-Tsuruta S, Hirabayashi J, Goda S, Unno H (2011) Galactose recognition by a tetrameric C-type lectin, CEL-IV, containing the EPN carbohydrate recognition motif. J Biol Chem 286:10305–10315

Ishikawa E, Ishikawa T, Morita YS, Toyonaga K, Yamada H, Takeuchi O, Kinoshita T, Akira S, Yoshikai Y, Yamasaki S (2009) Direct recognition of the mycobacterial glycolipid, trehalose dimycolate, by C-type lectin mincle. J Exp Med 206:2879–2888

Jegouzo SA, Feinberg H, Dungarwalla T, Drickamer K, Weis WI, Taylor ME (2015) A novel mechanism for binding of galactose-terminated glycans by the C-type carbohydrate recognition domain in blood dendritic cell antigen 2. J Biol Chem 290:16759–16771

Kansas GS (1996) Selectins and their ligands: current concepts and controversies. Blood 88:3259–3287

Kolenko P, Rozbesky D, Vanek O, Kopecky V Jr, Hofbauerova K, Novak P, Pompach P, Hasek J, Skalova T, Bezouska K, Dohnalek J (2011) Molecular architecture of mouse activating NKR-P1 receptors. J Struct Biol 175:434–441

Leth-Larsen R, Garred P, Jensenius H, Meschi J, Hartshorn K, Madsen J, Tornoe I, Madsen HO, Sorensen G, Crouch E, Holmskov U (2005) A common polymorphism in the SFTPD gene influences assembly, function, and concentration of surfactant protein D. J Immunol 174:1532–1538

Li Y, Mariuzza RA (2014) Structural basis for recognition of cellular and viral ligands by NK cell receptors. Front Immunol 5:123

Liu Y, Eisenberg D (2002) 3D domain swapping: as domains continue to swap. Protein Sci 11:1285–1299

Maita N, Nishio K, Nishimoto E, Matsui T, Shikamoto Y, Morita T, Sadler JE, Mizuno H (2003) Crystal structure of von Willebrand factor A1 domain complexed with snake venom, bitiscetin: insight into glycoprotein Ibalpha binding mechanism induced by snake venom proteins. J Biol Chem 278:37777–37781

Malik S, Greenwood CM, Eguale T, Kifle A, Beyene J, Habte A, Tadesse A, Gebrexabher H, Britton S, Schurr E (2006) Variants of the SFTPA1 and SFTPA2 genes and susceptibility to tuberculosis in Ethiopia. Hum Genet 118:752–759

Matsunaga I, Moody DB (2009) Mincle is a long sought receptor for mycobacterial cord factor. J Exp Med 206:2865–2868

Mehta-D'souza P, Klopocki AG, Oganesyan V, Terzyan S, Mather T, Li Z, Panicker SR, Zhu C, McEver RP (2017) Glycan bound to the selectin low affinity state engages Glu-88 to stabilize the high affinity state under force. J Biol Chem 292:2510–2518

Nagae M, Yamanaka K, Hanashima S, Ikeda A, Morita-Matsumoto K, Satoh T, Matsumoto N, Yamamoto K, Yamaguchi Y (2013) Recognition of bisecting N-acetylglucosamine: structural basis for asymmetric interaction with the mouse lectin dendritic cell inhibitory receptor 2. J Biol Chem 288:33598–33610

Nagae M, Morita-Matsumoto K, Kato M, Kaneko MK, Kato Y, Yamaguchi Y (2014a) A platform of C-type lectin-like receptor CLEC-2 for binding *O*-glycosylated podoplanin and nonglycosylated rhodocytin. Structure 22:1711–1721

Nagae M, Ikeda A, Kitago Y, Matsumoto N, Yamamoto K, Yamaguchi Y (2014b) Crystal structures of carbohydrate recognition domain of blood dendritic cell antigen-2 (BDCA2) reveal a common domain-swapped dimer. Proteins 82:1512–1518

Nagae M, Ikeda A, Hanashima S, Kojima T, Matsumoto N, Yamamoto K, Yamaguchi Y (2016) Crystal structure of human dendritic cell inhibitory receptor C-type lectin domain reveals the binding mode with *N*-glycan. FEBS Lett 590:1280–1288

Ng KK, Weis WI (1997) Structure of a selectin-like mutant of mannose-binding protein complexed with sialylated and sulfated Lewis(x) oligosaccharides. Biochemistry 36:979–988

Ng KK, Drickamer K, Weis WI (1996) Structural analysis of monosaccharide recognition by rat liver mannose-binding protein. J Biol Chem 271:663–674

Ng KK, Kolatkar AR, Park-Snyder S, Feinberg H, Clark DA, Drickamer K, Weis WI (2002) Orientation of bound ligands in mannose-binding proteins. Implications for multivalent ligand recognition. J Biol Chem 277:16088–16095

Palma AS, Feizi T, Zhang Y, Stoll MS, Lawson AM, Diaz-Rodriguez E, Campanero-Rhodes MA, Costa J, Gordon S, Brown GD, Chai W (2006) Ligands for the β-glucan receptor, Dectin-1, assigned using "designer" microarrays of oligosaccharide probes (neoglycolipids) generated from glucan polysaccharides. J Biol Chem 281:5771–5779

Poget SF, Legge GB, Proctor MR, Butler PJ, Bycroft M, Williams RL (1999) The structure of a tunicate C-type lectin from *Polyandrocarpa misakiensis* complexed with D-galactose. J Mol Biol 290:867–879

Preston RC, Jakob RP, Binder FP, Sager CP, Ernst B, Maier T (2016) E-selectin ligand complexes adopt an extended high-affinity conformation. J Mol Cell Biol 8:62–72

Riboldi E, Daniele R, Parola C, Inforzato A, Arnold PL, Bosisio D, Fremont DH, Bastone A, Colonna M, Sozzani S (2011) Human C-type lectin domain family 4, member C (CLEC4C/BDCA-2/CD303) is a receptor for asialo-galactosyl-oligosaccharides. J Biol Chem 286:35329–35333

Rozbesky D, Adamek D, Pospisilova E, Novak P, Chmelik J (2016) Solution structure of the lymphocyte receptor Nkrp1a reveals a distinct conformation of the long loop region as compared to in the crystal structure. Proteins 84:1304–1311

Sancho D, Reis e Sousa C (2013) Sensing of cell death by myeloid C-type lectin receptors. Curr Opin Immunol 25:46–52

Sanhueza CA, Baksh MM, Thuma B, Roy MD, Dutta S, Preville C, Chrunyk BA, Beaumont K, Dullea R, Ammirati M, Liu S, Gebhard D, Finley JE, Salatto CT, King-Ahmad A, Stock I, Atkinson K, Reidich B, Lin W, Kumar R, Tu M, Menhaji-Klotz E, Price DA, Liras S, Finn MG, Mascitti V (2017) Efficient liver targeting by polyvalent display of a compact ligand for the asialoglycoprotein receptor. J Am Chem Soc 139:3528–3536

Shrive AK, Tharia HA, Strong P, Kishore U, Burns I, Rizkallah PJ, Reid KB, Greenhough TJ (2003) High-resolution structural insights into ligand binding and immune cell recognition by human lung surfactant protein D. J Mol Biol 331:509–523

Silva-Martin N, Bartual SG, Ramirez-Aportela E, Chacon P, Park CG, Hermoso JA (2014) Structural basis for selective recognition of endogenous and microbial polysaccharides by macrophage receptor SIGN-R1. Structure 22:1595–1606

Somers WS, Tang J, Shaw GD, Camphausen RT (2000) Insights into the molecular basis of leukocyte tethering and rolling revealed by structures of P- and E-selectin bound to SLe^X and PSGL-1. Cell 103:467–479

Sutkeviciute I, Thepaut M, Sattin S, Berzi A, McGeagh J, Grudinin S, Weiser J, Le Roy A, Reina JJ, Rojo J, Clerici M, Bernardi A, Ebel C, Fieschi F (2014) Unique DC-SIGN clustering activity of a small glycomimetic: a lesson for ligand design. ACS Chem Biol 9:1377–1385

Swaminathan GJ, Myszka DG, Katsamba PS, Ohnuki LE, Gleich GJ, Acharya KR (2005) Eosinophil-granule major basic protein, a C-type lectin, binds heparin. Biochemistry 44:14152–14158

Toyonaga K, Torigoe S, Motomura Y, Kamichi T, Hayashi JM, Morita YS, Noguchi N, Chuma Y, Kiyohara H, Matsuo K, Tanaka H, Nakagawa Y, Sakuma T, Ohmuraya M, Yamamoto T, Umemura M, Matsuzaki G, Yoshikai Y, Yano I, Miyamoto T, Yamasaki S (2016) C-type lectin receptor DCAR recognizes mycobacterial phosphatidyl-inositol mannosides to promote a Th1 response during infection. Immunity 45:1245–1257

Unno H, Itakura S, Higuchi S, Goda S, Yamaguchi K, Hatakeyama T (2019) Novel Ca(2+)-independent carbohydrate recognition of the C-type lectins, SPL-1 and SPL-2, from the bivalve Saxidomus purpuratus. Protein Sci 28:766–778

van Eijk M, Rynkiewicz MJ, Khatri K, Leymarie N, Zaia J, White MR, Hartshorn KL, Cafarella TR, van Die I, Hessing M, Seaton BA, Haagsman HP (2018) Lectin-mediated binding and sialoglycans of porcine surfactant protein D synergistically neutralize influenza A virus. J Biol Chem 293:10646–10662

Varki A (2017) Biological roles of glycans. Glycobiology 27:3–49

Varki A, Cummings RD, Aebi M, Packer NH, Seeberger PH, Esko JD, Stanley P, Hart G, Darvill A, Kinoshita T, Prestegard JJ, Schnaar RL, Freeze HH, Marth JD, Bertozzi CR, Etzler ME, Frank M, Vliegenthart JF, Lütteke T, Perez S, Bolton E, Rudd P, Paulson J, Kanehisa M, Toukach P, Aoki-Kinoshita KF, Dell A, Narimatsu H, York W, Taniguchi N, Kornfeld S (2015) Symbol nomenclature for graphical representations of glycans. Glycobiology 25:1323–1324

Verdijk P, Dijkman R, Plasmeijer EI, Mulder AA, Zoutman WH, Mieke Mommaas A, Tensen CP (2005) A lack of birbeck granules in langerhans cells is associated with a naturally occurring point mutation in the human langerin gene. J Invest Dermatol 124:714–717

Walker JR, Nagar B, Young NM, Hirama T, Rini JM (2004) X-ray crystal structure of a galactose-specific C-type lectin possessing a novel decameric quaternary structure. Biochemistry 43:3783–3792

Wang Y, Kuan PJ, Xing C, Cronkhite JT, Torres F, Rosenblatt RL, DiMaio JM, Kinch LN, Grishin NV, Garcia CK (2009) Genetic defects in surfactant protein A2 are associated with pulmonary fibrosis and lung cancer. Am J Hum Genet 84:52–59

Ward EM, Stambach NS, Drickamer K, Taylor ME (2006) Polymorphisms in human langerin affect stability and sugar binding activity. J Biol Chem 281:15450–15456

Wedepohl S, Dernedde J, Vahedi-Faridi A, Tauber R, Saenger W, Bulut H (2017) Reducing Macro- and microheterogeneity of N-glycans enables the crystal structure of the lectin and EGF-like domains of human L-selectin to be solved at 1.9 Å resolution. ChemBioChem 18:1338–1345

Weis WI, Drickamer K, Hendrickson WA (1992) Structure of a C-type mannose-binding protein complexed with an oligosaccharide. Nature 360:127–134

Xia L, Ramachandran V, McDaniel JM, Nguyen KN, Cummings RD, McEver RP (2003) N-terminal residues in murine P-selectin glycoprotein ligand-1 required for binding to murine P-selectin. Blood 101:552–559

Zelensky AN, Gready JE (2005) The C-type lectin-like domain superfamily. FEBS J 272:6179–6217

Physiological and Pathological Functions of CARD9 Signaling in the Innate Immune System

Larsen Vornholz and Jürgen Ruland

Contents

1 Introduction ... 178
2 CARD9 Structure, Expression and Binding Partners ... 179
3 CARD9 Activation and Innate CBM Complex Assembly ... 179
4 Effector Functions of CARD9 Signaling ... 181
5 Negative Regulation of CARD9 Signaling ... 185
6 Physiological Functions of CARD9 Signaling in Host Protection ... 186
6.1 CARD9 Signaling in Antifungal Immunity ... 186
6.2 CARD9 in Antibacterial Defense ... 188
6.3 CARD9 in Immunity to Parasites ... 189
6.4 CARD9 Signaling upon Viral Recognition ... 189
7 CARD9 Signaling in Noninfectious Inflammation and Cancer Models ... 190
8 Genetic Alterations of *CARD9* in Human Disease ... 192
9 Conclusions and Therapeutic Perspective ... 194
References ... 195

L. Vornholz · J. Ruland (✉)
TUM School of Medicine, Institute of Clinical Chemistry and Pathobiochemistry, Technical University of Munich, 81675 Munich, Germany
e-mail: j.ruland@tum.de

TranslaTUM, Center for Translational Cancer Research, Technical University of Munich, 81675 Munich, Germany

J. Ruland
German Cancer Consortium (DKTK), Heidelberg, Germany

German Center for Infection Research (DZIF), Munich Partner Site, Munich, Germany

Current Topics in Microbiology and Immunology (2020) 429: 177–203
https://doi.org/10.1007/82_2020_211

Published Online: 16 May 2020

Abstract Caspase recruitment domain protein 9 (CARD9) forms essential signaling complexes in the innate immune system that integrate cues from C-type lectin receptors and specific intracellular pattern recognition receptors. These CARD9-mediated signals are pivotal for host defense against fungi, and they mediate immunity against certain bacteria, viruses and parasites. Furthermore, CARD9-regulated pathways are involved in sterile inflammatory responses critical for immune homeostasis and can control pro- and antitumor immunity in cancer microenvironments. Consequently, multiple genetic alterations of human *CARD9* are connected to primary immunodeficiencies or prevalent inflammatory disorders in patients. This review will summarize our current understanding of CARD9 signaling in the innate immune system, its physiological and pathological functions and their implications for human immune-mediated diseases.

1 Introduction

The cells of the innate immune system make up our first line of defense against microbial pathogens, and they mediate homeostasis of the organism upon tissue damage. To mediate these effects, innate immune cells utilize germline-encoded pattern recognition receptors (PRRs) such as C-type lectin receptors (CLRs) to sense pathogen-associated molecular patterns (PAMPs) or endogenous danger-associated molecular patterns (DAMPs) (Gong et al. 2020; Rivera et al. 2016). The controlled activation of PRR-associated intracellular signaling pathways is critical for immune physiology and for host protection, and pathological dysfunctions of the same pathways are frequent causes of human immunodeficiencies and inflammatory disorders (Cao 2016; Huber-Lang et al. 2018; Zhang and Cao 2019).

Caspase recruitment domain protein 9 (CARD9) is a central signaling adapter molecule in the innate immune system. This protein integrates signals from the C-type lectin receptor and plays a very prominent role downstream of the antifungal CLRs Dectin-1, Dectin-2 and Mincle (Gross et al. 2006; Robinson et al. 2009; Shenderov et al. 2013). In addition, CARD9 can be activated by several intracellular nucleic acid-sensing PRRs and NOD-like receptors (Hsu et al. 2007; Poeck et al. 2010; Roth et al. 2014), and it controls innate immune responses to fungi, certain bacteria, viruses and parasites and is also involved in sterile danger pathways upon cellular damage (Brown et al. 2018). These CARD9-controlled pathways are highly relevant to human immune pathology, as genetic alterations of *CARD9* are causal for primary immunodeficiencies (Drummond and Lionakis 2016) or risk factors for prevalent human inflammatory disorders (Drummond et al. 2018). CARD9 has also been reported to play critical roles in tumor immune microenvironments (Bergmann et al. 2017; Wang et al. 2018; Zhong et al. 2019). In the next chapters, we discuss our current understanding of the molecular regulation of CARD9 signaling with a focus on CLR pathways and outline the known physiological and pathological functions of CARD9 signaling in human disease.

2 CARD9 Structure, Expression and Binding Partners

The CARD9 protein has a bipartite structure consisting of an amino-terminal caspase recruitment domain (CARD) (Bertin et al. 2000), which represents a homotypic protein–protein interaction module that can associate with other CARDs, and a carboxy-terminal coiled-coil region that serves as an oligomerization site (Bertin et al. 2000) (Fig. 1a). CARD9 is highly conserved between mice and humans, sharing approximately 87% similarity and 83% identity. CARD9 is specifically expressed in myeloid cells of the innate immune system, such as dendritic cells, macrophages and neutrophils, and is found in various tissues, including the spleen, liver, brain, placenta, peripheral blood and bone marrow (Bertin et al. 2000; Hsu et al. 2007; Roth and Ruland 2013). Initial studies have demonstrated that CARD9 can utilize its CARD to directly interact with the ubiquitously expressed CARD-containing adapter protein B-cell lymphoma 10 (BCL10) (Bertin et al. 2000). BCL10 in turn is constitutively associated with the multifunctional protein mucosa-associated lymphoid tissue lymphoma translocation protein 1 (MALT1), which is composed of a death domain, three immunoglobulin (Ig)-like domains and a catalytic caspase-like domain (Fig. 1a), which has proteolytic function. MALT1 is therefore also referred to as a paracaspase and is unique to the mammalian proteome (Uren et al. 2000). Activation of CARD9 by innate immune signals also leads to rapid activation of the BCL10-MALT1 module and the formation of a CARD9-BCL10-MALT1 complex (also called the CBM complex), which then mediates context-dependent inflammatory responses (Fig. 1b, c). The BCL10-MALT1 module can also be activated by a series of additional immune and inflammatory receptors in other cell types, including antigen receptors in T and B lymphocytes, via CARD11 (Ruland and Hartjes 2019).

3 CARD9 Activation and Innate CBM Complex Assembly

The activation of CARD9 is a highly regulated process that can be triggered by multiple and potentially all SYK-coupled CLRs, including Dectin-1 (Gross et al. 2006), Dectin-2 (Robinson et al. 2009) and Mincle (also known as CLEC4E) (Shenderov et al. 2013) (Fig. 1b). CARD9 is also activated by other ITAM-containing receptors on myeloid cells, including the Fcγ receptor FcγRIII, myeloid-associated immunoglobulin-like receptor II (MAIR-II) (also known as CD300D), osteoclast-associated immunoglobulin-like receptor (OSCAR) and triggering receptor expressed on myeloid cells 1 (TREM1) (Hara et al. 2007) (Fig. 1b). Furthermore, the cytosolic nucleic acid sensors retinoic acid-inducible gene 1 (RIG-I), melanoma differentiation-associated protein 5 (MDA5, also known as IFIH1) (Poeck et al. 2010) and RAD50 (Roth et al. 2014) as well as nucleotide-binding oligomerization domain-containing protein 2 (NOD2) (Hsu et al. 2007) can utilize CARD9 for signal transduction (Fig. 1c).

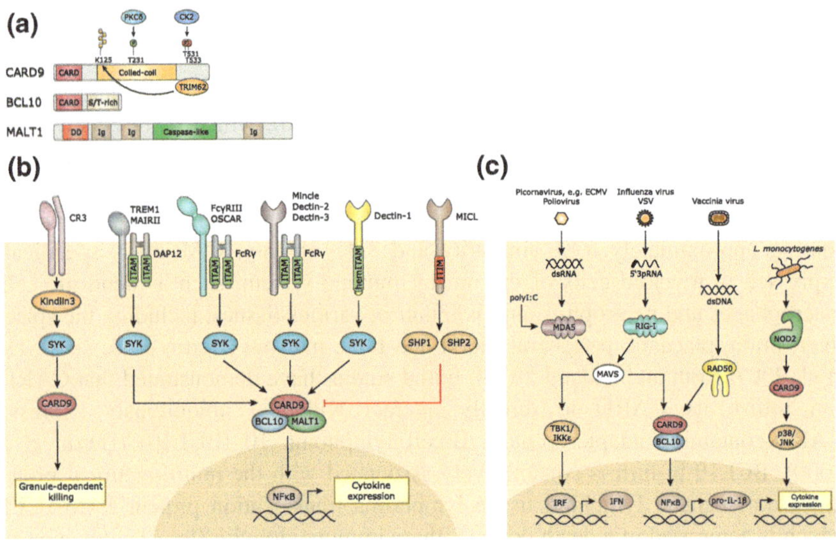

Fig. 1 CARD9-BCL10-MALT1 complex and signaling. **a** In myeloid cells, the CBM signalosome consists of CARD9-BCL10-MALT1. CARD9 has a bipartite structure with a CARD domain at the amino-terminal end and a coiled-coil domain at the carboxy-terminal end. TRIM62 binds CARD9 at its C-terminal end and is required for K125 ubiquitylation. Activating phosphorylation at T231 is mediated by PKCd, whereas inhibitory phosphorylation at T531/T533 is mediated by CK2. BCL10 consists of amino-terminal CARD and a serine-/threonine-rich carboxy-terminal region. MALT1 comprises an amino-terminal death domain (DD), three immunoglobulin-like (Ig) domains and a caspase-like catalytic domain. **b** Upon extracellular DAMPs/PAMPs (e.g., fungi), CARD9 relays signals from SYK-coupled immunoreceptor tyrosine-based activation motif (ITAM)-containing CLRs such as Dectin-1 and ITAM-associated CLRs such as Dectin-2, Dectin-3 and Mincle, which lack a cytoplasmic ITAM domain and thus require ITAM-containing adaptor proteins such as DAP12 or FcRγ for signal transduction. Further, ITAM-associated receptors engaging CARD9 are the Fcγ receptor FcγRIII, myeloid-associated immunoglobulin-like receptor II (MAIR-II, CD300D), osteoclast-associated immunoglobulin-like receptor (OSCAR) and triggering receptor expressed on myeloid cells 1 (TREM1). In contrast, the immunoreceptor tyrosine-based inhibition motif (ITIM)-bearing myeloid inhibitory C-type lectin-like receptors (MICLs) generally suppress inflammatory pathways by recruiting tyrosine phosphatases such as SHP1 and SHP2. Alternatively, unopsonized fungi can be sensed via complement receptor 3 (CR3), which leads to granule-dependent fungal killing via Kindlin3/SYK/CARD9. **c** CARD9 is also activated by intracellular pathogens of bacterial or viral origin. The cytosolic nucleic acid sensors RIG-I and MDA5 and the DNA double-strand break repair protein RAD50 promote NF-κB signaling via CARD9, which leads to inflammatory cytokine expression (e.g., IL-1β), cellular activation and selective adaptive immunity. Furthermore, CARD9 mediates p38/ERK activation upon NOD2 engagement of intracellular bacteria

To date, the activation of CARD9 in the CLR pathways in the context of fungal recognition is best understood (Fig. 2). The early activation of CLRs, such as Dectin-1, Dectin-2 or Mincle, leads to Src family tyrosine kinase-mediated phosphorylation of immunoreceptor tyrosine-based activation motifs (ITAMs),

ITAM-like regions within Dectin-1 or Dectin-2, or Mincle at associated transmembrane ITAM adapters, such as the Fc receptor γ-chain or DAP10 (Geijtenbeek and Gringhuis 2009; Rogers et al. 2005; Sato et al. 2006; Yamasaki et al. 2008). These receptor proximal events result in the recruitment of SYK to receptors and their activation, leading to the activation of PKCδ (Strasser et al. 2012) in cooperation with VAV proteins (Roth et al. 2016). Active PKCδ then triggers direct phosphorylation of CARD9 within its coiled-coil domain at position T231, and this modification is critical to induce assembly of the CARD9-BCL10-MALT1 complex (Strasser et al. 2012). Structural analysis revealed that CBM assembly upon CARD9 activation requires disruption of its autoinhibitory state, which leads to a conformational change that exposes CARD9-CARD for BCL10 binding (Holliday et al. 2019). CARD9-BCL10 interactions are then mediated via CARDs, whereby CARD9-CARD forms a helical assembly that serves as a template for unidirectional filamentous BCL10 polymerization, presumably dependent on Zn^{2+} (Holliday et al. 2018).

The assembled CBM complex in innate immune cells serves subsequently as a scaffold structure that likely places ubiquitin modifiers and protein kinases such as TAK1 in proximity for downstream signaling (Strasser et al. 2012). Although the precise collection of interaction partners in the CARD9-controlled innate immune pathways remains to be identified, the BCL10-MALT1 module can in principle recruit the E2 ubiquitin-conjugating enzymes UBC13 and UEV1a (Deng et al. 2000; Sun et al. 2004), the E3 ubiquitin ligases tumor necrosis factor (TNF) receptor-associated factor 2 (TRAF2) and TRAF6 (Oeckinghaus et al. 2007; Sun et al. 2004), cellular inhibitor of apoptosis protein 1 (cIAP1) and cIAP2 (Yang et al. 2016a), and the linear ubiquitin chain assembly complex (LUBAC) (Yang et al. 2016b). These factors have been reported to regulate BCL10-MALT1-dependent lymphocyte signaling (Ruland and Hartjes 2019). It is therefore conceivable that similar mechanisms would also control CARD9-BCL10-MALT1 signaling for innate immunity. Ubiquitylation reactions at CARD9-BCL10-MALT1 signalosomes are expected to recruit ubiquitin-binding proteins such as TAK1-binding protein 2 (TAB2) and TAB3 and for recruitment of TGFβ-associated kinase 1 (TAK1) (Sun et al. 2004; Wang et al. 2001) to activate downstream pathways. Nevertheless, the exact regulation of individual CARD9 effectors upon CARD9 activation by distinct PRRs is still insufficiently resolved.

4 Effector Functions of CARD9 Signaling

One key output of activated CARD9-BCL10-MALT1 signalosomes is activation of the canonical NF-κB signaling pathway (Roth and Ruland 2013) involving TAK1 and activation of the IκB kinase-β (IKKβ) (Fig. 2). This mechanism is particularly important for the control of the transcriptionally active NF-κB subunits RelA/p65 and c-Rel (Hayden and Ghosh 2012; Liu et al. 2017), which are essential for orchestrating the expression of a large series of inflammatory cytokines, such as

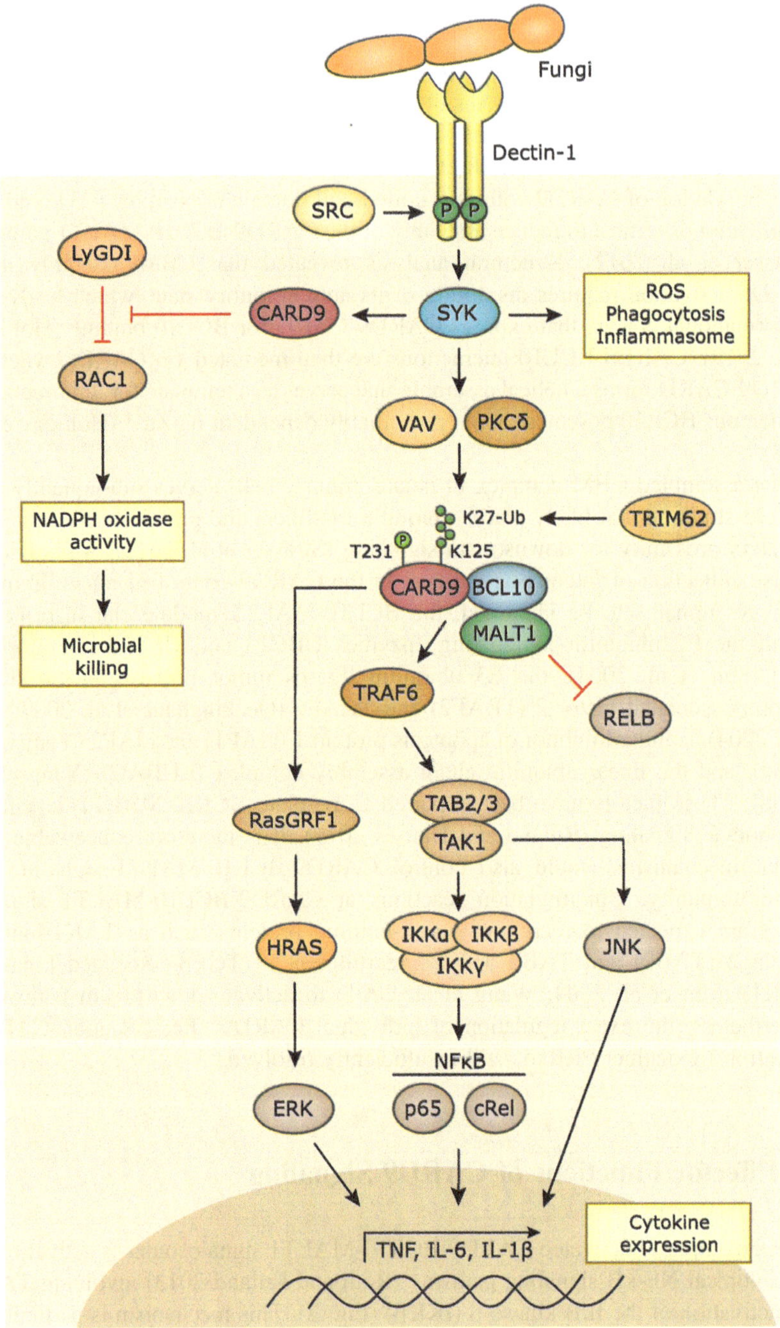
Fungi
Dectin-1
SRC
P
P
LyGDI
CARD9
SYK
ROS
Phagocytosis
Inflammasome
RAC1
VAV
PKCδ
NADPH oxidase
activity
K27-Ub
TRIM62
T231
P
K125
CARD9
BCL10
MALT1
Microbial
killing
TRAF6
RELB
TAB2/3
TAK1
RasGRF1
HRAS
IKKα
IKKβ
IKKγ
JNK
NFκB
ERK
p65
cRel
TNF, IL-6, IL-1β
Cytokine
expression

◂Fig. 2 CARD9 effector functions. After fungal engagement of CLRs (such as Dectin-1), the Src family kinase member Src phosphorylates the intracellular ITAM domain, allowing the recruitment of SYK. SYK subsequently activates the protein kinase C delta (PKCδ), which phosphorylates CARD9 at T231 in the coiled-coil domain in cooperation with VAV proteins, allowing CBM complex assembly. Subsequent K27-linked polyubiquitination of CARD9 at K125 by TRIM62 is required for NF-κB activation. Additionally, TRAF6 is recruited to the CBM signalosome upon assembly to mediate its ubiquitination. Then, the IKK complex is recruited to the CBM complex and activated by TAK1 together with TAB2 and TAB3. Activated IKK phosphorylates IκBα to induce proteasomal degradation, releasing the NF-kB subunits p65 and c-Rel for nuclear translocation and subsequent inflammatory cytokine expression. Additionally, TAK1 activates JNK. The NF-κB subunit RELB is counter-regulated by MALT1-mediated cleavage. Independent of BCL10/MALT1, Dectin-1 engagement initiates SYK-dependent Ras-GRF1 phosphorylation, which complexes with CARD9 and subsequently activates the H-Ras/ERK pathway. Moreover, CARD9 inhibits LyGDI-mediated suppression of RAC1, allowing NADPH oxidase activity, which is needed for microbial killing in the phagolysosome. For details, see text

TNF, IL-1β, IL-6, IL-10, IL-12, IL-23 and GM-CSF, and chemokines, such as CCL3, CXCL1, CXCL2 and CXCL8, which can be induced upon CARD9 engagement (Drummond et al. 2015; Gavino et al. 2016; Gross et al. 2006; Jhingran et al. 2015; LeibundGut-Landmann et al. 2007; Németh et al. 2016; Poeck et al. 2010; Roth et al. 2014). To drive full NF-κB activation, CARD9 recruits the ubiquitin ligase TRIM62, which binds to the C-terminal tail of CARD9 and mediates K27-linked ubiquitination of CARD9 at K125 (Cao et al. 2015). This ubiquitin-conjugating modification sterically disrupts the autoinhibitory state of the CARD9 CARD-coiled-coil interface, enabling BCL10 polymerization and subsequent NF-κB activation (Holliday et al. 2019). CARD9 signaling is also implicated in CLR-induced ERK (Jia et al. 2014) and in NOD2-induced p38 and JNK signaling (Hara et al. 2007), which further contribute to fully inducing the inflammatory cascade.

Triggering of the CARD9-BCL10-MALT1 complex induces activation of MALT1 protease function (Yu et al. 2015). Although the specific role of MALT1-mediated proteolysis in CARD9-dependent innate immunity is still ill-defined, nine MALT1 substrates have been identified thus far, largely in activated T cells (Ruland and Hartjes 2019). The currently known MALT1 substrates include MALT1 itself and BCL10, the cleavage of which is thought to serve as a feedback loop to support optimal gene transcription (Baens et al. 2014). Furthermore, MALT1 can cleave and thereby inactivate the negative NF-κB and MAPK regulators A20 (Coornaert et al. 2008) and CYLD (Staal et al. 2018) as well as HOIL1 (Douanne et al. 2016; Elton et al. 2016; Klein et al. 2015) to amplify the NF-κB and JNK pathways. Similarly, upon lymphocyte activation, MALT1 can cleave the ribonuclease Regnase 1 (also known as ZC3H12A) (Uehata et al. 2013) or the mRNA decay regulators Roquin 1 (also known as RC3H1) and Roquin 2 (also known as RC3H2) (Jeltsch et al. 2014), which are important for controlling the stability of inflammatory mRNAs by binding to common stem-loop elements in their 3′ untranslated region (3′-UTR) (Mino et al. 2015). Although A20 (Coornaert et al. 2009), CYLD (Zhang et al. 2011), HOIL1 (MacDuff et al. 2017), Regnase 1 (Mao et al. 2017) or Roquin 1/2 (Athanasopoulos et al. 2016) are also involved in the regulation of innate immune responses, it is currently unknown whether and under what conditions MALT1 may inactivate these factors during CARD9-controlled innate immunity. Nevertheless, the fact that MALT1 cleaves overlapping substrates in innate immune cells and in activated lymphocytes is demonstrated in the case of the NF-κB subunit RELB that is inactivated in stimulated T cells (Hailfinger et al. 2011) as well as in myeloid lineage cells upon CARD9-mediated Dectin-1 signaling (Xu et al. 2018). The NF-κB RELB subunit acts primarily in the noncanonical NF–κB pathway and controls upon Dectin-1 stimulation the expression of the Th-2 cytokine IL-5, which is an important example of how CARD9-triggered MALT1 proteolysis tunes the inflammatory response.

In addition to the activation of CARD9-BCL10-MALT1 complexes, CARD9 has also been reported to engage other immune effector pathways. In response to Dectin-1-initiated SYK-mediated Ras-GRF1 phosphorylation, CARD9 can associate with Ras-GRF1 to activate the H-Ras/ERK pathway for the full induction of inflammatory cytokines such as IL-6, TNF and IL-1β (Jia et al. 2014). Moreover, CARD9 can control the production of ROS under certain conditions. Upon phagocytosis of *Listeria monocytogenes*, CARD9 is recruited to bacteria-containing phagolysosomes and suppresses the LyGDI-mediated inhibition of Rac1, enabling guanine nucleotide exchange factors (GEFs) to fuel Rac1 with GTP, which in turn drives NADPH complex-dependent ROS production required for antimicrobial killing (Wu et al. 2009). Furthermore, CARD9 can mediate granule-dependent killing of unopsonized fungal species that are recognized by complement receptor 3 (CR3) (Gazendam et al. 2014, 2016) through mechanisms that remain to be further defined.

In response to cytosolic PRR engagement, CARD9 plays additional selective roles. After the activation of NOD2, CARD9 selectively activates the MAP kinases p38 and JNK without affecting the NF-κB pathway (Hara et al. 2007). Furthermore, in response to cytosolic RNA or DNA detection by the nucleic acid receptors RIG-I, MDA5 (Poeck et al. 2010) or RAD50 (Roth et al. 2014), CARD9 engages BCL10 presumably without MALT1 to drive canonical NF-κB signaling for the production of inflammatory cytokines, particularly IL-1β.

To induce the production of IL-1β during innate immunity, CARD9 frequently collaborates with the activation of inflammasomes (Gross et al. 2009; Poeck et al. 2010), which are multiprotein complexes that induce caspase-1 activation to cleave pro-IL-1β and related cytokines into their bioactive form (Guo et al. 2015). While CARD9 is not directly involved in the assembly of inflammasome complexes, it mediates pro-IL-1β expression as well as NF-κB-dependent inflammasome priming in response to fungal cell wall components (e.g., β-glucan) (Gross et al. 2009), mycobacterial glycolipids (e.g., TDM) (Schweneker et al. 2013; Shenderov et al. 2013) and viral nucleic acids (e.g., 3pRNA) (Poeck et al. 2010). Recent work has also reported that CARD9 can associate with necroptosis-inducing kinases RIPK1 and RIPK3 to regulate myeloid cell necroptosis upon fungal infection (Cao et al.

2019). While these noncanonical functions of CARD9 are currently emerging, putative BCL10-MALT1-independent mechanisms of CARD9 during cytosolic PRR signaling, ROS or necroptosis control are still insufficiently defined.

5 Negative Regulation of CARD9 Signaling

To dampen or downregulate CARD9 signaling to counter-regulate inflammation, the innate immune system has developed several mechanisms that operate at different levels of the signaling cascade (Fig. 3). In addition to ITAM-containing activating CLRs that activate CARD9, myeloid cells also express inhibitory CLRs that contain immunoreceptor tyrosine-based inhibition motives (ITIMs), such as the myeloid inhibitory C-type lectin-like receptor MICL (Redelinghuys and Brown 2011). These ITIM CLRs can recruit protein tyrosine phosphatases such as SHP1 and SHP2 to negatively regulate SYK-induced cellular activation (Marshall et al. 2004) and, consequently, CARD9 engagement. An additional mechanism that counteracts activating CLR signaling involves the E3 ubiquitin ligase CBLB (Wirnsberger et al. 2016; Xiao et al. 2016), which is active during antifungal immunity. CBLB mediates ubiquitin-dependent degradation of Dectin-1 and Dectin-2 as well as SYK, resulting in the negative regulation of CARD9 signaling (Wirnsberger et al. 2016; Xiao et al. 2016).

Direct negative regulation of the CARD9-BCL10-MALT1 complex itself is mediated by the regulator of autophagosome maturation and endocytosis called Rubicon (Yang et al. 2012). In response to fungal PAMPs or RNA viruses, Rubicon switches its interaction from the phosphoserine-/phosphothreonine-binding protein 14-3-3β to CARD9 and thereby induces the disassembly of CARD9-BCL10-MALT1 complexes as an inhibitory feedback for CBM signaling in response to CLR and RIG-I ligands (Yang et al. 2012). Here, the S_{248} phosphorylation status of Rubicon in its ST-N region is critical for selective 14-3-3β or CARD9 binding.

CARD9 is also negatively regulated by phosphorylation-dependent events. In this context, casein kinase 2 (CK2) can directly phosphorylate the c-terminus of CARD9 at T531/T533 and mediate subsequent negative regulation of the NF-κB pathway via the von Hippel–Lindau tumor suppressor protein (pVHL) (Yang et al. 2007). However, whether pVHL can directly bind CARD9 or requires an accessory protein remains unclear. Downstream of kinase 3 (DOK3) can induce protein phosphatase (PP)1-dependent dephosphorylation of CARD9 at threonine residues, which subsequently attenuates CARD9-dependent NF-κB and JNK activation, and the subsequent reduction in cytokine production results in negatively regulated neutrophil antifungal immunity (Loh et al. 2019). Moreover, it is conceivable that additional inhibitory phosphorylation or the inhibition of activating phosphorylation and physical interference with activating binding partners or selective degradation of CARD9 complex components could further negatively regulate this pathway, although these hypotheses have been less studied.

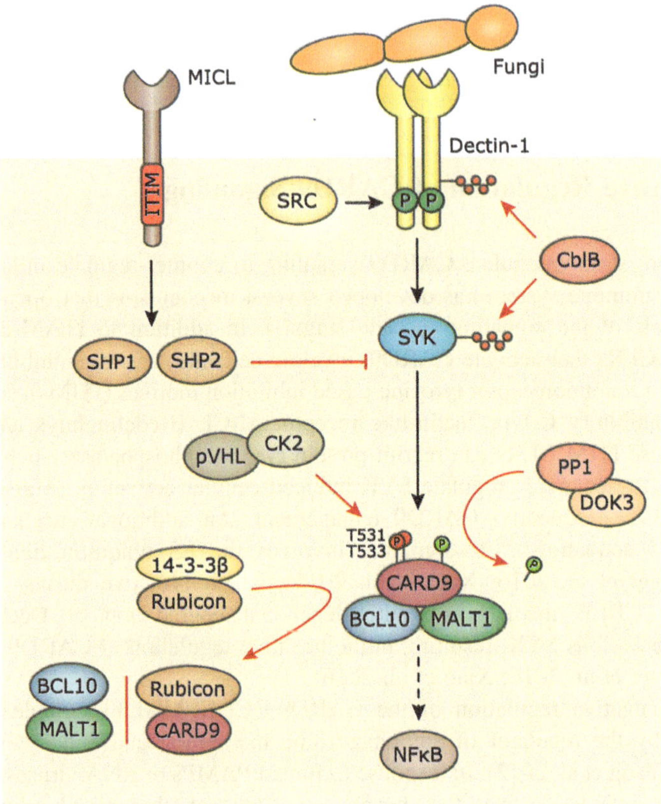

Fig. 3 Negative regulation of CARD9 signaling. ITIM-containing CLRs such as MICL recruit SHP1/SHP2 to generally inhibit SYK-mediated activating receptor signaling. Serving as inhibitory feedback, Rubicon switches its interaction from 14-3-3β to CARD9 upon stimulation and thereby induces disassembly of the CBM complex. Via pVHL, CK2 can directly phosphorylate the c-terminus of CARD9 at T531/T533 and mediate subsequent negative regulation of NF-κB. Dephosphorylation of CARD9 by DOK3-induced PP1 activity can further attenuate NF-κB signaling. CARD9 signaling during antifungal immunity is further counter-regulated by CBLB-mediated degradation of Dectin-1 and Dectin-2 and their associated SYK kinase

6 Physiological Functions of CARD9 Signaling in Host Protection

6.1 CARD9 Signaling in Antifungal Immunity

The nonredundant functions of CARD9 in innate immunity and antifungal defense were originally discovered in studies of CARD9-deficient mice using *Candida albicans* infections as a model (Gross et al. 2006). Subsequent investigations demonstrated that CARD9 signaling also mediates host protection against

Aspergillus fumigatus (Jhingran et al. 2012), *Coccidioides posadasii* (Hung et al. 2016), *Microsporum canis* (Mao et al. 2014), *Phialophora verrucosa* (Wu et al. 2016) and *Cryptococcus neoformans* (Yamamoto et al. 2014), which together establish CARD9 as one of the most important regulators of the mammalian antifungal immune response. These functions can be explained by the nonredundant role of CARD9 downstream of the SYK-coupled CLRs Dectin-1, Dectin-2, Dectin-3 and Mincle and the role of these receptors in sensing the fungal PAMPs β-glucan (Dectin-1) (Taylor et al. 2007), α-mannans (Dectin-2 and Dectin-3) (Saijo et al. 2010; Zhu et al. 2013) and α-mannose (Mincle) (Bugarcic et al. 2008).

To mediate antifungal defense, CARD9 triggers at least two important immune effector functions (Fig. 4). On the one hand, CARD9 signaling in monocytes and microglia induces the production of GM-CSF and CXCL1, respectively, which initiates the recruitment of neutrophils to the site of fungal infections (Drummond et al. 2015, 2019; Gavino et al. 2016). Neutrophils are required for fungal phagocytosis and killing and for the subsequent clearance of pathogens. The failure of CARD9-deficient mice to produce GM-CSF results in impaired neutrophil recruitment and allows invasive fungal growth (Drummond et al. 2015; Gavino et al. 2016; Jhingran et al. 2015). In addition to these immediate innate immune effects, CARD9 signaling in antigen-presenting cells also couples innate immune sensing to the activation of T cell-mediated adaptive immunity and controls, in particular, the differentiation of Th1 and Th17 cells (LeibundGut-Landmann et al. 2007; Robinson et al. 2009; Saijo et al. 2010; Sokol et al. 2013). CARD9 drives the

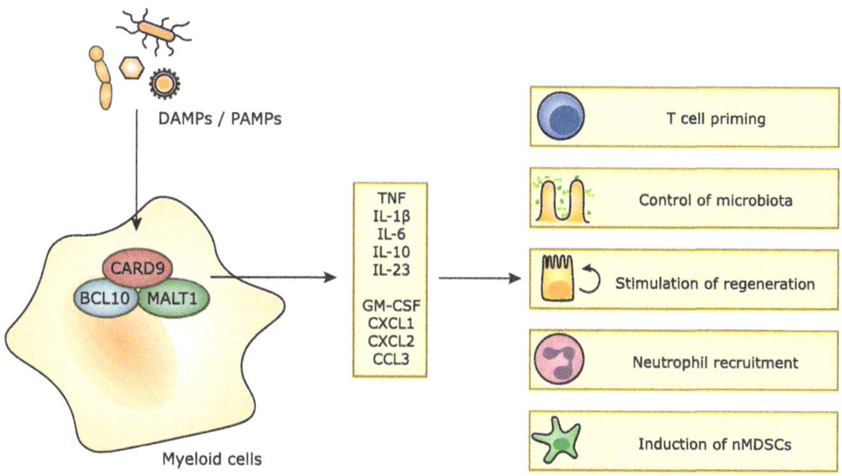

Fig. 4 Physiological functions of CARD9 in host defense and homeostasis. CARD9 signaling upon sensing DAMPs or PAMPs by myeloid cells leads to the expression of inflammatory cytokines and chemokines and presumably other effector molecules that orchestrate homeostatic and host defense function including T cell priming, control of the microbiota, stimulation of regeneration of epithelial cells, recruitment of neutrophils and the induction of neutrophilic myeloid-derived suppressor cells (nMDSCs). For details, see text

production of a specific set of cytokines, including IL-1β, IL-6 and IL-23, which provide a milieu that induces Th17 differentiation of antigen-stimulated naïve T cells (LeibundGut-Landmann et al. 2007; Saijo et al. 2010). Th17 cells are crucial for host defense against extracellular pathogens, including fungi such as *Candida albicans* (LeibundGut-Landmann et al. 2007) and *Cryptococcus neoformans* (Yamamoto et al. 2014). The failure of CARD9-deficient mice to mount adequate Th17 immunity is an additional cause of defective antifungal defense. During later phases of fungal infection, CARD9 regulates the development of neutrophilic myeloid-derived suppressor cells (nMDSCs) in a manner that depends on ROS, caspase-8 and IL-1β (Rieber et al. 2015). In general, nMDSCs can suppress T lymphocytes via ROS, arginase and NO to prevent pathological T cell and NK cell responses (Gabrilovich and Nagaraj 2009), and CARD9-mediated nMDSC generation after fungal recognition prevents tissue damage at later stages of infection.

6.2 CARD9 in Antibacterial Defense

Since CARD9-coupled CLRs sense not only fungi but also bacteria, it is not surprising that CARD9-deficient animals also exhibit pathological deficiencies in certain bacterial infection models. The mycobacterial cord factor trehalose-6,6-dimycolate (TMD) from *M. tuberculosis* is a potent proinflammatory PAMP detected by Mincle, and it induces potent CARD9-BCL10-MALT1 signaling (Shenderov et al. 2013; Werninghaus et al. 2009). This mechanism is particularly important for promoting the production of IL-10 during *M. tuberculosis* infection, which prevents neutrophil-mediated systemic hyperinflammation (Dorhoi et al. 2010). Furthermore, Mincle can mediate host protection against group A *streptococcus* (GAS) after recognition of the GAS-produced anchor of lipoteichoic acid monoglucosyldiacylglycerol (MDGD) by inducing the production of inducible nitric oxidase synthase (iNOS), ROS and inflammatory cytokines such as TNF, MIP-2 and IL-6, the latter of which is directly regulated by CARD9 (Imai et al. 2018). In the context of *Vibrio cholerae* infections, FcRg-CARD9 signaling induces inflammatory responses upon stimulation with the nontoxic cholera toxin subunit CTB (Phongsisay et al. 2015). Independent of CLRs, the ITAM-like motif bearing the CEACAM3 receptor engages the CARD9 pathway upon infection with *Moraxella catarrhalis* to induce the production of chemokines such as CXCL8 and CCL3 (Heinrich et al. 2016). The function of CARD9 in response to NOD2 stimulation has been implicated in the antibacterial defense against intracellular pathogens such as *Listeria monocytogenes* (Hsu et al. 2007). Moreover, CARD9-deficient animals are also hypersusceptible to bacterial infections with *Citrobacter rodentium*, which is commonly used to study experimental colitis (Lamas et al. 2018; Sokol et al. 2013). Linking innate to adaptive immunity, SYK-CARD9 signaling in APCs is also important for the optimal activation of T cells specific for bacterial proteins such as flagellin (Atif et al. 2015).

6.3 CARD9 in Immunity to Parasites

Parasites such as *Schistosoma mansoni,* which are causative for human schistosomiasis (hepatic and intestinal granulomatous inflammation), are also detected by the SYK-coupled CLRs Dectin-2, Mincle and SIGNR5, which synergistically activate CARD9 for IL-1β and IL-23 production and subsequent Th17-mediated immunity (Kalantari et al. 2018; Ritter et al. 2010). The function of CARD9 is also required for host defense against the intracellular protozoan parasite *Toxoplasma gondii* by mediating IL-1β production in primary human peripheral blood monocytes (Pandori et al. 2019), presumably upon activation of Dectin-1 and Mincle (He et al. 2016; Yan et al. 2014). Furthermore, CARD9-mediated IL-1β production contributes to malaria-associated acute respiratory distress syndrome (ARDS) in response to *Plasmodium falciparum*, the most virulent species of malaria parasites (Maknitikul et al. 2017, 2018).

6.4 CARD9 Signaling upon Viral Recognition

Several SYK-coupled CLRs, such as DC-SIGN, langerin, MMR, DCIR, MDL-1 and LSECtin, have evolved to sense viruses such as HIV-1, HBV, influenza virus, measles virus, dengue virus and SARS corona virus (Geijtenbeek and Gringhuis 2009; Monteiro and Lepenies 2017), and it is therefore conceivable that these could signal via CARD9 during viral recognition. Furthermore, in vivo experiments with influenza virus (IFV) have demonstrated that CARD9-mediated innate activation plays a critical role in severe influenza pneumonia (Uematsu et al. 2015), presumably via DC-SIGN receptor engagement (Wang et al. 2008). In CARD9-deficient mice, pneumonia was drastically attenuated concomitant with reduced inflammatory cytokines and chemokines in the influenza-infected lungs. However, the type I interferon response and virus-specific adaptive lymphocyte immunity were not altered (Uematsu et al. 2015). Interestingly, CARD9 deficiency selectively altered the cytokine response to IFV in DCs in vitro but not in macrophages. Similarly, the Mincle-CARD9 signaling axis in DCs also controls the early antiviral response phase against the mosquito-transmitted La Crosse virus (LACV), whereas viral clearance was unaffected (Monteiro et al. 2019). Moreover, as indicated above, RNA and DNA viruses, which are detected by cytosolic nucleic acid sensors RIG-I, MDA5 (Poeck et al. 2010) and RAD50 (Roth et al. 2014), engage CARD9 for proinflammatory cytokine production. Consistently, RNA viruses such as vesicular stomatitis viruses (VSV) and DNA viruses such as vaccinia viruses (VV) activate CARD9 signaling in vitro, and deficiency of CARD9 results in impaired IL-1β, TNF and IL-6 responses to these pathogens in vivo (Poeck et al. 2010). While these CARD9-dependent signals are important in shaping the T cell response to viruses at later stages of infection, they are not involved in the immediate early type I interferon responses that mediate innate antiviral protection.

7 CARD9 Signaling in Noninfectious Inflammation and Cancer Models

More recent experiments with CARD9-deficient mice have demonstrated critical functions for CARD9 signaling in the regulation of noninfectious immune homeostasis (Fig. 4). Using the classical inflammatory bowel disease (IBD) model of chemically induced colitis with dextran sodium sulfate (DSS) (Chassaing et al. 2014), CARD9-deficient animals were found to be hypersusceptible compared with wild-type animals (Bergmann et al. 2017). Mechanistic studies have demonstrated that CARD9 signaling in innate immune cells regulates the production of cytokines such as IL-6, TNF and IFNγ and the Th17- and innate lymphoid cell (ILC)-related cytokines IL-17A, IL-17F and IL-22 during intestinal inflammation. These CARD9-induced cytokines are necessary to stimulate optimal IL-22 production by ILCs, which subsequently activate the IL-22 receptor on intestinal epithelial cells (IECs) to promote an epithelial regenerative response via STAT3 activation (Bergmann et al. 2017). Furthermore, CARD9 signaling shapes the composition of the gut microbiome (Lamas et al. 2016, 2018). Consequently, CARD9-deficient mice at baseline already exhibit an underrepresentation of bacterial species from the *Adlercreutzia* genus and *Lactobacillus reuteri*, which metabolize tryptophan into aryl hydrocarbon receptor (AHR) ligands, which are also crucial for the expression of IL-22 by lymphoid cells (Lamas et al. 2016). The impaired AHR ligand production is another reason for impaired IL-22 production and delayed epithelial regeneration in CARD9-deficient mice (Lamas et al. 2016). In line with the function of CARD9 in the antifungal response, the intestinal flora of CARD9-deficient mice also exhibits an aberrant fungal microbiome, which includes *Candida tropicalis* species that can increase the susceptibility to intestinal inflammation (Wang et al. 2018).

The effects of CARD9 signaling during intestinal inflammation have further consequences in models of colitis-associated cancer (CAC) (Bergmann et al. 2017; Hartjes and Ruland 2019; Malik et al. 2018; Wang et al. 2018). On the one hand, the CARD9-triggered IL–22-mediated proliferative IEC response can stimulate the growth of transformed colon epithelial cells and thereby enhance colon cancer growth (Bergmann et al. 2017). On the other hand, presumably dependent on the microbiota, CARD9 signaling can enhance granulocytic MDSC accumulation and activation in the lamina propria, which inhibits T cell-dependent antitumor immunity in the colon (Wang et al. 2018). Moreover, CARD9 signaling in the gut controls IL-18 production. This effect supports antitumor immunity during CAC development by promoting epithelial barrier integrity and by stimulating intestinal CD8 + T cells (Malik et al. 2018).

While these recent studies have demonstrated that CARD9 signaling can mediate pro- and antitumor immune effects during CAC (Hartjes and Ruland 2019), CARD9 signaling not only affects colon cancer growth at the primary side but also contributes to liver metastasis of colon carcinoma cells by promoting metastasis-associated macrophage polarization in response to VEGF CARD9 engagement and by inducing the production of tumor-promoting cytokines such as IL-10 and IL-1α (Yang et al. 2014). In addition to colon cancer, CARD9 has been implicated in the pathogenesis of renal cell carcinoma (RCC), whereby loss of the RCC tumor suppressor molecule pVHL leads to CARD9-BCL10-TRAF6-mediated signaling, which subsequently leads to oncogenic hyperactive JNK activation (An et al. 2013; Yang et al. 2007). In Lewis lung carcinoma, CARD9 has been implicated in regulating the accumulation of MDSCs in tumor tissues and subsequent IDO production, which can negatively regulate tumor growth (Qu et al. 2019). CARD9 has thus emerged as an important regulator of tumor immune microenvironments, although the cancer-derived ligands that can engage CARD9 signaling remain largely unclear.

Endogenous DAMPs that could in principle activate CARD9 include cholesterol crystals, SAP130 (a subunit of the histone deacetylase complex) and β-glucosylceramide (a ubiquitous intracellular metabolite that indicates massive tissue damage if elevated), which can all trigger Mincle (cholesterol crystals, β-glucosylceramide, SAP130) (Kiyotake et al. 2015; Nagata et al. 2017; Seifert et al. 2016) and vimentin (an intermediate filament protein), which activates Dectin-1 (vimentin) (Thiagarajan et al. 2013). While the in vivo scenarios under which these specific DAMPs activate CARD9 signaling are not well explored, there is emerging evidence that CARD9 signaling contributes to noninfectious inflammatory responses under conditions beyond the cancer microenvironment. In acute pancreatitis, the severity of pancreatic injury is correlated with overexpression of CARD9 in mononuclear cells, which leads to enhanced NF-κB and p38 MAPK signaling-mediated cytokine expression that promotes inflammatory pathology (Yang et al. 2016c, 2017). In the context of ischemic heart disease, CARD9 signaling promotes neutrophil infiltration following the acute ischemia/reperfusion (I/R) phase by inducing inflammatory cytokines such as TNF, IL-6, CXCL1 and MCP-1, thereby damaging the myocardium and promoting I/R injury (Qin et al. 2018). Similarly, after cerebral ischemic injury, which leads to an upregulation of triggering receptor expressed on myeloid cells (TREM)-1, the activation of CARD9 together with inflammasomes leads to pyroptosis and the release of inflammatory factors, which results in neuroinflammatory injury (Xu et al. 2019). Likewise, in a rat model of brain injury after subarachnoid hemorrhage (SAH), the sensing of necrotic cells by Mincle was linked to the activation of CARD9 signaling for inflammatory cytokine production, promoting neuroinflammation-induced brain damage following cerebral ischemia and reperfusion injury in the central nervous system (He et al. 2015). In contrast, in a murine model of atherosclerosis induced by a Western-type diet using *Ldlr*$^{-/-}$ mice, CARD9 deficiency increased the atherosclerotic lesion size and severity, indicating that CARD9 signaling may be protective for the development of atherosclerosis (Thiem et al. 2019). Two additional studies demonstrated further critical functions of CARD9 during neutrophil-mediated sterile inflammatory pathology. These reports demonstrated that CARD9 signaling can drive autoantibody-induced rheumatoid arthritis and dermatitis by controlling inflammatory chemokine/cytokine release in neutrophils (Németh et al. 2016) and that CARD9 engages in cross talk with SHP-1 to regulate IL-1-driven neutrophilic footpad inflammation in an animal model of human neutrophilic dermatosis (Tartey et al. 2018). Thus, these studies demonstrate that CARD9 signaling can mediate critical pathophysiological responses during sterile inflammatory stress in several organ systems.

8 Genetic Alterations of *CARD9* in Human Disease

Following the discovery of CARD9 as a nonredundant regulator of innate immunity and inflammation (Gross et al. 2006; Hara et al. 2007; Hsu et al. 2007) pathological *CARD9* loss-of-function (LOF) mutations and *CARD9* risk polymorphisms have been identified in human patients (Fig. 5). In line with the high susceptibility of CARD9-deficient mice to fungal infections (Gross et al. 2006), familial homozygous *CARD9* LOF mutations in patients also result in a high susceptibility to fungal infections (Glocker et al. 2009). Since the report of autosomal recessive *CARD9* LOF variants due to the Q295X mutation (Glocker et al. 2009), multiple *CARD9* LOF mutations have been identified in patients (Drummond and Lionakis 2016), including the homozygous *CARD9* missense mutations R18W (Lanternier et al. 2015a), R35Q (Lanternier et al. 2015b), R57H (Drummond et al. 2015), R70W (Lanternier et al. 2015b), Y91H (Gavino et al. 2014), R101C (Lanternier et al. 2013) and R101L (Grumach et al. 2015). All these alterations lead to impaired CARD9 signaling functions and a high susceptibility to fungal pathogens, including *Candida albicans, Aspergillus fumigatus, Phialophora verrucosa, Corynespora cassiicola,* several dermatophytes and some *Trichophyton* and *Exophilia* species, some of which cause lethal infections (Drummond and Lionakis 2016; Drummond et al. 2018). *Candida* infections are predominantly localized to the oral mucosa, subcutaneous tissues or the central nervous system (CNS), while *Aspergillus*, *Exophilia* and *Phialophora* species are recurrently detected in skin, CNS, abdominal organs or bone tissues (Drummond et al. 2018). Similar to CARD9-deficient mice, CARD9 signaling defects in patients lead to impaired Th17 cell responses (Glocker et al. 2009; LeibundGut-Landmann et al. 2007; Wang et al. 2014) and defective monocytic GM-CSF and C-X-C chemokine (e.g., CXCL1) production with impaired neutrophil recruitment to the site of infections (Drummond et al. 2015; Gavino et al. 2016).

In addition to *CARD9* LOF mutations, multiple human *CARD9* single-nucleotide polymorphisms (SNPs) have been associated with prevalent inflammatory disorders. These include IBDs, such as ulcerative colitis and Crohn's disease (Jostins et al. 2012; Rivas et al. 2011; Zhernakova et al. 2008) (McGovern et al. 2010) (Franke et al. 2010); ankylosing spondylitis, a severe form of arthritis affecting the

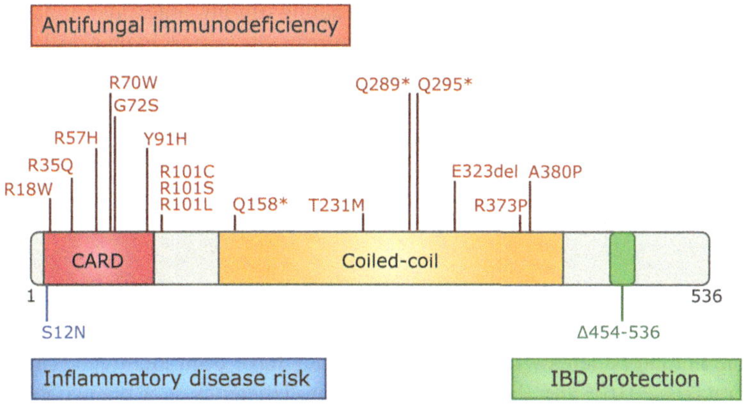

Fig. 5 Genetic alterations of CARD9 in human disease. While several CARD9 loss-of-function mutations lead to antifungal immunodeficiencies (red), the CARD9^{S12N} SNP leads to increased inflammatory disease risk (blue). Conversely, the CARD9$^{\Delta 11}$ variant, which results from a skipping of exon 11, has protective functions in IBD (green)

spine (Evans et al. 2011; Pointon et al. 2010); primary sclerosing cholangitis, a complex inflammatory disorder gradually resulting in liver cirrhosis (Janse et al. 2011); and IgA nephropathy, in which IgA deposits trigger glomerulonephritis leading to renal failure (Kiryluk et al. 2014). While most of the *CARD9* disease SNPs do not affect the coding regions of *CARD9*, they presumably influence the level of *CARD9* expression and thereby modulate CARD9 signaling strength. One prevalent *CARD9* SNP, which is correlated with an increased risk for IBD, encodes a missense variant at position 12 where the asparagine (N) is replaced by a serine (S) residue (Franke et al. 2010; McGovern et al. 2010). Experiments in knock-in mice that carry the respective CARD9^{S12N} allele demonstrated that this specific CARD9 variant can deviate CARD9 signaling to induce activation of the RELB NF-κB subunit for IL-5 production in response to CLR stimulation, resulting in a deviated T cell response toward Th2 differentiation (Xu et al. 2018). Consistently, human individuals with the CARD9^{S12N} variant exhibit an increased risk of Th2-mediated allergic bronchopulmonary fungal infections (Xu et al. 2018). Therefore, it is conceivable that related pathways might contribute to the inflammatory pathology in IBD patients with CARD9^{S12N} alteration.

Another human CARD9 IBD risk variant that has been functionally studied is the protective CARD9 isoform that decreases the risk of inflammatory bowel diseases and results from the exclusion of exon 11 within *CARD9*, creating a CARD9$^{\Delta 11}$ protein (Rivas et al. 2011). Mechanistically, CARD9$^{\Delta 11}$ can assemble trimeric signaling complexes with BCL10 and MALT1, but these CARD9$^{\Delta 11}$-BCL10-MALT1 complexes are defective in their ability to recruit TRIM62 for NF-κB activation. Therefore, it has been suggested that a reduction in TRIM62-mediated inflammation underlies the protective function of CARD9$^{\Delta 11}$ in the intestinal immune system during IBD pathogenesis. While these investigations into the function of aberrant CARD9 signaling during intestinal inflammation provide the first insights into the role of these pathways in human inflammatory disorders, mechanistic studies on the function of CARD9 in ankylosing spondylitis, primary sclerosing cholangitis or IgA nephropathy have not yet been carried out.

9 Conclusions and Therapeutic Perspective

Over the last decade, the myeloid cell-specific adapter molecule CARD9 has emerged as a key regulator of innate immunity and inflammatory responses. Multiple human patients with CARD9 deficiency have been identified (Drummond and Lionakis 2016), and their primary immunodeficiency is a phenocopy of that observed in CARD9-deficient mice (Gross et al. 2006), confirming that CARD9 function is highly conserved between humans and mice. The recently developed mechanistic understanding of the CARD9 signaling pathways has paved the way for rational strategies for the treatment of CARD9-dependent innate immune disorders. On the one hand, patients with inherited CARD9 deficiencies and defective immune responses have been successfully reconstituted with allogeneic stem cell transplantation to reconstitute leukocyte-mediated CARD9 immunity, resulting in complete clinical remissions (Queiroz-Telles et al. 2019). Other patients with CARD9 mutations have been successfully treated with GM-CSF, because this CARD9-controlled cytokine is critical for neutrophil recruitment and control of CNS candidiasis (Gavino et al. 2014). One additional strategy to enforce CLR-CARD9 signaling for the potential treatment of disseminated fungal infections could involve inhibition of the negative regulator CBLB, which has been experimentally tested in preclinical models (Wirnsberger et al. 2016; Xiao et al. 2016). Further strategies have been explored to boost CARD9-mediated immunity in vaccination regimens against mycobacterium tuberculosis (Ostrop et al. 2015), where selective triggering of the CLR-CARD9 signaling pathways with defined adjuvants can generate profound protective Th1 cell and Th17 cell responses (Werninghaus et al. 2009).

Conversely, in scenarios where aberrant CARD9 signaling might drive inflammatory pathologies, strategies are being developed to dampen the CARD9 pathway. Based on the knowledge that the IBD-protective $CARD9^{\Delta 11}$ protein exhibits impairment of TRIM62 recruitment and subsequent NF-κB activation, small molecule inhibitors that disrupt the proinflammatory CARD9-TRIM62 interaction have been developed as candidate IBD therapeutics (Leshchiner et al. 2017). Other strategies to attenuate pathological CARD9 activity could potentially utilize MALT1 paracaspase inhibitors, which were initially developed for the treatment of malignant lymphomas that are driven by pathological BCL10-MALT1 signaling in transformed B lymphocytes (Ferch et al. 2009; Fontan et al. 2012; Hailfinger et al. 2011; Nagel et al. 2012). Although the functional identity and biological effects of MALT1 substrates in innate immune responses remain largely unclear, the activation of CARD9-BCL10-MALT1 signalosomes induces proteolytic activity of

MALT1 (Gringhuis et al. 2011), which likely controls important inflammatory pathways. Therefore, it is necessary to further study the innate immune function of the MALT1 protease in vivo and to explore whether its pharmacological inhibition could also be utilized to target CARD9-mediated immune pathology. Therefore, further mechanistic knowledge on the regulation of CARD9-BCL10-MALT1 signaling complexes in innate immunity and inflammation is required; as such, knowledge will open new avenues for therapeutic strategies that target inflammatory pathways during human pathology.

Acknowledgements The work performed in the authors' laboratory was supported by research grants from the Deutsche Forschungsgemeinschaft (DFG, German Research Foundation) (SFB 1054/B01, Projektnummer 360372040 – SFB 1335/P01 and P08, Projektnummer 395357507 – SFB 1371/P05, TRR 237/A10, RU 695/9-1) and the European Research Council (ERC) under the European Union's Horizon 2020 research and innovation program (grant agreement No 834154). We thank Lara Hartjes for her help developing the graphics.

Declaration of Interests

The authors declare no competing interests.

References

An J, Liu H, Magyar CE, Guo Y, Veena MS, Srivatsan ES, Huang J, Rettig MB (2013) Hyperactivated JNK is a therapeutic target in pVHL-deficient renal cell carcinoma. Cancer Res 73:1374–1385

Athanasopoulos V, Ramiscal RR, Vinuesa CG (2016) ROQUIN signalling pathways in innate and adaptive immunity. Eur J Immunol 46:1082–1090

Atif SM, Lee S-J, Li L-X, Uematsu S, Akira S, Gorjestani S, Lin X, Schweighoffer E, Tybulewicz VLJ, McSorley SJ (2015) Rapid CD4+ T-cell responses to bacterial flagellin require dendritic cell expression of Syk and CARD9. Eur J Immunol 45:513–524

Baens M, Bonsignore L, Somers R, Vanderheydt C, Weeks SD, Gunnarsson J, Nilsson E, Roth RG, Thome M, Marynen P (2014) MALT1 auto-proteolysis is essential for NF-κB-dependent gene transcription in activated lymphocytes. PLoS ONE 9:e103774

Bergmann H, Roth S, Pechloff K, Kiss EA, Kuhn S, Heikenwälder M, Diefenbach A, Greten FR, Ruland J (2017) Card9-dependent IL-1β regulates IL-22 production from group 3 innate lymphoid cells and promotes colitis-associated cancer. Eur J Immunol 47:1342–1353

Bertin J, Guo Y, Wang L, Srinivasula SM, Jacobson MD, Poyet JL, Merriam S, Du MQ, Dyer MJ, Robison KE et al (2000) CARD9 is a novel caspase recruitment domain-containing protein that interacts with BCL10/CLAP and activates NF-kappa B. J Biol Chem 275:41082–41086

Brown GD, Willment JA, Whitehead L (2018) C-type lectins in immunity and homeostasis. Nat Rev Immunol 18:374–389

Bugarcic A, Hitchens K, Beckhouse AG, Wells CA, Ashman RB, Blanchard H (2008) Human and mouse macrophage-inducible C-type lectin (Mincle) bind Candida albicans. Glycobiology 18:679–685

Cao X (2016) Self-regulation and cross-regulation of pattern-recognition receptor signalling in health and disease. Nat Rev Immunol 16:35–50

Cao Z, Conway KL, Heath RJ, Rush JS, Leshchiner ES, Ramirez-Ortiz ZG, Nedelsky NB, Huang H, Ng A, Gardet A et al (2015) Ubiquitin Ligase TRIM62 regulates CARD9-mediated anti-fungal immunity and intestinal inflammation. Immunity 43:715–726

Cao M, Wu Z, Lou Q, Lu W, Zhang J, Li Q, Zhang Y, Yao Y, Zhao Q, Li M et al (2019) Dectin-1-induced RIPK1 and RIPK3 activation protects host against *Candida albicans* infection. Cell Death Differ

Chassaing B., Aitken JD, Malleshappa M, Vijay-Kumar M (2014) Dextran sulfate sodium (DSS)-induced colitis in mice. Curr Protoc Immunol (Ed. John E Coligan Al 104, Unit-15.25)

Coornaert B, Baens M, Heyninck K, Bekaert T, Haegman M, Staal J, Sun L, Chen ZJ, Marynen P, Beyaert R (2008) T cell antigen receptor stimulation induces MALT1 paracaspase-mediated cleavage of the NF-kappaB inhibitor A20. Nat Immunol 9:263–271

Coornaert B, Carpentier I, Beyaert R (2009) A20: central gatekeeper in inflammation and immunity. J Biol Chem 284:8217–8221

Deng L, Wang C, Spencer E, Yang L, Braun A, You J, Slaughter C, Pickart C, Chen ZJ (2000) Activation of the IkappaB kinase complex by TRAF6 requires a dimeric ubiquitin-conjugating enzyme complex and a unique polyubiquitin chain. Cell 103:351–361

Dorhoi A, Desel C, Yeremeev V, Pradl L, Brinkmann V, Mollenkopf H-J, Hanke K, Gross O, Ruland J, Kaufmann SHE (2010) The adaptor molecule CARD9 is essential for tuberculosis control. J Exp Med 207:777–792

Douanne T, Gavard J, Bidère N (2016) The paracaspase MALT1 cleaves the LUBAC subunit HOIL1 during antigen receptor signaling. J Cell Sci 129:1775–1780

Drummond RA, Lionakis MS (2016) Mechanistic insights into the role of C-type lectin receptor/CARD9 signaling in human antifungal immunity. Front Cell Infect Microbiol 6:39

Drummond RA, Collar AL, Swamydas M, Rodriguez CA, Lim JK, Mendez LM, Fink DL, Hsu AP, Zhai B, Karauzum H et al (2015) CARD9-dependent neutrophil recruitment protects against fungal invasion of the central nervous system. PLoS Pathog 11:e1005293

Drummond RA, Franco LM, Lionakis MS (2018) Human CARD9: a critical molecule of fungal immune surveillance. Front Immunol 9:1836

Drummond RA, Swamydas M, Oikonomou V, Zhai B, Dambuza IM, Schaefer BC, Bohrer AC, Mayer-Barber KD, Lira SA, Iwakura Y et al (2019) CARD9+ microglia promote antifungal immunity via IL-1β- and CXCL1-mediated neutrophil recruitment. Nat Immunol 20:559–570

Elton L, Carpentier I, Staal J, Driege Y, Haegman M, Beyaert R (2016) MALT1 cleaves the E3 ubiquitin ligase HOIL-1 in activated T cells, generating a dominant negative inhibitor of LUBAC-induced NF-κB signaling. FEBS J 283:403–412

Evans DM, Spencer CCA, Pointon JJ, Su Z, Harvey D, Kochan G, Oppermann U, Opperman U, Dilthey A, Pirinen M et al (2011) Interaction between ERAP1 and HLA-B27 in ankylosing spondylitis implicates peptide handling in the mechanism for HLA-B27 in disease susceptibility. Nat Genet 43:761–767

Ferch U, Kloo B, Gewies A, Pfänder V, Düwel M, Peschel C, Krappmann D, Ruland J (2009) Inhibition of MALT1 protease activity is selectively toxic for activated B cell-like diffuse large B cell lymphoma cells. J Exp Med 206:2313–2320

Fontan L, Yang C, Kabaleeswaran V, Volpon L, Osborne MJ, Beltran E, Garcia M, Cerchietti L, Shaknovich R, Yang SN et al (2012) MALT1 small molecule inhibitors specifically suppress ABC-DLBCL in vitro and in vivo. Cancer Cell 22:812–824

Franke A, McGovern DPB, Barrett JC, Wang K, Radford-Smith GL, Ahmad T, Lees CW, Balschun T, Lee J, Roberts R et al (2010) Genome-wide meta-analysis increases to 71 the number of confirmed Crohn's disease susceptibility loci. Nat Genet 42:1118–1125

Gabrilovich DI, Nagaraj S (2009) Myeloid-derived suppressor cells as regulators of the immune system. Nat Rev Immunol 9:162–174

Gavino C, Cotter A, Lichtenstein D, Lejtenyi D, Fortin C, Legault C, Alirezaie N, Majewski J, Sheppard DC, Behr MA et al (2014) CARD9 deficiency and spontaneous central nervous system candidiasis: complete clinical remission with GM-CSF therapy. Clin Infect Dis Off Publ Infect Dis Soc Am 59:81–84

Gavino C, Hamel N, Zeng JB, Legault C, Guiot M-C, Chankowsky J, Lejtenyi D, Lemire M, Alarie I, Dufresne S et al (2016) Impaired RASGRF1/ERK-mediated GM-CSF response characterizes CARD9 deficiency in French-Canadians. J Allergy Clin Immunol 137:1178–1188.e7

Gazendam RP, van Hamme JL, Tool ATJ, van Houdt M, Verkuijlen PJJH, Herbst M, Liese JG, van de Veerdonk FL, Roos D, van den Berg TK et al (2014) Two independent killing mechanisms of *Candida albicans* by human neutrophils: evidence from innate immunity defects. Blood 124:590–597

Gazendam RP, van de Geer A, Roos D, van den Berg TK, Kuijpers TW (2016) How neutrophils kill fungi. Immunol Rev 273:299–311

Geijtenbeek TBH, Gringhuis SI (2009) Signalling through C-type lectin receptors: shaping immune responses. Nat Rev Immunol 9:465–479

Glocker E-O, Hennigs A, Nabavi M, Schäffer AA, Woellner C, Salzer U, Pfeifer D, Veelken H, Warnatz K, Tahami F et al (2009) A homozygous CARD9 mutation in a family with susceptibility to fungal infections. N Engl J Med 361:1727–1735

Gong T, Liu L, Jiang W, Zhou R (2020) DAMP-sensing receptors in sterile inflammation and inflammatory diseases. Nat Rev Immunol 20:95–112

Gringhuis SI, Wevers BA, Kaptein TM, van Capel TMM, Theelen B, Boekhout T, de Jong EC, Geijtenbeek TBH (2011) Selective C-Rel activation via Malt1 controls anti-fungal T(H)-17 immunity by dectin-1 and dectin-2. PLoS Pathog 7:e1001259

Gross O, Gewies A, Finger K, Schäfer M, Sparwasser T, Peschel C, Förster I, Ruland J (2006) Card9 controls a non-TLR signalling pathway for innate anti-fungal immunity. Nature 442:651–656

Gross O, Poeck H, Bscheider M, Dostert C, Hannesschläger N, Endres S, Hartmann G, Tardivel A, Schweighoffer E, Tybulewicz V et al (2009) Syk kinase signalling couples to the Nlrp3 inflammasome for anti-fungal host defence. Nature 459:433–436

Grumach AS, de Queiroz-Telles F, Migaud M, Lanternier F, Filho NR, Palma SMU, Constantino-Silva RN, Casanova JL, Puel A (2015) A homozygous CARD9 mutation in a Brazilian patient with deep dermatophytosis. J Clin Immunol 35:486–490

Guo H, Callaway JB, Ting JP-Y (2015) Inflammasomes: mechanism of action, role in disease, and therapeutics. Nat Med 21:677–687

Hailfinger S, Nogai H, Pelzer C, Jaworski M, Cabalzar K, Charton J-E, Guzzardi M, Décaillet C, Grau M, Dörken B et al (2011) Malt1-dependent RelB cleavage promotes canonical NF-κB activation in lymphocytes and lymphoma cell lines. Proc Natl Acad Sci 108:14596–14601

Hara H, Ishihara C, Takeuchi A, Imanishi T, Xue L, Morris SW, Inui M, Takai T, Shibuya A, Saijo S et al (2007) The adaptor protein CARD9 is essential for the activation of myeloid cells through ITAM-associated and Toll-like receptors. Nat Immunol 8:619–629

Hartjes L, Ruland J (2019) CARD9 signaling in intestinal immune homeostasis and oncogenesis. Front Immunol 10:419

Hayden MS, Ghosh S (2012) NF-κB, the first quarter-century: remarkable progress and outstanding questions. Genes Dev 26:203–234

He Y, Xu L, Li B, Guo Z-N, Hu Q, Guo Z, Tang J, Chen Y, Zhang Y, Tang J et al (2015) Macrophage-inducible C-type lectin/spleen tyrosine kinase signaling pathway contributes to neuroinflammation after subarachnoid hemorrhage in rats. Stroke 46:2277–2286

He J-J, Ma J, Song H-Q, Zhou D-H, Wang J-L, Huang S-Y, Zhu X-Q (2016) Transcriptomic analysis of global changes in cytokine expression in mouse spleens following acute Toxoplasma gondii infection. Parasitol Res 115:703–712

Heinrich A, Heyl KA, Klaile E, Müller MM, Klassert TE, Wiessner A, Fischer K, Schumann RR, Seifert U, Riesbeck K et al (2016) *Moraxella catarrhalis* induces CEACAM3-Syk-CARD9-dependent activation of human granulocytes. Cell Microbiol 18:1570–1582

Holliday MJ, Ferrao R, de Leon Boenig G, Estevez A, Helgason E, Rohou A, Dueber EC, Fairbrother WJ (2018) Picomolar zinc binding modulates formation of Bcl10-nucleating assemblies of the caspase recruitment domain (CARD) of CARD9. J Biol Chem 293:16803–16817

Holliday MJ, Witt A, Rodríguez Gama A, Walters BT, Arthur CP, Halfmann R, Rohou A, Dueber EC, Fairbrother WJ (2019) Structures of autoinhibited and polymerized forms of CARD9 reveal mechanisms of CARD9 and CARD11 activation. Nat Commun 10:3070

Hsu Y-MS, Zhang Y, You Y, Wang D, Li H, Duramad O, Qin X-F, Dong C, Lin X (2007) The adaptor protein CARD9 is required for innate immune responses to intracellular pathogens. Nat Immunol 8:198–205

Huber-Lang M, Lambris JD, Ward PA (2018) Innate immune responses to trauma. Nat Immunol 19:327–341

Hung C-Y, Castro-Lopez N, Cole GT (2016) Card9- and MyD88-mediated gamma interferon and nitric oxide production is essential for resistance to subcutaneous *Coccidioides posadasii* infection. Infect Immun 84:1166–1175

Imai T, Matsumura T, Mayer-Lambertz S, Wells CA, Ishikawa E, Butcher SK, Barnett TC, Walker MJ, Imamura A, Ishida H et al (2018) Lipoteichoic acid anchor triggers Mincle to drive protective immunity against invasive group A Streptococcus infection. Proc Natl Acad Sci USA 115:E10662–E10671

Janse M, Lamberts LE, Franke L, Raychaudhuri S, Ellinghaus E, Muri Boberg K, Melum E, Folseraas T, Schrumpf E, Bergquist A et al (2011) Three ulcerative colitis susceptibility loci are associated with primary sclerosing cholangitis and indicate a role for IL2, REL, and CARD9. Hepatol Baltim Md 53:1977–1985

Jeltsch KM, Hu D, Brenner S, Zöller J, Heinz GA, Nagel D, Vogel KU, Rehage N, Warth SC, Edelmann SL et al (2014) Cleavage of roquin and regnase-1 by the paracaspase MALT1 releases their cooperatively repressed targets to promote T(H)17 differentiation. Nat Immunol 15:1079–1089

Jhingran A, Mar KB, Kumasaka DK, Knoblaugh SE, Ngo LY, Segal BH, Iwakura Y, Lowell CA, Hamerman JA, Lin X et al (2012) Tracing conidial fate and measuring host cell antifungal activity using a reporter of microbial viability in the lung. Cell Rep. 2:1762–1773

Jhingran A, Kasahara S, Shepardson KM, Junecko BAF, Heung LJ, Kumasaka DK, Knoblaugh SE, Lin X, Kazmierczak BI, Reinhart TA et al (2015) Compartment-specific and sequential role of MyD88 and CARD9 in chemokine induction and innate defense during respiratory fungal infection. PLoS Pathog 11:e1004589

Jia X-M, Tang B, Zhu L-L, Liu Y-H, Zhao X-Q, Gorjestani S, Hsu Y-MS, Yang L, Guan J-H, Xu G-T et al (2014) CARD9 mediates Dectin-1-induced ERK activation by linking Ras-GRF1 to H-Ras for antifungal immunity. J Exp Med 211:2307–2321

Jostins L, Ripke S, Weersma RK, Duerr RH, McGovern DP, Hui KY, Lee JC, Philip Schumm L, Sharma Y, Anderson CA et al (2012) Host–microbe interactions have shaped the genetic architecture of inflammatory bowel disease. Nature 491:119–124

Kalantari P, Morales Y, Miller EA, Jaramillo LD, Ponichtera HE, Wuethrich MA, Cheong C, Seminario MC, Russo JM, Bunnell SC et al (2018) CD209a synergizes with dectin-2 and mincle to drive severe Th17 cell-mediated schistosome egg-induced immunopathology. Cell Rep 22:1288–1300

Kiryluk K, Li Y, Scolari F, Sanna-Cherchi S, Choi M, Verbitsky M, Fasel D, Lata S, Prakash S, Shapiro S et al (2014) Discovery of new risk loci for IgA nephropathy implicates genes involved in immunity against intestinal pathogens. Nat Genet 46:1187–1196

Kiyotake R, Oh-hora M, Ishikawa E, Miyamoto T, Ishibashi T, Yamasaki S (2015) Human Mincle binds to cholesterol crystals and triggers innate immune responses. J Biol Chem 290:25322–25332

Klein T, Fung S-Y, Renner F, Blank MA, Dufour A, Kang S, Bolger-Munro M, Scurll JM, Priatel JJ, Schweigler P et al (2015) The paracaspase MALT1 cleaves HOIL1 reducing linear ubiquitination by LUBAC to dampen lymphocyte NF-κB signalling. Nat Commun 6:8777

Lamas B, Richard ML, Leducq V, Pham H-P, Michel M-L, Da Costa G, Bridonneau C, Jegou S, Hoffmann TW, Natividad JM et al (2016) CARD9 impacts colitis by altering gut microbiota metabolism of tryptophan into aryl hydrocarbon receptor ligands. Nat Med 22:598–605

Lamas B, Michel M-L, Waldschmitt N, Pham H-P, Zacharioudaki V, Dupraz L, Delacre M, Natividad JM, Costa GD, Planchais J et al (2018) Card9 mediates susceptibility to intestinal pathogens through microbiota modulation and control of bacterial virulence. Gut 67:1836–1844

Lanternier F, Cypowyj S, Picard C, Bustamante J, Lortholary O, Casanova J-L, Puel A (2013) Primary immunodeficiencies underlying fungal infections. Curr Opin Pediatr 25:736–747
Lanternier F, Barbati E, Meinzer U, Liu L, Pedergnana V, Migaud M, Héritier S, Chomton M, Frémond M-L, Gonzales E et al (2015a) Inherited CARD9 deficiency in 2 unrelated patients with invasive exophiala infection. J Infect Dis 211:1241–1250
Lanternier F, Mahdaviani SA, Barbati E, Chaussade H, Koumar Y, Levy R, Denis B, Brunel A-S, Martin S, Loop M et al (2015b) Inherited CARD9 deficiency in otherwise healthy children and adults with Candida species-induced meningoencephalitis, colitis, or both. J Allergy Clin Immunol 135:1558–1568.e2
LeibundGut-Landmann S, Gross O, Robinson MJ, Osorio F, Slack EC, Tsoni SV, Schweighoffer E, Tybulewicz V, Brown GD, Ruland J et al (2007) Syk- and CARD9-dependent coupling of innate immunity to the induction of T helper cells that produce interleukin 17. Nat Immunol 8:630–638
Leshchiner ES, Rush JS, Durney MA, Cao Z, Dančík V, Chittick B, Wu H, Petrone A, Bittker JA, Phillips A et al (2017) Small-molecule inhibitors directly target CARD9 and mimic its protective variant in inflammatory bowel disease. Proc Natl Acad Sci USA 114:11392–11397
Liu T, Zhang L, Joo D, Sun S-C (2017) NF-κB signaling in inflammation. Signal Transduct Target Ther 2:1–9
Loh JT, Xu S, Huo JX, Kim SS-Y, Wang Y, Lam K-P (2019) Dok3-protein phosphatase 1 interaction attenuates Card9 signaling and neutrophil-dependent antifungal immunity. J Clin Invest 130:2717–2729
MacDuff DA, Reese TA, Baldridge MT, Kimmey JM, Weiss LA, Song C, Nice TJ, Carrero J, Colonna M, Edelson BT et al (2017) Regulation of inflammation, innate immunity and intestinal homeostasis by HOIL1. FASEB J 31:622.1–622.1
Maknitikul S, Luplertlop N, Grau GER, Ampawong S (2017) Dysregulation of pulmonary endothelial protein C receptor and thrombomodulin in severe falciparum malaria-associated ARDS relevant to hemozoin. PLoS ONE 12:e0181674
Maknitikul S, Luplertlop N, Chaisri U, Maneerat Y, Ampawong S (2018) Featured article: immunomodulatory effect of hemozoin on pneumocyte apoptosis via CARD9 pathway, a possibly retarding pulmonary resolution. Exp Biol Med Maywood NJ 243:395–407
Malik A, Sharma D, Malireddi RKS, Guy CS, Chang T-C, Olsen SR, Neale G, Vogel P, Kanneganti T-D (2018) SYK-CARD9 signaling axis promotes gut fungi-mediated inflammasome activation to restrict colitis and colon cancer. Immunity 49:515–530.e5
Mao L, Zhang L, Li H, Chen W, Wang H, Wu S, Guo C, Lu A, Yang G, An L et al (2014) Pathogenic fungus *Microsporum canis* activates the NLRP3 inflammasome. Infect Immun 82:882–892
Mao R, Yang R, Chen X, Harhaj EW, Wang X, Fan Y (2017) Regnase-1, a rapid response ribonuclease regulating inflammation and stress responses. Cell Mol Immunol 14:412–422
Marshall ASJ, Willment JA, Lin H-H, Williams DL, Gordon S, Brown GD (2004) Identification and characterization of a novel human myeloid inhibitory C-type lectin-like receptor (MICL) that is predominantly expressed on granulocytes and monocytes. J Biol Chem 279:14792–14802
McGovern DPB, Gardet A, Törkvist L, Goyette P, Essers J, Taylor KD, Neale BM, Ong RTH, Lagacé C, Li C et al (2010) Genome-wide association identifies multiple ulcerative colitis susceptibility loci. Nat Genet 42:332–337
Mino T, Murakawa Y, Fukao A, Vandenbon A, Wessels H-H, Ori D, Uehata T, Tartey S, Akira S, Suzuki Y et al (2015) Regnase-1 and Roquin regulate a common element in inflammatory mRNAs by spatiotemporally distinct mechanisms. Cell 161:1058–1073
Monteiro J, Lepenies B (2017) Myeloid C-Type lectin receptors in viral recognition and antiviral immunity. Viruses 9:59
Monteiro JT, Schön K, Ebbecke T, Goethe R, Ruland J, Baumgärtner W, Becker SC, Lepenies B (2019) The CARD9-associated C-type lectin, mincle, recognizes La Crosse Virus (LACV) but plays a limited role in early antiviral responses against LACV. Viruses 11

Nagata M, Izumi Y, Ishikawa E, Kiyotake R, Doi R, Iwai S, Omahdi Z, Yamaji T, Miyamoto T, Bamba T et al (2017) Intracellular metabolite β-glucosylceramide is an endogenous Mincle ligand possessing immunostimulatory activity. Proc Natl Acad Sci 114:E3285–E3294

Nagel D, Spranger S, Vincendeau M, Grau M, Raffegerst S, Kloo B, Hlahla D, Neuenschwander M, Peter von Kries J, Hadian K et al (2012) Pharmacologic inhibition of MALT1 protease by phenothiazines as a therapeutic approach for the treatment of aggressive ABC-DLBCL. Cancer Cell 22:825–837

Németh T, Futosi K, Sitaru C, Ruland J, Mócsai A (2016) Neutrophil-specific deletion of the CARD9 gene expression regulator suppresses autoantibody-induced inflammation in vivo. Nat Commun 7:11004

Oeckinghaus A, Wegener E, Welteke V, Ferch U, Arslan SC, Ruland J, Scheidereit C, Krappmann D (2007) Malt1 ubiquitination triggers NF-kappaB signaling upon T-cell activation. EMBO J 26:4634–4645

Ostrop J, Jozefowski K, Zimmermann S, Hofmann K, Strasser E, Lepenies B, Lang R (2015) Contribution of MINCLE-SYK signaling to activation of primary human APCs by mycobacterial cord factor and the novel adjuvant TDB. J Immunol Baltim Md 1950 (195):2417–2428

Pandori WJ, Lima TS, Mallya S, Kao TH, Gov L, Lodoen MB (2019) Toxoplasma gondii activates a Syk-CARD9-NF-κB signaling axis and gasdermin D-independent release of IL-1β during infection of primary human monocytes. PLoS Pathog 15:e1007923

Phongsisay V, Iizasa E, Hara H, Yoshida H (2015) Evidence for TLR4 and FcRγ-CARD9 activation by cholera toxin B subunit and its direct bindings to TREM2 and LMIR5 receptors. Mol Immunol 66:463–471

Poeck H, Bscheider M, Gross O, Finger K, Roth S, Rebsamen M, Hannesschläger N, Schlee M, Rothenfusser S, Barchet W et al (2010) Recognition of RNA virus by RIG-I results in activation of CARD9 and inflammasome signaling for interleukin 1β production. Nat Immunol 11:63–69

Pointon JJ, Harvey D, Karaderi T, Appleton LH, Farrar C, Stone MA, Sturrock RD, Brown MA, Wordsworth BP (2010) Elucidating the chromosome 9 association with AS; CARD9 is a candidate gene. Genes Immun 11:490–496

Qin X, Peterson MR, Haller SE, Cao L, Thomas DP, He G (2018) Caspase recruitment domain-containing protein 9 (CARD9) knockout reduces regional ischemia/reperfusion injury through an attenuated inflammatory response. PLoS ONE 13:e0199711

Qu J, Liu L, Xu Q, Ren J, Xu Z, Dou H, Shen S, Hou Y, Mou Y, Wang T (2019) CARD9 prevents lung cancer development by suppressing the expansion of myeloid-derived suppressor cells and IDO production. Int J Cancer

Queiroz-Telles F, Mercier T, Maertens J, Sola CBS, Bonfim C, Lortholary O, Constantino-Silva RMN, Schrijvers R, Hagen F, Meis JF et al (2019) Successful allogenic stem cell transplantation in patients with inherited CARD9 deficiency. J Clin Immunol 39:462–469

Redelinghuys P, Brown GD (2011) Inhibitory C-type lectin receptors in myeloid cells. Immunol Lett 136:1–12

Rieber N, Singh A, Öz H, Carevic M, Bouzani M, Amich J, Ost M, Ye Z, Ballbach M, Schäfer I et al (2015) Pathogenic fungi regulate immunity by inducing neutrophilic myeloid-derived suppressor cells. Cell Host Microbe 17:507–514

Ritter M, Gross O, Kays S, Ruland J, Nimmerjahn F, Saijo S, Tschopp J, Layland LE, Prazeres da Costa C (2010) *Schistosoma mansoni* triggers Dectin-2, which activates the Nlrp3 inflammasome and alters adaptive immune responses. Proc Natl Acad Sci 107:20459–20464

Rivas MA, Beaudoin M, Gardet A, Stevens C, Sharma Y, Zhang CK, Boucher G, Ripke S, Ellinghaus D, Burtt N et al (2011) Deep resequencing of GWAS loci identifies independent rare variants associated with inflammatory bowel disease. Nat Genet 43:1066–1073

Rivera A, Siracusa MC, Yap GS, Gause WC (2016) Innate cell communication kick-starts pathogen-specific immunity. Nat Immunol 17:356–363

Robinson MJ, Osorio F, Rosas M, Freitas RP, Schweighoffer E, Gross O, Verbeek JS, Ruland J, Tybulewicz V, Brown GD et al (2009) Dectin-2 is a Syk-coupled pattern recognition receptor crucial for Th17 responses to fungal infection. J Exp Med 206:2037–2051

Rogers NC, Slack EC, Edwards AD, Nolte MA, Schulz O, Schweighoffer E, Williams DL, Gordon S, Tybulewicz VL, Brown GD et al (2005) Syk-dependent cytokine induction by Dectin-1 reveals a novel pattern recognition pathway for C type lectins. Immunity 22:507–517

Roth S, Ruland J (2013) Caspase recruitment domain-containing protein 9 signaling in innate immunity and inflammation. Trends Immunol 34:243–250

Roth S, Rottach A, Lotz-Havla AS, Laux V, Muschaweckh A, Gersting SW, Muntau AC, Hopfner K-P, Jin L, Vanness K et al (2014) Rad50-CARD9 interactions link cytosolic DNA sensing to IL-1β production. Nat Immunol 15:538–545

Roth S, Bergmann H, Jaeger M, Yeroslaviz A, Neumann K, Koenig P-A, Prazeres da Costa C, Vanes L, Kumar V, Johnson M et al (2016) Vav proteins are key regulators of Card9 signaling for innate antifungal immunity. Cell Rep 17:2572–2583

Ruland J, Hartjes L (2019) CARD–BCL-10–MALT1 signalling in protective and pathological immunity. Nat Rev Immunol 19:118–134

Saijo S, Ikeda S, Yamabe K, Kakuta S, Ishigame H, Akitsu A, Fujikado N, Kusaka T, Kubo S, Chung S et al (2010) Dectin-2 recognition of alpha-mannans and induction of Th17 cell differentiation is essential for host defense against *Candida albicans*. Immunity 32:681–691

Sato K, Yang X, Yudate T, Chung J-S, Wu J, Luby-Phelps K, Kimberly RP, Underhill D, Cruz PD, Ariizumi K (2006) Dectin-2 is a pattern recognition receptor for fungi that couples with the Fc receptor gamma chain to induce innate immune responses. J Biol Chem 281:38854–38866

Schweneker K, Gorka O, Schweneker M, Poeck H, Tschopp J, Peschel C, Ruland J, Gross O (2013) The mycobacterial cord factor adjuvant analogue trehalose-6,6'-dibehenate (TDB) activates the Nlrp3 inflammasome. Immunobiology 218:664–673

Seifert L, Werba G, Tiwari S, Giao Ly NN, Alothman S, Alqunaibit D, Avanzi A, Barilla R, Daley D, Greco SH et al (2016) The necrosome promotes pancreatic oncogenesis via CXCL1 and Mincle-induced immune suppression. Nature 532:245–249

Shenderov K, Barber DL, Mayer-Barber KD, Gurcha SS, Jankovic D, Feng CG, Oland S, Hieny S, Caspar P, Yamasaki S et al (2013) Cord factor and peptidoglycan recapitulate the Th17-promoting adjuvant activity of mycobacteria through mincle/CARD9 signaling and the inflammasome. J Immunol Baltim Md 1950(190):5722–5730

Sokol H, Conway KL, Zhang M, Choi M, Morin B, Cao Z, Villablanca EJ, Li C, Wijmenga C, Yun SH et al (2013) Card9 mediates intestinal epithelial cell restitution, T-helper 17 responses, and control of bacterial infection in mice. Gastroenterology 145:591–601.e3

Staal J, Driege Y, Haegman M, Borghi A, Hulpiau P, Lievens L, Gul IS, Sundararaman S, Gonçalves A, Dhondt I et al (2018) Ancient origin of the CARD-coiled coil/Bcl10/MALT1-like paracaspase signaling complex indicates unknown critical functions. Front Immunol 9:1136

Strasser D, Neumann K, Bergmann H, Marakalala MJ, Guler R, Rojowska A, Hopfner K-P, Brombacher F, Urlaub H, Baier G et al (2012) Syk kinase-coupled C-type lectin receptors engage protein kinase C-σ to elicit Card9 adaptor-mediated innate immunity. Immunity 36:32–42

Sun L, Deng L, Ea C-K, Xia Z-P, Chen ZJ (2004) The TRAF6 ubiquitin ligase and TAK1 kinase mediate IKK activation by BCL10 and MALT1 in T lymphocytes. Mol Cell 14:289–301

Tartey S, Gurung P, Samir P, Burton A, Kanneganti T-D (2018) Cutting edge: dysregulated CARD9 signaling in neutrophils drives inflammation in a mouse model of neutrophilic dermatoses. J Immunol Baltim Md 1950(201):1639–1644

Taylor PR, Tsoni SV, Willment JA, Dennehy KM, Rosas M, Findon H, Haynes K, Steele C, Botto M, Gordon S et al (2007) Dectin-1 is required for beta-glucan recognition and control of fungal infection. Nat Immunol 8:31–38

Thiagarajan PS, Yakubenko VP, Elsori DH, Yadav SP, Willard B, Tan CD, René Rodriguez E, Febbraio M, Cathcart MK (2013) Vimentin is an endogenous ligand for the pattern recognition receptor Dectin-1. Cardiovasc Res 99:494–504

Thiem K, Hoeke G, van den Berg S, Hijmans A, Jacobs CWM, Zhou E, Mol IM, Mouktaroudi M, Bussink J, Kanneganti TD et al (2019) Deletion of hematopoietic Dectin-2 or CARD9 does not protect against atherosclerotic plaque formation in hyperlipidemic mice. Sci Rep 9:4337

Uehata T, Iwasaki H, Vandenbon A, Matsushita K, Hernandez-Cuellar E, Kuniyoshi K, Satoh T, Mino T, Suzuki Y, Standley DM et al (2013) Malt1-induced cleavage of regnase-1 in CD4(+) helper T cells regulates immune activation. Cell 153:1036–1049

Uematsu T, Iizasa E, Kobayashi N, Yoshida H, Hara H (2015) Loss of CARD9-mediated innate activation attenuates severe influenza pneumonia without compromising host viral immunity. Sci Rep 5:17577

Uren AG, O'Rourke K, Aravind LA, Pisabarro MT, Seshagiri S, Koonin EV, Dixit VM (2000) Identification of paracaspases and metacaspases: two ancient families of caspase-like proteins, one of which plays a key role in MALT lymphoma. Mol Cell 6:961–967

Wang C, Deng L, Hong M, Akkaraju GR, Inoue J, Chen ZJ (2001) TAK1 is a ubiquitin-dependent kinase of MKK and IKK. Nature 412:346–351

Wang S-F, Huang JC, Lee Y-M, Liu S-J, Chan Y-J, Chau Y-P, Chong P, Chen Y-MA (2008) DC-SIGN mediates avian H5N1 influenza virus infection in cis and in trans. Biochem Biophys Res Commun 373:561–566

Wang X, Wang W, Lin Z, Wang X, Li T, Yu J, Liu W, Tong Z, Xu Y, Zhang J et al (2014) CARD9 mutations linked to subcutaneous phaeohyphomycosis and TH17 cell deficiencies. J Allergy Clin Immunol 133:905–908.e3

Wang T, Fan C, Yao A, Xu X, Zheng G, You Y, Jiang C, Zhao X, Hou Y, Hung M-C et al (2018) The adaptor protein CARD9 protects against colon cancer by restricting mycobiota-mediated expansion of myeloid-derived suppressor cells. Immunity 49:504–514.e4

Werninghaus K, Babiak A, Gross O, Hölscher C, Dietrich H, Agger EM, Mages J, Mocsai A, Schoenen H, Finger K et al (2009) Adjuvanticity of a synthetic cord factor analogue for subunit *Mycobacterium tuberculosis* vaccination requires FcRgamma-Syk-Card9-dependent innate immune activation. J Exp Med 206:89–97

Wirnsberger G, Zwolanek F, Asaoka T, Kozieradzki I, Tortola L, Wimmer RA, Kavirayani A, Fresser F, Baier G, Langdon WY et al (2016) Inhibition of CBLB protects from lethal *Candida albicans* sepsis. Nat Med 22:915–923

Wu W, Hsu Y-MS, Bi L, Songyang Z, Lin X (2009) CARD9 facilitates microbe-elicited production of reactive oxygen species by regulating the LyGDI-Rac1 complex. Nat Immunol 10:1208–1214

Wu W, Zhang R, Wang X, Song Y, Liu Z, Han W, Li R (2016) Impairment of immune response against dematiaceous fungi in Card9 knockout mice. Mycopathologia 181:631–642

Xiao Y, Tang J, Guo H, Zhao Y, Tang R, Ouyang S, Zeng Q, Rappleye CA, Rajaram MVS, Schlesinger LS et al (2016) Targeting CBLB as a potential therapeutic approach for disseminated candidiasis. Nat Med 22:906–914

Xu X, Xu J-F, Zheng G, Lu H-W, Duan J-L, Rui W, Guan J-H, Cheng L-Q, Yang D-D, Wang M-C et al (2018) CARD9S12N facilitates the production of IL-5 by alveolar macrophages for the induction of type 2 immune responses. Nat Immunol 19:547–560

Xu P, Zhang X, Liu Q, Xie Y, Shi X, Chen J, Li Y, Guo H, Sun R, Hong Y et al (2019) Microglial TREM-1 receptor mediates neuroinflammatory injury via interaction with SYK in experimental ischemic stroke. Cell Death Dis 10:555

Yamamoto H, Nakamura Y, Sato K, Takahashi Y, Nomura T, Miyasaka T, Ishii K, Hara H, Yamamoto N, Kanno E et al (2014) Defect of CARD9 leads to impaired accumulation of gamma interferon-producing memory phenotype T cells in lungs and increased susceptibility to pulmonary infection with *Cryptococcus neoformans*. Infect Immun 82:1606–1615

Yamasaki S, Ishikawa E, Sakuma M, Hara H, Ogata K, Saito T (2008) Mincle is an ITAM-coupled activating receptor that senses damaged cells. Nat Immunol 9:1179–1188

Yan J, Wu B, Huang B, Huang S, Jiang S, Lu F (2014) Dectin-1-CD37 association regulates IL-6 expression during Toxoplasma gondii infection. Parasitol Res 113:2851–2860

Yang H, Minamishima YA, Yan Q, Schlisio S, Ebert BL, Zhang X, Zhang L, Kim WY, Olumi AF, Kaelin WG (2007) pVHL acts as an adaptor to promote the inhibitory phosphorylation of the NF-kappaB agonist Card9 by CK2. Mol Cell 28:15–27

Yang C-S, Rodgers M, Min C-K, Lee J-S, Kingeter L, Lee J-Y, Jong A, Kramnik I, Lin X, Jung JU (2012) The autophagy regulator Rubicon is a feedback inhibitor of CARD9-mediated host innate immunity. Cell Host Microbe 11:277–289

Yang M, Shao J-H, Miao Y-J, Cui W, Qi Y-F, Han J-H, Lin X, Du J (2014) Tumor cell-activated CARD9 signaling contributes to metastasis-associated macrophage polarization. Cell Death Differ 21:1290–1302

Yang Y, Kelly P, Shaffer AL, Schmitz R, Yoo HM, Liu X, Huang DW, Webster D, Young RM, Nakagawa M et al (2016a) Targeting non-proteolytic protein ubiquitination for the treatment of diffuse large B cell lymphoma. Cancer Cell 29:494–507

Yang Y-K, Yang C, Chan W, Wang Z, Deibel KE, Pomerantz JL (2016b) Molecular determinants of scaffold-induced linear ubiquitinylation of B cell lymphoma/leukemia 10 (Bcl10) during T cell receptor and oncogenic caspase recruitment domain-containing protein 11 (CARD11) signaling. J Biol Chem 291:25921–25936

Yang Z, Weng C, Wang J, Xu P (2016c) The role of Card9 overexpression in peripheral blood mononuclear cells from patients with aseptic acute pancreatitis. J Cell Mol Med 20:441–449

Yang Z-W, Meng X-X, Zhang C, Xu P (2017) CARD9 gene silencing with siRNA protects rats against severe acute pancreatitis: CARD9-dependent NF-κB and P38MAPKs pathway. J Cell Mol Med 21:1085–1093

Yu JW, Hoffman S, Beal AM, Dykon A, Ringenberg MA, Hughes AC, Dare L, Anderson AD, Finger J, Kasparcova V et al (2015) MALT1 protease activity is required for innate and adaptive immune responses. PLoS ONE 10:e0127083

Zhang Q, Cao X (2019) Epigenetic regulation of the innate immune response to infection. Nat Rev Immunol 19:417–432

Zhang M, Lee AJ, Wu X, Sun S-C (2011) Regulation of antiviral innate immunity by deubiquitinase CYLD. Cell Mol Immunol 8:502–504

Zhernakova A, Festen EM, Franke L, Trynka G, van Diemen CC, Monsuur AJ, Bevova M, Nijmeijer RM, van't Slot R, Heijmans R et al (2008) Genetic analysis of innate immunity in Crohn's disease and ulcerative colitis identifies two susceptibility loci harboring CARD9 and IL18RAP. Am J Hum Genet 82:1202–1210

Zhong X, Chen B, Yang L, Yang Z (2019) Card9 as a critical regulator of tumor development. Cancer Lett 451:150–155

Zhu L-L, Zhao X-Q, Jiang C, You Y, Chen X-P, Jiang Y-Y, Jia X-M, Lin X (2013) C-type lectin receptors dectin-3 and dectin-2 form a heterodimeric pattern-recognition receptor for host defense against fungal infection. Immunity 39:324–334

The manufacturer's authorised representative in the EU is Springer Nature Customer Service Centre GmbH, Europaplatz 3, 69115 Heidelberg, Germany. If you have any concerns regarding our products, please contact ProductSafety@springernature.com

Printed and bound by CPI Group (UK) Ltd, Croydon, CR0 4YY
28/09/2026
02224074-0012